THE AMERICAN CITIES
AND TECHNOLOGY READER

Cities and Technology, a series of three textbooks and three readers, explores one of the most fundamental changes in the history of human society: the transition from predominantly rural to urban ways of living. This series presents a new social history of technology, using primarily urban settings as a source of historical evidence and a focus for the interpretation of the historical relations of technology and society.

Drawing on perspectives and writings from across a number of disciplines involved in urban historical studies – including archaeology, urban history, historical geography and architectural history – the books in this series explore: how towns and cities have been shaped by applications of a range of technologies; and how such technological applications have been influenced by their social contexts, including politics, economics, culture and the natural environment.

The American Cities and Technology Reader is designed to be used on its own or as a companion volume to the accompanying *American Cities and Technology* textbook. Chronologically, this volume ranges from the earliest technological dimensions of Amerindian settlements to the 'wired city' concept of the 1960s and internet communications of the 1990s. Geographically, its focus is the continental United States, and the majority of the readings selected deal with American cases. However, the context for the study of modern electronic communications in relation to cities transcends national boundaries just as the technologies themselves do; consequently, the contents of some of the later texts in the volume extend to global coverage. Among the issues discussed are the rise of the skyscraper, the coming of the automobile age, relations between private and public transport, the development of infrastructural technologies and systems, the implications of electronic communications and the emergence of city planning.

Gerrylynn K. Roberts is Senior Lecturer in the History of Science and Technology at the Open University.

THE AMERICAN CITIES AND TECHNOLOGY READER

Wilderness to wired city

Edited by Gerrylynn K. Roberts

 in association with

The Open University

London and New York

First published 1999
by Routledge
11 New Fetter Lane, London EC4P 4EE

Simultaneously published in the USA and Canada
by Routledge
29 West 35th Street, New York, NY 10001

Typeset in Palatino by J&L Composition Ltd, Filey, North Yorkshire
Printed and bound in Great Britain by
TJ International Ltd, Padstow, Cornwall

British Library Cataloguing in Publication Data
A catalogue record for this book is available from the British Library

Library of Congress Cataloging in Publication Data
The American cities and technology reader: wilderness to wired city/
edited by Gerrylynn K. Roberts.
p.   cm. - (Cities and technology)
1. Technology-Social aspects-United States-History.
2. Urbanization-United States-History.
I. Roberts, Gerrylynn K.   II. Series.
T14.5.A45   1999   98-42028
303.48'3-dc21

ISBN 0-415-20085-7 (hbk)
ISBN 0-415-20086-5 (pbk)

CONTENTS

LIST OF FIGURES AND TABLES

Tables

INTRODUCTION

The Cities and Technology Series

This reader is the third in a series about the technological dimension of one of the most fundamental changes in the history of human society: the transition from predominantly rural to urban ways of living. It is first and foremost a series in the social history of technology; the urban setting serves above all as a repository of historical evidence with which to interpret the historical relations of technology and society. The main focus, though not an exclusive one, is on the social relations of technology as exhibited in the physical form and fabric of towns and cities.

The main aims of the series are twofold. The first is to investigate the extent to which major changes in the physical form and fabric of towns and cities have been stimulated by technological developments (and conversely how far urban development has been constrained by the existing state of technology). The second aim is to explore within the urban setting the social origins and contexts of technology. The series draws upon a number of disciplines involved in urban historical studies (urban archaeology, urban history, urban historical geography, architectural history and environmental history) in order to correct any illusion of perspective that all major changes in urban form and fabric might be sufficiently explained by technological innovations. In brief, the series attempts to show not only how towns and cities have been shaped by applications of technology, but also how such applications have been influenced by, for example, politics, economics, culture and the natural environment.

The wide chronological and geographical compass of the series serves to bring out the general features of urban form which differentiate particular civilizations and economic orders. Attention to these differences shows how civilizations and societies can be characterized both by their use of certain complexes of technologies, and also by the peculiar political, social and economic pathways through which the potentials of these technologies are channelled and shaped. Despite its wide sweep, the series does not sacrifice depth for breadth: studies of technologies in particular urban settings form the bulk of the three collections of existing texts in the series.

The American Cities and Technology Reader

This book is designed to stand on its own, but readers might note that there is a companion, specially written illustrated textbook,[1] which the contents of this reader have been selected to complement. Chronologically, this volume ranges from the earliest technological dimensions of settlements of Amerindians to the "wired city" concept of the 1960s and internet communications of the 1990s. Geographically, its focus is the continental United States and the majority of the readings selected deal with American cases. However, the context for the study of modern electronic communications in relation to cities transcends national boundaries just as the technologies themselves do; consequently, the contents of some of the later texts in the volume range more widely around the globe.

Readers should also note the policy that has guided the editing of the selected texts. With a general and undergraduate readership in mind, scholarly apparatus has been kept to a minimum. This means in many instances the omission of carefully compiled primary and secondary source references. Despite this pruning, in almost all cases notes of sources have been retained where they would be expected at any level of scholarly writing, including undergraduate essays; in particular, where there is direct quotation, explicit reference to another author or citation of statistical data. In practice, an unarguably consistent policy has been difficult to implement. In the end, I would as editor encourage readers especially interested in any text to follow up the reference to the original source, which is always given. I should also like to record my gratitude to the many authors who have responded generously to my request that their work be made available

in this edited form; some, of course, have been unavailable for comment.

It should be made clear that in a collection like this, a product of that distinctive collaborative unit, the Open University Course Team, editorship signifies rather less individual input than in other such volumes with a single name on the cover. It means more that the named individual has taken on responsibility for corresponding with authors and publishers, and with preparing the handover manuscript. In particular, Philip Steadman, co-author of the companion textbook, has played a substantial role in the preparation of this volume. And in all cases, the selection and editing has been subject to Course Team discussion and approval. Finally, the sheer amount of coordination and administration involved in the preparation of such a volume, particularly with regard to the procurement of illustrations, must not go unremarked and I wish to thank Denise Hall for her part in the preparation of the manuscript.

Note

1 G.K. Roberts and J.P. Steadman (eds), *American Cities and Technology: wilderness to wired city*, London, Routledge, 1999

I

THE INDIAN LEGACY IN THE AMERICAN LANDSCAPE

by Karl W. Butzer

Source: Karl W. Butzer, "The Indian Legacy in the American Landscape", in M.P. Conzen (ed.), *The Making of the American Landscape*, London, Routledge, 1990, pp. 27–50

North America was not a sparsely populated "virgin land" when the French and English first settled Québec, Plymouth Rock, and the James River estuary in the early 1600s. As generations of colonists slashed their way through the eastern forests and pushed back the "savages," their introspective and ethnocentric view excluded native Americans from the cherished image of a new European landscape. Frontiersmen and later frontier historians saw Indians as outsiders, people without legitimate claim to the land they lived on and, not surprisingly, Indians were excluded from the new society that emerged. The Spanish, who came earlier, had a very different vision. The De Soto expedition, pillaging through the Southeast in 1539–42, noted mortuary temples as a potential source of loot, and Coronado, who explored the Southwest in the same years, described pueblos such as Cibola. Whatever their motives, Spaniards "saw" the indigenous cultural landscape, and they ultimately sought to assimilate its people into their own world.

These very different visions of North America are also reflected in two traditions of cultural and historical geography, one emphasizing the indigenous roots, the other the European contributions. But America did not begin on the banks of the James River, rather, when Asian peoples crossed the Bering Straits about 15,000 years ago. Their descendants settled the continent and, over many millennia, adapted their hunting and foraging ways of life to different combinations of resources, reflecting North American environmental diversity. They created farming towns, following an independent trajectory of agricultural origins during what in Europe were the so-called Dark Ages. The farming frontier in most areas was pushed to its ecological limits, while on the west coast, alternative ways of life were developed that could support surprisingly large populations by fishing and intensified plant collecting. In the period when Gothic cathedrals were erected in medieval Europe, many thousands of native Americans built impressive towns in the Southwest and Mississippi Basin, sites now visited by tourists from both continents.

There is, then, a pre-European cultural landscape, one that represented the trial and error as well as the achievement of countless human generations. It is upon this imprint that the more familiar Euro-American landscape was grafted, rather than created anew.

Adapting to new environments

The first peopling of the New World remains the subject of controversy. The earliest immigrants arrived from Asia via the Bering Straits, to confront the problems of an inhospitable environment, a cold water body, bleak mountain ranges, and oscillating glaciers. The persistently sparse archeological record of Siberia, Alaska, and northwestern Canada, however, hinders our interpretation of this movement. [. . .]

It was technically possible for prehistoric hunters to pass from Asia into more productive regions of the New World for tens of millennia prior to 30,000 BP [before present]. But the coeval record of prehistoric settlement in eastern and northern Asia is poor, and there still is no convincing record of such antiquity in Canada or the United States. The earliest documented site in Alaska is from about 14,000 BP, and in the United States the oldest is Meadowcroft Rockshelter near Pittsburgh, of about the same age. [. . .] Dating from shortly after 12,000 BP, there is a veritable explosion of archeological sites in the continental United States (Fig. 1.1) and to a lesser degree in Alaska and South America. This dramatic influx of Paleoindians represents a highly successful human adaptation to big-game hunting.[1] [. . .] Within 2,500 years, the Paleoindian people had settled much of the United States, and not long thereafter they appeared at the other end of South America, near Tierra del Fuego.

The Paleoindians evidently were highly mobile, efficient, and adaptable. But within the United States their site concentrations suggest a preference for relatively open environments with a high animal population: the pine-grass parklands of the High Plains and incipient

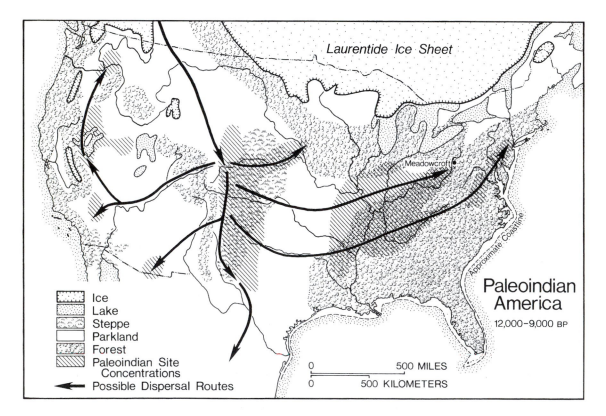

Figure 1.1 The Paleoindian entry into North America after about 12,000 BP. Ice margins and proglacial or pluvial lakes represent their maximum extent about 12,000–11,000 BP.

Prairie Peninsula, the pine–sagebrush parkland of the western Great Basin, and the then assembling deciduous woodlands of the east-central and mid-Atlantic United States (Fig. 1.1).[2] The classic Paleoindian sites on the High Plains represent the ephemeral encampments of bison hunters. [. . .]

Most Paleoindian sites are small, with comparatively few artifacts, even where large numbers of animals had been killed, but the fine projectile points were carefully husbanded in the course of a mobile, seasonal schedule. A millennium or two after Paleoindian dispersal into most American environments, changes in shape and size of projectile points become apparent in different areas, reflecting an adaptation to specific prey as well as the emergence of regional stylistic differences. [. . .]

After 5000 BP, finding food in the Late Archaic period focused more specifically on exploiting a limited range of resources, a trend apparent in different environments of North America. In the Pacific Northwest, finds of barbed antler harpoons point to the increasingly effective use of marine and river derived food such as salmon, while the existence of larger and more numerous documented sites may imply seasonal settlement. In the Mississippi, Ohio, and Tennessee drainage, manipulation of weedy seeds gradually led to domestication of marsh elder (sumpweed) and maygrass by 4000 BP.[3] The native squash was domesticated and generally available about 3000 to 2000 BP, while the bottle gourd, a tropical cultigen of Mexican origin, was introduced before 4300 BP and was widely cultivated by 2500 BP. In the Southwest, domesticated maize of Mesoamerican origin indicates the presence of supplementary agriculture about 3000 to 2500 BP, but sites are limited to some caves near the Mexican border.

All in all, there are parallels between American Archaic and European Mesolithic developments. They were periods of environment-specific specialization and diversification, in which increased labor was devoted to raising the calorific or protein yield of food. [. . .]

Towards an agricultural landscape in the East

About 3000 to 1500 BP, economic trends that emerged during the Late Archaic period crystallized into more definite patterns. Mesoamerican cultivated crops (cultigens), pottery technology, and cultural ideas became important in the Southwest and the Mississippi Basin. The bow and arrow, a major improvement in hunting technology and warfare, were introduced from the North. Trade in food, raw materials such as copper and marine shells, and ornaments accelerated and affected economic life in the back country of the coasts and river valleys. New forms of social organization and ideology appeared and were reflected in large ritual centers in the Mississippi Basin and a general increase almost everywhere in the clarity of the picture archeologists were able to build up.

In the Mississippi, Ohio, and Tennessee Basins the period from 1000 BC to AD 900 represents the Woodland culture complex, a disparate group of proto-agricultural tribes that was interconnected by an active, long-distance trade network. Intensive gathering of wild plant and aquatic foods continued, but the array of local cultigens was increased by the addition of sunflower (an oil plant) and goosefoot (a starchy seed), while eight-rowed "flint" and twelve-rowed "dent" maize were introduced from the Southwest and Mexico respectively.[4] Maize of both types has been verified in Tennessee about 350 BC, in the Ohio drainage after 300 BC, and in the Illinois valley by AD 650. Tobacco was also introduced from Mesoamerica about AD 200, while pottery traditions of similar origin were established in the Ohio and Tennessee basins by 900 BC, spreading to the northern High Plains by AD 500.[5]

In effect, the Woodland phenomenon represents a 2,000-year period of diffusion, innovation, and development: regions where humans could live expanded and productivity increased; populations grew significantly and settlements became semi-permanent. The rôle of domesticated foods also increased progressively. [. . .] By this time one can speak of "supplementary" agriculture within an intensified gathering economy. But even prior to the dissemination of maize, sizeable towns with great burial mounds sprang up (Fig. 1.2). The largest of these is Poverty Point, Louisiana, a complex of artificial earth mounds and geometrical earthworks that contain nearly 1 million cubic yards of material, begun about 1200 BC. A

cluster of such sites in the middle Ohio valley around Adena and Hopewell includes towns with up to 38 burial and effigy mounds about 100 acres in size, that date from between 500 BC and AD 400.[6] Trade goods are prominent in such centers, indicating a far-flung exchange system that actively linked a multitude of small villages (50–100 inhabitants) and raw material sources across the Midwest, while maintaining indirect contacts with towns in the Mississippi valley and the Southeast. Presumably trade also assured complementary food supplies, at least during years with average crop yields.

Although overall population density was low, perhaps as low as one person per square mile, the persistence of some towns with

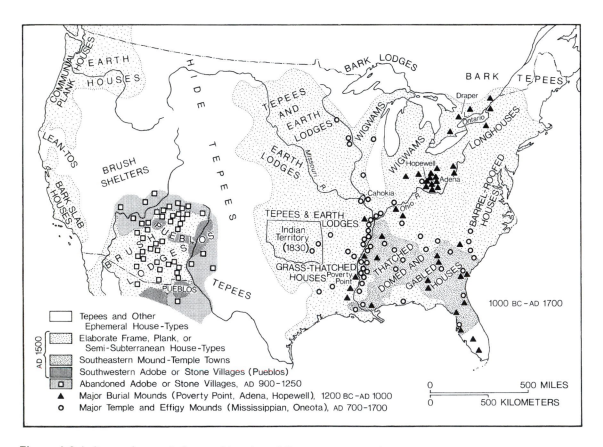

Figure 1.2 Indian settlement in late prehistoric and European contact times.

several thousand inhabitants over four and five centuries – without a true agricultural base–has no parallels in Old World prehistory. [. . .]

The Mississippian phenomenon refers to the agricultural high point of native American settlement in the Mississippi and Ohio basins. Geographically, at any one time this Mississippian phase represents a dozen or so settlement clusters along different floodplain segments (Fig. 1.2). Many such clusters were short-lived, perhaps enduring a mere 75 years, while others spanned most of the 600 or so years represented by the Mississippian period (about AD 900–1500). The designation again encompassed different tribes, with varying sociopolitical complexity, but each geographical and temporal component was concentrated around one or more ceremonial centers, with conspicuous "temple mounds," that also served economic and political functions. The hierarchical nature of settlement size, function, and arrangement seems to have been paralleled by some degree of social hierarchy ("chiefdoms"). The unifying elements of the different regional components, spread from the Gulf coast to the Great Lakes watersheds, appear to have included riverine trade networks, a common system of agriculture, and a broadly shared body of ideas and beliefs.

The Mississippian phase developed from indigenous, Woodland roots, with some infusion of cultural and ideological elements, from the Gulf coast and from Mesoamerica, in part via the Southwest. The configurations emerging through archeological research took form over some two centuries, attained their maximum visibility between AD 1100 and 1300, and subsequently show evidence of decline and regional abandonment. [. . .]

The Mississippian agricultural system was therefore highly diversified, rather than specialized, but invariably dependent on proximity to floodplains for both their fertile alluvial soils and natural pulses of energy. Nothing is known about the scheduling of fallow periods, and

manure was not used, but the simple hoe and digging stick technology would have been unable to provide sustained yields on sandier soil without long fallow intervals. Overall, this method of agriculture was extensive, rather than intensive. Allowing for the absence of domesticated animals, the closest European analogy was with simple Neolithic farming.

Excavated site residues suggest several categories of settlement: (a) short-term, special-purpose sites used in hunting, plant collecting, or processing; (b) homesteads of one or several families; (c) hamlets of perhaps 10 or 20 houses; (d) villages, with an area of 1/2–3 acres and from 30 to over 300 houses, enclosed by a palisade or earthworks; and (e) ceremonial towns, ranging from 12 to over 200 acres in size and including anything from 200 to 1,000 houses.[7] Houses enclosed space of 30 to 60 square yards and were roughly rectangular, with numerous post impressions in the soil indicating permanence but frequent rebuilding with perishable materials, in pole and thatch style; there was a central hearth, with storage pits inside or outside. They are thought to have been inhabited by extended nuclear families of seven or eight people. Such structures were commonly arranged in rough rows, at a density of 12 to 28 per acre. A typical hamlet had about 100 people, a village between 700 and 1,300 inhabitants, and a ceremonial town 2,500 or more.

The largest settlement of the time and region was Cahokia, located on the former levee of a cut-off Mississippi River meander, near East St. Louis (Fig. 1.3). [. . .] The site was occupied by a large settlement from at least AD 1100 to 1350, but enjoyed its heyday during the 13th century.[8] Over 100 mounds have been identified in the area illustrated by Figure 1.3. [. . .] Most served as platforms for public buildings or the residence of prominent people, although at other sites mounds were often still used for mortuary rites or burials. The Cahokia mounds were primarily oriented along the crest of the

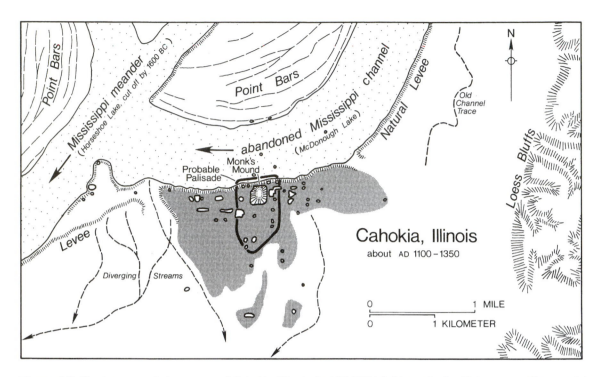

Figure 1.3 The large population center of Cahokia, Illinois, had 30,000 inhabitants in the 13th century. Ceremonial mounds visible in the 5-foot contours of the 1:24,000 topographic map are shown by strong outlines of various shapes.

levee, centered on the four-tiered Monk's Mound (13.5 acres, with an intact relief of 112 feet); further lines of mounds were arranged in perpendicular fashion, probably with large open "plazas" adjacent. A central area of 200 acres was once enclosed by a log palisade, with watchtowers and gates set at regular intervals. Rebuilt four times, this palisade may have served to enclose a defended refuge as well as a high-status area.

Residential land use in Cahokia was concentrated in a roughly 2,000-acre area, with several adjacent satellite clusters of houses, and an estimated total population of 30,000 people about AD 1250.[9] Goods found within such residences indicate strong differentiation according to wealth, as well as between craftsmen and farmers. Several other large ceremonial

towns of 120 to 300 acres surrounded Cahokia, at least during its early stages, as did dozens of villages, suggesting some form of central place hierarchy. Cahokia was a major center, the largest settlement in the United States until it was surpassed by Philadelphia in AD 1800, and it remains prominent in the landscape today.

The demise of the Mississippian settlement clusters is poorly understood; however, the cemetery skeletal record of the 13th century indicates poor nutrition, widespread infectious disease, and high numbers of births per woman.[10] Since many potentially productive areas remained unsettled, this implicit subsistence crisis was apparently compounded by social constraints on dispersal and by unequal access to resources. [. . .]

The Mississippian economic network stimulated agricultural development and village agglomerations well beyond the direct influence of this cultural sphere, in the Northeast and on the Great Plains. In upstate New York, the Iroquois, a peripheral offshoot of the Woodland tradition, shifted from small, oval houses to great longhouses during the 13th century, indicating a change from nuclear to extended residences, with up to two dozen units; from then until about AD 1500 they congregated into increasingly large villages (Fig. 1.4), supported by relatively intensive agriculture and by hunting, fishing, and plant gathering within a large territorial radius.[11]

To the west, Woodland groups first penetrated river valleys of the eastern Plains about 2000 BP, building countless small river-bluff mounds. After AD 700 semi-agricultural villages began to appear along the central Plains rivers where maize, squash, and sunflower were cultivated on the major floodplains, complemented by bison hunting.[12] These villages frequently shifted their location, and consisted of some 20 to 30 multifamily lodges of rectangular, semi-subterranean type. These Plains Village Indians competed with the established, mobile bison hunters and berry foragers of the region, but they began to abandon some valleys by AD 1300, partly in response to recurrent droughts and erratic floods. This withdrawal, recalling that in the Ohio–Mississippi drainage, continued over several centuries and was accompanied by social changes, reflected in a shift to circular or oval lodges, larger villages with at least 30-100 houses, and stout palisades. Oñate visited Wichita Indians at Quivira on the middle Arkansas in 1601, estimating the number of houses in this large but otherwise unremarkable town to be about 1,200. Further retraction of these communities on the ecological limits of extensive farming ensued when both they and the neighboring Plains hunters adapted to horseback riding during the early 18th century. [. . .]

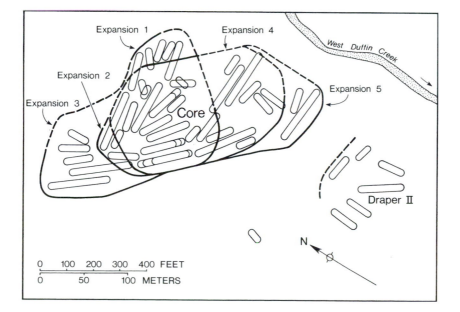

Figure 1.4 The Draper site, Ontario, represents a dense cluster of Huron longhouses within an expanding stockade, during the 16th century.

Pueblo and irrigation agriculture in the Southwest

The agricultural transition in the Southwest was also gradational. Eight-row maize, squash, bottle gourd, and beans were all introduced between 3000 and 2000 BP, the timespan of the San Pedro stage, for which available sites are limited to rockshelters in the mountains of southern New Mexico and Arizona.[13] Plant gathering and hunting were, however, the staple food practices. Proto-agricultural settlements soon began to spread to the Mogollon Rim and onto the Colorado Plateau (about 200 BC) with hamlets or small villages and increasing use of semi-subterranean houses, and the

gradual appearance of two different pottery traditions (Mogollon and Pueblo or Anasazi). Simple villages with a third pottery tradition appeared in the arid Gila and Salt River lowlands after AD 1 where, by AD 500, there was a progressive introduction of several new beans, Mesoamerican cotton (and loom weaving), and grain amaranths, their cultivation made possible by irrigation. This Hohokam tradition supported larger agricultural settlements around AD 550 to 700, and new varieties of drought-resistant maize were developed to increase the dependability of the food supply.

Eventually two distinctive settlement styles, linked to different ecologies, emerged after AD 950. In the high country, increasingly large

Figure 1.5 The masonry structures of the Pueblo Bonito ceremonial and population center, Chaco Canyon, New Mexico, illustrate the durability of 11th- and 12th-century settlements. The arroyo in the background was incised after AD 1100 but before abandonment, probably impeding floodplain cultivation.

settlements were constructed of multiroom, multistory, and flat-roofed, dry-masonry houses, arranged around large, circular, masonry-lined, ceremonial pit-houses, known as "kivas."[14] These pueblos have a strikingly urban appearance (Figs 1.5 & 1.6), whether they are situated in open valleys, at canyon heads, or in immense rockshelters in or below the canyon walls. Supported by cultivation of maize, wild foods such as pinyon nuts and juniper berries, as well as jackrabbits and domesticated turkey, such towns sometimes housed several thousand people. Cultivation depended on rainfall and the diversion and control of sporadic flood waters, with successive checkdams slowing the runoff of small upland streams. It also relied on rock lines along the lower borders of cultivated fields to prevent soil erosion. The best known emergent towns with large apartment

Figure 1.6 Masonry construction and pine crossbeams, Pueblo del Arroyo, Chaco Canyon, New Mexico.

complexes date after the period AD 1150 to 1175, when defensive situations were generally selected and satellite hamlets increasingly abandoned. At some point between AD 1290 and 1450 these settlements were either totally abandoned, or abruptly reduced to very modest proportions.

In the lowlands, the Hohokam of the Gila–Salt drainage developed a complex irrigation network around modern Phoenix that is the largest (over 250 square miles) and most elaborate of the New World (Fig. 1.7). Some of the canals were 15 to 18 miles and more in length by the time that this system achieved its maximum development (around AD 1400), and flows of up to 237 cubic feet per second have been estimated for trunk channels.[15] Feeders appear to have taken off directly at the Salt River banks, presumably when rainfall was more regular and the present erratic flooding was not a factor, and without the use of the mortared, masonry diversion dams characteristic of Spanish irrigation. Hohokam canals were not "lined," although centuries of flowing water have impregnated many with hard lime, and sluice gates were simple arrangements and involved backfilling and removal of earth, unlike the mortared counterparts in Spain, with wood or iron traps.[16] The prehistoric Salt River system remained sufficiently visible and logical in its arrangement that in 1878 Mormon settlers hired Pima Indians to reconstitute the 300 miles of major Hohokam canals. Interspersed within this network are at least 80 Hohokam settlement sites that have been classified into several size categories, some of which were larger than 250 acres and many of which remained occupied over a span of 500–800 years. The settlement surfaces of the Salt River south bank, roughly half of the total, add up to nearly 5,000 acres,[17] suggesting a maximum possible population of 75,000 to 100,000. By any reckoning, this was one of the largest ever traditional irrigation systems in human history. [. . .]

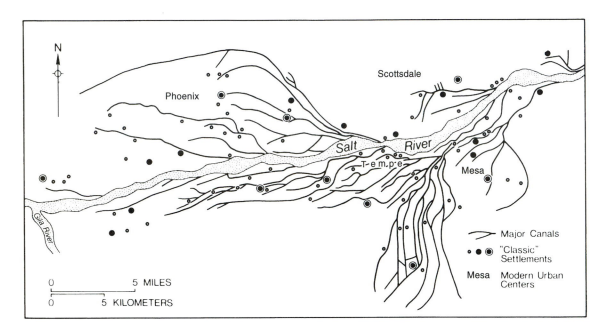

Figure 1.7 The irrigation and settlement network along the lower Salt River, Arizona, in Classic Hohokam times (AD 1150–1400), was the largest in the prehistoric New World.

House and town construction by the Hohokam was less permanent than in the masonry pueblos of the plateau. Puddled adobe was the basic building material, poured in regular courses of calcic mud that hardened to the consistency of a low grade concrete to allow the raising of multistoried, rectangular structures. Casa Grande, near the Gila River, had four floors and walls three feet thick (Fig. 1.8), and has remained a prominent, if derelict landmark since it was described in that state by Kino in 1694. Other ruins have generally fared less well, "melted-down" slowly by rain or quarried as a source of soil in the late 19th century. But the mass of adobe accumulated in Hohokam villages over centuries of occupation has created conspicuous mounds similar to Near Eastern tells. Roofs were flat and supported by large wooden beams (that allow tree-ring dating), covered by a bed of stout reeds and then several layers of adobe. Windows were small and rare. The so-called

Spanish domestic adobe architecture in the Southwest and northern Mexico is in fact indigenous, with the exception that Spaniards substituted preformed, sun-dried adobe bricks for puddled adobe and then added drain spouts from the roof. Nowhere in Spain is adobe plastered on masonry walls, as is the custom in many surviving southwestern pueblos (Fig. 1.9) although Spanish walls may be surfaced with a thin coat of cement before whitewashing.

The cycle of demographic growth, settlement groupings, and eventual abandonment evident in the late prehistoric site clusters of the Southwest paralleled that of counterparts in the Mississippi–Ohio Basin. It suggests a latent instability in such settlement systems that needs further exploration.

Significantly, the southwestern site clusters that appeared about AD 950 to 1050 showed a parallel but not strictly synchronous development. Such clustering peaked as early as

Figure 1.8 The Casa Grande, near the Gila River of Arizona, was built about AD 1300 and abandoned during the 15th century. Measuring 41 by 62 feet, this adobe structure was 33 feet high and had some 60 rooms. The walls taper upwards from a base of 53 inches to 21 inches, and were constructed with regular 25-inch courses of puddled adobe, poured between some sort of formwork. The lower wall surfaces were restored in 1891, the shelter built in 1932.

AD 1075 and as late as 1325, and partial or total abandonment took place in some areas during the late 1200s, in others during the mid-1400s. Maximum population tended to accompany or precede congregation in large settlements, suggesting social changes, possibly a switch from an inter-community exchange system to one of centralized redistribution. Abandonment sometimes followed droughts evidenced in long tree rings, or local floodplain downcutting (with lower water tables and loss of irrigation "head"), but more commonly coincided with periods of wildly erratic rainfall. [. . .]

The basic problem in the Southwest, therefore, seems to have been that productivity could not be sustained in the face of demographic growth, given a relatively static technology. Thus, the social systems appear to have been too rigid to adjust, and wholesale abandonment ensued.

The displaced populations subsequently relocated to existing pueblo centers, where a dramatic upsurge of population occurred between 1250 and 1400. At Zuñi, a cluster of six or seven archeologically documented villages was described as the "seven cities of Cibola" at the time of Coronado (1540), and in 1582 the Spanish estimated 130,000 inhabitants in 61 pueblos for the Southwest.[18] Perhaps the Old World diseases introduced by the Spaniards headed off further crises of sustainability and Malthusian "overshoot."

While agricultural economies with large,

Figure 1.9 The multistoried, flat-roofed and adobe-faced masonry houses of Taos, New Mexico, are representative of surviving pueblos in the Southwest.

permanent settlements evolved in the East and Southwest, the productive environments of the west coast became the scene of highly successful foraging societies. In California, a vast range of wild plant foods was utilized, with much emphasis on acorns that were ground into bread meal, while tobacco was the sole cultivated crop. Freshwater and marine fish were equally important, and exchange networks bound together people of the coast and the interior.[19] Prior to European contact, a population of over 300,000 included at least a dozen centers with more than 1,200 people.

In the Pacific Northwest, by 1500 BP equally large communities lived in fortified, seashore villages of communal plank houses that were supported by salmon, halibut, and cod fishing from boats, with harpoons and nets. Small, curly-haired dogs and mountain sheep provided a form of wool for blanket weaving, while some groups planted and tended gardens of clover roots and other plants.[20] Further inland, smaller villages consisted of large, circular pithouses sunk into the ground, the roofs formed by heavy, sloping rafters covered with bark and earth. Wild, starchy roots and bulbs were roasted in earthen ovens; spawning salmon were taken in the rivers and lakes, along with beaver and mussels, while moose, deer, bear, and mountain sheep were sought farther afield on seasonal hunting forays. Because they were built of perishable materials, there is little visible evidence of the Northwest coast settlements today, other than an occasional totem pole. But early travelers left vivid accounts of their strange charm, teeming populations, and industrious bustle. [. . .]

The surviving legacy

The most obvious imprint of native Americans on the landscape are the Indian place names. Of the 48 coterminous states, 25 carry Indian names, as do 13 percent of some 1300 counties, hundreds of rivers and mountains, and thousands of towns and cities. So familiar to the average Anglo-American as to be unrecognized, these toponyms serve as a constant reminder that the landscape had been humanized by the first Americans. [. . .]

Physical configurations of the Indian landscape [. . .] survive directly. Apart from the abandoned or living pueblos of the Southwest, thousands of mounds in the East remain conspicuous landmarks of an earlier civilization, despite road building and mechanized plowing. The 19th-century Mississippi boatmen returned upstream to Tennessee by the Natchez Trace, previously the Chickasaw Trail, and still visible today. The Angeleno who drives over the Cajon Pass towards a Las Vegas weekend follows an Indian trail already adapted by the Spaniard. The modern irrigation system around Phoenix, Arizona, is largely a recreation of its Hohokam counterpart. The flat-roofed adobe house of the Borderlands, and its gentrified application to new architectural designs, is basically an Indian form, not a Spanish introduction of a Berber house type.[21] French fur trade posts and Anglo-American forts were located at Indian communication or population nodes, and served as nuclei for civilian settlement: Kingston, Ontario; Albany, New York; Pittsburgh, Pennsylvania; Detroit, Michigan; Fort Wayne, Indiana; Peoria, Illinois; Green Bay, Wisconsin; Des Moines, Iowa; Fort Smith, Arkansas; Fort Worth, Texas; Missoula, Montana; or Walla Walla, Washington, provide some examples. Spanish presidios and missions were located next to Indian settlements or ceremonial centers in the Southwest and California, to become centers like San Antonio, Texas; Santa Fe, New Mexico; Tucson, Arizona; and in California San Diego, Los Angeles, or San Francisco.

Thousands of years of Indian settlements influenced the Anglo-American landscape in many other subtle ways. The quality of land had already been determined by generations of Indian use, a realization that may help to explain the insatiable greed of the homesteader and rancher for Indian core territory. Indian expertise in countless facets of forest and prairie living greatly facilitated British colonization and American westward expansion, preventing much costly trial and error. Determined Indian resistance by the Comanche, Sioux, Apache, and other tribes probably affected rates and patterns of settlement as much in a negative way as passive tribes or thinly settled lands did in a positive way. Although the average American might well not appreciate this legacy, cultural and historical geographers have no excuse for lacking a deeper appreciation of the American roots of the American landscape.

Notes

I am deeply indebted to William E. Doolittle (Austin), for sharing his library and experience with me. Fred Eggan (Chicago), B.L. Turner II (Worcester), as well as Stephen A. Hall, Robert A. Ricklis and Michael D. Blum (Austin) provided discussion, suggestions, or information. The maps were ably drawn by John V. Cotter (Austin).

1 West 1983
2 Porter 1983; Bryant and Holloway 1985
3 Ford 1985; Delcourt *et al*. 1986
4 Ford 1985; Delcourt *et al*. 1986
5 Frison 1978
6 Brose and Greber 1979
7 Smith 1978
8 Fowler 1978
9 Gregg 1975
10 Cohen and Armelagos 1984
11 Ritchie 1980; Ricklis n.d.
12 Caldwell and Henning 1978
13 Rohn 1978; Ford 1985

14 Mortar appears to have been first introduced to Mexico and the Southwest by the Spanish

15 Nicholas and Neitzel 1984; flow data from Masse (1981)

16 However, precarious segments of Hohokam canals were sometimes strengthened by applying fire to a fresh adobe lining, to "bake" it, in lieu of cement (Haury 1976)

17 Nicholas and Neitzel 1984

18 The Zuñi archeological evidence is presented in Fish and Fish (1984); on the Spanish estimates, see Sauer (1980)

19 Hornbeck 1982

20 Fladmark 1986

21 Mudbrick housing is limited to a very few, out-of-the-way locations in southeastern and east-central Spain, and flat roofs are only characteristics in parts of Granada, Murcia and the island of Ibiza. None of these areas contributed to the stream of emigrants to the New World (Butzer 1988)

References

BROSE, D.S. and GREBER, N. (1979) *Hopewell Archaeology*, Kent, OH, Kent State University Press

BRYANT, V.M. and HOLLOWAY, R.G. (eds) (1985) *Pollen Records of Late Quaternary North American Sediments*, Dallas, American Association of Stratigraphic Palynologists Foundation

BUTZER, K.W. (1988) "Cattle and Sheep from Old to New Spain: Historical Antecedents", *Annals of the Association of American Geographers* 78, pp. 29-56

CALDWELL, W.W. and HENNING, D.R. (1978) "North American Plains", in TAYLOR, R.E. and MEIGHAN, C.W. (eds), *Chronologies in New World Archaeology*, New York, Academic Press, pp. 113-46

COHEN, M.N. and ARMELAGOS, G.J. (eds) (1984) *Paleopathology and the Origins of Agriculture*, Orlando, FL, Academic Press

DELACOURT, P.A. *et al.* (1986) "Holocene Ethnobotanical and Paleoecological Record of Human Impact on Vegetation in the Little Tennessee River Valley, Tennessee", *Quarternary Research* 25, pp. 330-49

FISH, S.K. and FISH, P.R. (eds) (1984) *Prehistoric Agricultural Strategies in the Southwest*, Anthropological Research Papers 33, Tempe, Arizona State University

FLADMARK, K.R. (1986) *British Columbia Prehistory*, Ottawa, National Museums of Canada

FORD, R.I. (ed.) (1985) *Prehistoric Food Production in North America*, Anthropological Paper 75, University of Michigan, Museum of Anthropology

FOWLER, M.L. (1978) "Cahokia and the American Bottom: Settlement Archaeology", in SMITH, B.D. (ed.), *Mississippian Settlement Patterns*, New York, Academic Press, pp. 455-78

FRISON, G.C. (1978) *Prehistoric Hunters of the High Plains*, New York, Academic Press

GREGG, M.L. (1975) "A Population Estimate for Cahokia", in *Perspectives in Cahokia Archaeology*, Illinois Archaeological Survey Bulletin 10, Urbana, University of Illinois, pp. 126-136

HAURY, E.W. (1976) *The Hohokam: Desert Farmers and Craftsmen*, Tucson, University of Arizona Press.

HORNBECK, D. (1982) "The California Indian before European Contact", *Journal of Cultural Geography* 2, pp. 23-39

MASSE, W.B. (1981) "Prehistoric Irrigation Systems in the Salt River Valley, Arizona", *Science*, 214, pp. 408-15

NICHOLAS, L. and NEITZEL, J. (1984) "Canal Irrigation and Sociopolitical Organization in the Lower Salt River Valley: A Diachronic Analysis", in FISH, S.K. and FISH, P.R. (eds), *Prehistoric Agricultural Strategies in the Southwest*, Anthropological Research Papers 33, Tempe, Arizona State University, pp. 161-78

PORTER, S.C. (ed.) (1983) *Late Quaternary Environments of the United States*, Minneapolis, University of Minnesota Press

RICKLIS, R.A. "Iroquois Cultural Development on the Colonial Frontier", in press

RITCHIE, W.A. (1980) *The Archaeology of New York State*, Harrison, NY, Harbor Hill Books

ROHN, A.H. (1978) "American Southwest", in TAYLOR, R.E. and MEIGHAN, C.W. (eds), *Chronologies in New World Archaeology*, New York, Academic Press, pp. 201-22

SAUER, C.O. (1980) *Seventeenth Century North America*, Berkeley, CA, Turtle Island Press

SMITH, B.D. (ed.) (1978) *Mississippian Settlement Patterns*, New York, Academic Press

WEST, F.H. (1983) "The Antiquity of Man in America", in *Late Quaternary Environments of the United States*, Vol. 1: PORTER, S.C. (ed.), *The Late Pleistocene*, Minneapolis, University of Minnesota Press, pp. 354-84

2

SPANISH LEGACY IN THE BORDERLANDS

by David Hornbeck

Source: David Hornbeck, "Spanish Legacy in the Borderlands", in M.P. Conzen (ed.), *The Making of the American Landscape*, London, Routledge, 1990, pp. 51–62

[. . .] Before English colonists settled the eastern seaboard, [. . .] Spain had explored and occupied much of the present day southeastern and southwestern parts of the United States.

Spain's influence on the United States has both geographical and institutional foundations. Today the names of six states – Florida, Colorado, Nevada, California, New Mexico, Texas, and Arizona – have their origins in the Spanish language, as do those of scores of rivers, mountains, and towns. [. . .]

For almost 300 years Spain occupied the southwestern part of the United States. Between 1762 and 1800, Spain possessed the entire trans-Mississippi West, granting lands, conducting trade in furs, and building trading and military posts as far north as Minnesota. Florida was in Spanish hands from 1526 until 1821, during which time military outposts and missions were established as far north as Port Royal, South Carolina; Spain even briefly occupied the Chesapeake Bay. Today, the areas once settled by the Spanish are usually referred to as the Spanish Borderlands. [. . .]

Spain began her search for new territories in North America from two established areas of settlement. The first push was from the Caribbean into Florida, along both the Gulf and Atlantic coasts. The second area from which Spain began to explore North America was central Mexico northward into the trans-Mississippi West and along the Pacific coast (Fig. 2.1).

[. . .]

Populating the land

Spain did not simply explore and then leave an area; rather, Spanish explorers established settlements in most of the areas they explored. In 1559, Spanish settlers founded Pensacola and six years later established St. Augustine. The first of many Jesuit missions along the South Atlantic coast (from southern Florida to Chesapeake Bay) was founded in Florida in 1566. By the beginning of the 17th century, Spain had placed permanent colonies in New Mexico and had established missions in the Hopi area of Arizona.

Settlement during the 16th century was for the most part driven by economic and religious motives. Mines, stock ranches, towns, and missions were established to exploit or convert local Indian populations. But with intrusions from other European powers, Spanish settlement began to be driven by a new factor – defense. During the 17th century, defensive settlements north of the Rio Grande and the Gulf of Mexico were established in response to French and British threats of incursion. Of Spain's settlements in North America, only

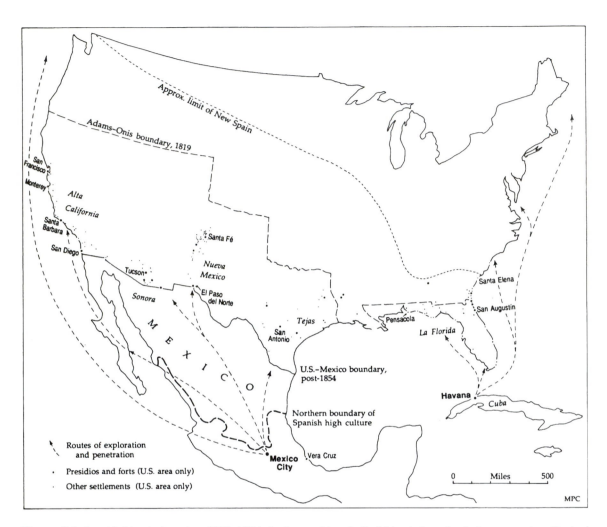

Figure 2.1 Spanish North America, 1600–1854. Spain considered all of North America in its possession. Dashed lines show the origin and direction of Spanish explorations of the continent.

New Mexico was not initially settled to create a buffer against encroachment, instead it was colonized to christianize the Indians. For the most part, however, new settlements throughout Spain's northern frontier during the 17th century were constructed primarily for defensive reasons. Even during the 18th century California was not occupied for economic reasons, but rather to thwart Russian expansion southward along the Pacific coast.

Spain's strategy was to protect the more heavily settled areas of the Caribbean and central Mexico from foreign intrusion by using the area north of the Gulf coast and west of the Mississippi River as a buffer zone. After the French arrived in force at the mouth of the Mississippi during the 1720s, Spain retreated and turned Texas into a buffer province. By 1750, the geopolitical maneuverings between Spain, England, and France began to have an effect on the Spanish Borderlands, causing Spain steadily to lose territory.

To carry out its settlement strategy, Spain employed three frontier institutions: the mission, the presidio and the pueblo. The missionary and the military were the primary means by which settlement was achieved, with small civil colonies established later. Short of both manpower and civilian colonists, Spain depended upon a settlement strategy that absorbed the indigenous population. To effect settlement, Spain employed a network of Catholic mission stations that were to convert the local Indians to Christianity and teach them to become loyal Spanish subjects (Fig. 2.2).[1] The type of mission most frequently used was

Figure 2.2 Mission Concepción, just southeast of San Antonio, Texas, built between 1731 and 1755, is the oldest unrestored church still used for religious services in the United States. Twin towers and an elaborately carved entrance typify Spanish mission architecture in the Borderlands.

the *reduccion* or *congregracion*. Its purpose was to attract natives who lived in small, dispersed villages, congregate them in the mission, and "reduce" them from their heathen way of life to that of Christians. After they had been successfully weaned off their native culture, the mission was to be turned over to secular clergy, with the missionaries moving on to another frontier to repeat the process. At one time Spanish missions extended from Florida and Georgia through Texas, to New Mexico and Arizona and into California. Today the remnants of these early missions serve as one of the most visible landscape elements of Spanish occupancy.

Presidios formed the defensive arm of Spanish settlement. As agents of the government, they were responsible for defending the area, subduing hostile Indians, maintaining peaceful relationships with friendly Indians, and acting as the secular authority until a civil government could be established. Presidios were scattered along a wide arc extending from Georgia and Florida on the Atlantic coast to four strung along the California coast (Figs 2.3 & 2.4).

Pueblos – civil communities – were usually a later addition to the Spanish colonization scheme, after missionary efforts were completed. They were established to supplement the soldiers with agricultural products, engage in trade when feasible, set examples of Spanish life for the Indians to follow, and, in times of emergency, to act as a reserve militia for the military (Fig. 2.4).

The ultimate goal of the presidio-mission-pueblo settlement strategy was to ensure Spain's claim to a vast area extending from Florida to California. The choice of settlement sites, therefore, was an important consideration and in large measure was predetermined by the specific rôle each institution played out on the frontier. As military outposts, the presidios were located in areas that would provide maximum advantage against foreign

Figure 2.3 Presidio at Santa Barbara, California, founded 1781, as seen in a 19th-century lithograph. Mission Santa Barbara can be seen at some distance from the military town.

intrusions and hostile Indian attacks. In contrast, pueblos were founded with an eye toward permanent settlement and agricultural development. Mission sites were no less planned than the presidio and pueblo but were more flexible in their location. Missions were found primarily in areas that contained large numbers of Indians and were allowed to take up and use as much land as was necessary to care properly for Indian neophytes, or converts. So the missions were able to take advantage of good sites and Indian labor to expand into large, well-developed settlements (Fig. 2.5).

Shaping the borders

Spanish settlement was mainly for defensive purposes and thus institutionally organized. Individualism was not encouraged in Spain's settlements as it was on the American frontier. [. . .]

As the American frontier moved farther west it ran into an uneven but nevertheless defined line of occupation that stretched from Texas through New Mexico and Arizona to California. These areas were the Spanish Borderlands, the outer rim of Spanish colonization, containing a population of almost 100,000. The Borderlands, however,

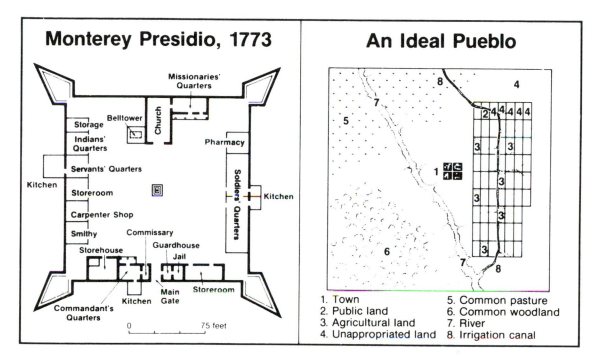

Figure 2.4 The building plan for the presidio of Monterey, founded in 1770, shows the internal arrangement of Spain's military fortresses in North America. The idealized layout of a pueblo is drawn from the evidence of the pueblo of Los Angeles, founded in 1782.

were not a wilderness; rivers had been mapped, towns founded, roads completed, agriculture developed, and trade routes established. [. . .]

Spanish legacy

Half of the land in the present-day contiguous United States was once under Spanish control, and the most recognizable area of Spanish influence is that extending from Texas to California. Here social, cultural, economic, and legal institutions derived from Spain remain a part of everyday life. The irrigation systems of both the small market gardeners of New Mexico and the large corporate farmers of California share in a common water-rights system that is a thinly disguised copy of Spanish water law. It was in the Spanish Borderlands

that Indian and Spanish culture came together, mingled and established a new pattern, a pattern that is only slightly altered today in many parts of the Southwest. The irregular land ownership patterns throughout the Borderlands remain as evidence of Spanish land tenure (Fig. 2.6). Spanish names of rivers, mountains, towns, and cities are the enduring witness in modern times to Spanish exploration and settlement that took place many centuries ago. [. . .]

One of the most obvious remains of Spanish occupation in the landscape is her architecture. The oldest standing dwelling today in the United States is not in Boston or Virginia but in Santa Fe, New Mexico. In addition to Santa Fe's historic buildings there is a trail of what were originally Spanish outposts composed of civic buildings, houses, missions,

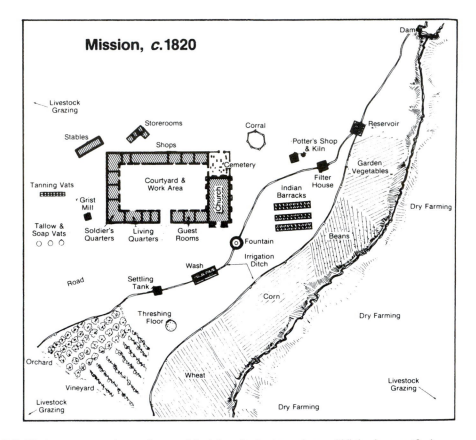

Figure 2.5 Missions were an integral part of Spain's colonization scheme. While the specific layout and design of each mission was different, the overall patterns were similar.

and military fortresses extending from Florida to California, whose construction was perfectly adapted to the climatic conditions of each region. Unlike their English counterparts, Spanish settlers did not disdain aboriginal architecture, but rather strove to mingle and assimilate all that could be used to good account, leaving us today a blend of Spanish and aboriginal buildings that are distinctive in their artistic design. Nowhere is this more evident than in the Spanish mission ruins of Texas, New Mexico, Arizona, and California (Fig. 2.7).

One of the more underplayed and least

noticed legacies of Spain in North America is her impact on modern urban patterns. As suggested earlier, Spain employed institutions to occupy new areas and peopled its land with three types of communities. Today, many of these communities have taken root and become major cities along the Gulf coast and throughout the Southwest. The major cities of New Mexico and Arizona were built upon Spanish foundations. Nowhere in the Spanish Borderlands, however, has Spanish settlement had a greater impact on the urban structure than in California. To settle and occupy that state, Spain established 21 missions, four pre-

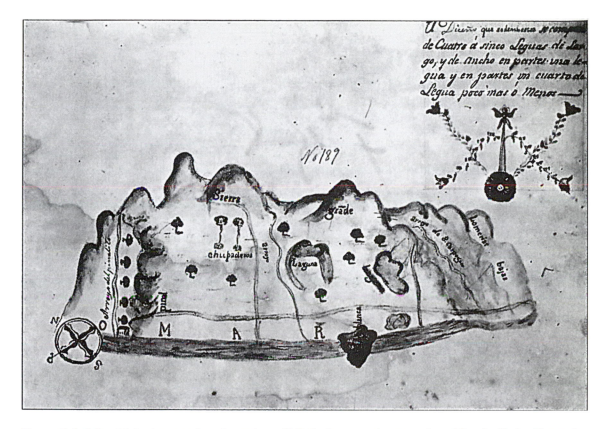

Figure 2.6 A Spanish land concession, shown in an 1840 *diseño*, or crude estate plan, of Rancho Piedra Blanca, San Luis Obispo County, California. Each citizen requesting land had to prepare a sketch map, depicting the area requested. Such vernacular cartography produced the earliest maps of California.

sidios, and three pueblos along the California coast. Today, 72 percent of the state's population live in one of the 28 sites originally founded by Spain. Within these cities, many street names, roads, boundaries, neighborhoods, the orientation of street patterns, water rights, and land tenure are of Spanish origin, to the surprise of many who reside there.

[. . .]

Note

1 Bolton 1917

Reference

BOLTON, H.E. (1917) "The Mission as a Frontier Institution in the Spanish-American Colonies", *American Historical Review* 23, pp. 42–61

Figure 2.7 Mission Santa Barbara in 1895. Founded in 1786 by the Franciscan order, this is the only mission in California continuously occupied since it was founded. The mission is considered the "Queen of California Missions" due to its distinctive architectural style.

3

THE LAWS OF THE INDIES

by John W. Reps

Source: John W. Reps, *The Making of Urban America: a history of city planning in the United States*, Princeton, NJ, Princeton University Press, 1965, pp. 29–32.

[. . .]

Philip II proclaimed the royal ordinances governing the planning of new cities on July 3, 1573, but the real author of these regulations is not known. [. . .] They begin, as might be expected, with the selection of a suitable site. One on an elevation surrounded by good farming land and with a good water supply and available fuel and timber was favored. The plan was to be decided upon before any construction, and it was to be ample in scope. As the regulations stated:

> "The plan of the place, with its squares, streets and building lots is to be outlined by means of measuring by cord and ruler, beginning with the main square from which streets are to run to the gates and principal roads and leaving sufficient open space so that even if the town grows it can always spread in a symmetrical manner."[1]

Several paragraphs of the regulations dealt with the plaza, that distinctive element in all Spanish-American towns. For coastal cities the regulations prescribed a location for the plaza near the shore; for inland cities, in the center of the town. As to shape, the regulations specified that the length should be at least one and a half times the width since "this proportion is the best for festivals in which horses are used. . . ." Planners were instructed to consider the eventual size of the town in deciding on the dimensions of the main plaza. [. . .]

The main plaza was to be oriented so that its four corners pointed to the four cardinal points of the compass. This feature was designed to prevent exposure "to the four principal winds," which would otherwise result in "much inconvenience." In other parts of the towns smaller, "well proportioned" open spaces were to be provided as sites for churches and other religious buildings.

From the main plaza principal streets were to lead from the middle of each side, with two minor streets also diverging from each corner. The regulations called for another distinctive aspect of Spanish colonial towns:

> "The whole plaza and the four main streets diverging from it shall have arcades, for these are a great convenience for those who resort thither for trade. The eight streets which run into the plaza at its four corners are to do so freely without being obstructed by the arcades of the plaza. These arcades are to end at the corners in such a way that the sidewalks of the streets can evenly join those of the plaza."[2]

Other streets were to be located "consecutively around the plaza," and, although nowhere do the regulations so state, it is obvious that the laws envisaged a gridiron or checkerboard pattern of straight streets with intersections at right angles.

The regulations provided precise guides for the location of the important buildings of the town. The main church of a coastal city was to face on the plaza and set near the harbor, so constructed that it might be used as a defensive fortification in the event of attack. In inland towns, however, the church was to be at a distance from the plaza, separate from other buildings and if possible on an elevated site. Other sites around the plaza were to be assigned for the town hall, the customs house, arsenal, a hospital, and other public buildings. Remaining sites on the plaza were to be allotted for shops and dwellings for merchants.

As for the location of ordinary dwelling sites and for future expansion, the regulations stated:

> "The remaining building lots shall be distributed by lottery to those of the settlers who are entitled to build around the main plaza. Those left over are to be held for us to grant settlers who may come later or to dispose of at our pleasure."[3]

After the drawing for lots the regulations provided that a settler should erect a tent or temporary hut on his site and then join in the construction of a palisade around the plaza for immediate safety against Indian attack.

The town itself was to be but one element in the settlement unit. Surrounding the central or urban core,

> "A common shall be assigned to each town, of adequate size so that even though it should grow greatly there would always be sufficient space for its inhabitants to find recreation and for cattle to pasture without encroaching upon private property."[4]

Beyond the common were to be located the agricultural lands. There were to be as many parcels as there were town lots, and these were also to be distributed by drawing lots. If within the community territory there were lands capable of being irrigated these were also to be subdivided into farming tracts and similarly distributed. Remaining farm land was reserved to the crown for distribution to settlers who might come at a later time. [. . .]

These regulations stand out as one of the most important documents in the history of urban development. The fact that almost without exception they were followed in the construction of so many towns throughout the

Spanish possessions in the Americas makes them doubly significant. [. . .]

Many of the regulations doubtless grew out of the experience gained in the first city planning efforts by the Spanish. [. . .]

[T]wo comprehensive treatises on town planning and civil architecture existed before the beginning of Spanish colonization. The earliest and probably best known was the *Ten Books on Architecture* of Vitruvius, written about 30 BC and rediscovered early in the fifteenth century. On many points the regulations in the Laws of the Indies closely resemble the principles of city planning established by Vitruvius. To select but a few examples, Vitruvius suggested that the forum should be rectangular, that its dimensions should be in the ratio of three to two; the forum should be located near the harbor if in a coastal city or in the center of the town if inland; public buildings should be given sites around the forum; and religious buildings should be located on an elevation. The similarity is unmistakable, and we have every reason to believe from internal evidence that the Spanish planners and colonial administrators drew heavily on Vitruvius in formulating their own regulations for town development.[5]

Another possible source, and one closer in time to the period of Spanish colonization, was the great work on architecture by Alberti. In addition to his own recommendations for the planning of towns and the siting and design of buildings, Alberti's work contained a summary of the suggestions of Vitruvius and other Roman and Greek authorities on the main elements of civic design. Alberti also advocated a gridiron plan with straight streets and right-angle intersections and that certain noxious uses, like tanneries, be located apart and downwind from the city. Both points find their counterparts in the Spanish regulations.

There were in addition actual examples of city planning on which the Spanish might have based some of their regulations. In Spain itself were many Roman colonial cities planted on the Iberian Peninsula centuries before. Santa Fé, the siege town laid out near Granada, and the *bastide* type communities in northern Spain may also have been partial models. North of the Pyrenees in southern France lay dozens of planned communities. Most of them were planned on a gridiron pattern with central colonnaded squares on which the church and other principal buildings fronted. [. . .]

Then, too, almost on the scale of complete towns were the great monastic complexes throughout Europe. Many of them were planned on roughly rectangular lines and usually included a colonnaded cloister surrounding some open court or green. [. . .]

The Spanish also were aware of the results of the Renaissance in Italy. [. . .]

Finally, in view of the strong military character of most of the early Spanish settlements, the influence of writers on castrametation should not be underestimated. [. . .]

What has been said about the possible origins of the city planning principles given legal form in the Laws of the Indies indicates that many diverse sources were available during the development of Spanish colonial policy. The history of ideas is never a simple matter, particularly when dealing with a series of events so distant in time and so meagerly documented. One fact is clear: the Laws of the Indies represented a unique combination of town planning doctrines and prescribed practices. They stemmed from a number of diverse European sources, modified by experience in the islands and on the mainland of Hispanic-America, and represent the best efforts by Spanish authorities to provide detailed guides for the founders of future colonial towns. [. . .]

Notes

1 The translation used throughout is from Zelia Nuttall, "Royal Ordinances Concerning the Laying Out of New Towns," *The Hispanic American Historical Review*, v, 1922, 249–54. The original document is Archivo Nacional Ms. 3017, "Bulas y Cedulas para el Gobierno de las Indias"

2 *Ibid.*

3 *Ibid.*

4 *Ibid.*

5 Vitruvius, *The Ten Books on Architecture*, translated and edited by Morris Hicky Morgan, Cambridge, 1926, p. 27

4

ST. AUGUSTINE, FLORIDA

by John W. Reps

Source: John W. Reps, *The Making of Urban America: a history of city planning in the United States*, Princeton, NJ, Princeton University Press, 1965, pp. 32–35

[. . .]

Earlier efforts to establish mission settlements in Florida, Georgia, and the Carolinas had been only partially successful. [. . .] Spain was soon spurred to action by the unwelcome news that a French expedition [. . .] had sailed in 1561 to establish a fortress colony somewhere along the southeast coast of the American continent. Although this colony did not succeed, a second party set out in 1564 [. . .] . Not only did this group represent a rival European power but the French expedition consisted of Huguenots, thus posing a double threat to Catholic Spain. [. . .]

Philip II began a series of negotiations with the French in an effort to bring about the withdrawal of the colony from territory claimed by the Spanish; when these broke down he ordered Don Pedro Menéndez de Avilés, an experienced soldier and skilled mariner to remove the French and establish a Spanish base on the Florida coast. In return for military and trading privileges in Florida, Menéndez was to transport five hundred colonists, including a hundred soldiers and an equal number of sailors; three hundred artisans, craftsmen, and laborers, two hundred of which were to be married; four Jesuits and ten or twelve monks of some other order; and five hundred slaves for the purpose of establishing at least two towns and appropriate fortifications. In July 1565 the fleet sailed from Cadiz, and after calling in Caribbean ports for supplies arrived at what is now the site of our earliest city on St. Augustine's day, August 28. [. . .]

Plans of St. Augustine showing the city soon after its founding provide only limited details of its plan. The map reproduced in Figure 4.1 shows the harbor, fort, and town at the time of Francis Drake's attack in 1586. The fort was located to command the narrow entrance to the harbor from the sea. The town itself may be seen in the upper left corner as a little gridiron settlement of eleven blocks of various sizes. The open space where the twelfth block might be expected may indicate the plaza, which, according to the Laws of the Indies specifications for seaport towns, would have opened to the water.

A more detailed plan of the town nearly two centuries later appears in Figure 4.2. The generally rectangular street system is readily apparent, and, in the size and location of the plaza, the city generally conforms to the regulations established by the Laws of the Indies. But many irregularities in the street alignment and size of blocks appear, indicating either a more casual plan or laxity in guiding development not exhibited in later colonial settlements. Menéndez, in 1565, was of course not guided by the Laws of the Indies, and he had evidently been given almost complete freedom in deciding on the plan of his settlement.

One feature that can be noticed on the plan is the open character of the community.

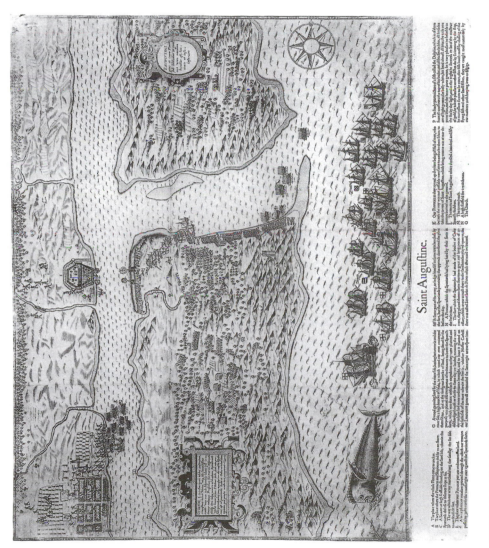

Figure 4.1 View of the harbor, fort and town of St. Augustine, Florida, 1586.

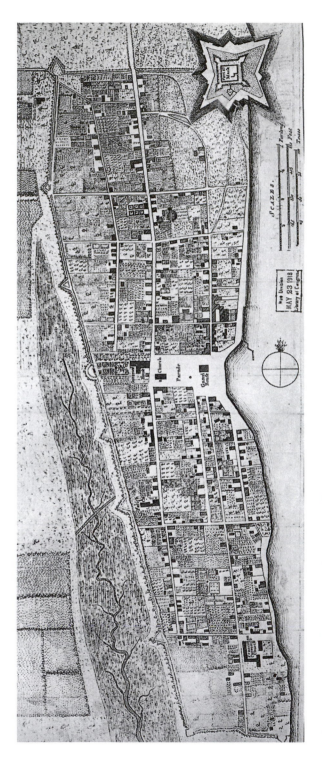

Figure 4.2 Plan of St. Augustine, Florida, c. 1770.

Houses fronted directly on the streets, leaving generous areas to the rear for lawns and gardens. As in most of the Spanish colonial towns, houses usually rose two stories high. The roofs were generally flat. Balconies at the upper levels over porticoed entrances added interest to the street façades.

Town life focused on the plaza. At its upper end stood the governor's house with a balcony on the front and galleries along its sides. The main church and guardhouse were on the plaza itself, while other churches were located along the principal streets. At the northern edge of the city stood the fortress surrounded by a ditch and earth bastions. The entire community was enclosed by a palisade, strengthened at intervals by watchtowers and minor fortifications.

St. Augustine combined three distinct functions in one community. The city was, first of all, a military post with its fort and military garrison. Secondly, it was designed as a civil settlement for trade, farming, and handicraft industry. And, finally, it was intended as a center from which religious orders would begin the work of converting the Indians to Christianity. Toward the end of the sixteenth century it became the policy of Spain to establish separate settlements for these three types of activity. [. . .]

5

MISSION, PRESIDIO AND PUEBLO *IN CALIFORNIA*

by John W. Reps

Source: John W. Reps, *The Making of Urban America: a history of city planning in the United States*, Princeton, NJ, Princeton University Press, 1965, pp. 46-54

Although Spanish explorers from Mexico had ventured north along the California coast as early as 1542, when Cabrillo entered San Diego Bay, attempts at settlement were not made until more than two hundred years later. In 1769 two expeditions, one by land and the other by sea, set out to establish three missions and two *presidios*. Thus began the final chapter of Spanish community planning within the area of the United States.

The first mission was founded at San Diego in July, the beginning of a chain of religious settlements that eventually stretched northward beyond San Francisco Bay as far as Sonoma. The mission communities followed the well-known pattern established over the years in New Spain. They are described by Eugène Duflot de Mofras, sent in 1840 by the French to report on conditions in California, then under Mexican rule. De Mofras pointed out that all of the missions resembled one another closely. He chose to describe one of the last to be founded and perhaps the most impressive – San Luis Rey, established about fifty miles north of San Diego and a few miles inland in 1798 [Figure 5.1]. [. . .]

"Mission San Luis Rey de Francia is built in the form of a quadrangle, 150 meters in width, with a chapel occupying one of the wings. Along the façade extends an ornamental cloister. The building, which is one story high, is raised a few feet above the ground. The interior, in the form of a court, is adorned with fountains and planted with trees. Off the surrounding cloister open doors lead into rooms occupied by priests, majordomos, and travelers, as well as the main rooms, schools, and shops. Infirmaries for men and women are placed in a secluded corner of the mission. Nearby the school is situated.

Surrounding the mission are the workshops, the huts of the neophytes, and the houses of a few white settlers. In addition to the main buildings, 15 or 20 subsidiary farms and some auxiliary chapels lie within a radius of 30 or 40 square leagues. Across from the mission are the quarters of the priests' bodyguard, an escort consisting of four cavalrymen and a sergeant."[1]

The Spanish government expected these Indian mission communities to become self-governing within a few years. This hope was ill-founded. The Indians failed to adjust to this changed mode of life, often they rebelled against what in reality was a system of conscript labor, and there is substantial evidence that the religious authorities resisted strongly the attempts by civil officials to secularize the missions. Not until 1833, by which time California had passed to Mexico, was a general decree passed for the secularization of the missions. Actual accomplishment of this order took several years. The results were quite different from what had been hoped. In most cases the Indians were quickly dispossessed of both lands and political power by white

Figure 5.1 View of the mission of San Luis Rey, California, 1840.

settlers from both Mexico and the United States.

Four California *presidios* or military communities were also founded: San Diego in 1769, Monterey in 1770, San Francisco in 1776, and Santa Barbara in 1782. These, too, were essentially similar in design, and de Mofras again furnishes the details:

> "*Presidios* were invariably built in the following uniform manner: After a suitable place had been chosen, a trench about 4 meters broad and 2 deep was then excavated. What earth was removed was used for an outer embankment. The presidio was then enclosed by a quadrangle that measured approximately 200 meters on each side. The rampart, or wall, constructed of adobes, was 4 or 5 meters high and a meter thick, with small bastions at each corner. The presidio had only two gates. These fortifications were never protected by more than eight bronze cannon, usually of 8, 12, or 16 pounds. Although incapable of resisting a serious attack by warships, these fortifications were adequate to repulse Indian raids. Not far from the presidios, at a point selected to conform with the local topography, stood the outpost batteries, inappropriately designated as the *castillo*, or fort. Within the presidio, the church, barracks for officers and soldiers, the houses of a few settlers, and the stores, shops, stables, wells, and cisterns were provided. Outside stood more groups of houses."[2]

Figure 5.2 shows the *presidio* of San Francisco as it existed in 1820. [. . .] The plan [. . .] is misleading in one respect. It does not show the houses and farm structures that sprang up beyond the walls of the *presidio* proper. These features of civil life, as has already been mentioned, tended to give these military communities a closer resemblance to the purely civil towns. [. . .]

The missions and *presidios* were numerically more important in California, but the *pueblo* towns and their land system are of greater interest. The California *pueblos* owe their founding to Philip de Neve, governor of the province. In 1775 the Spanish king ordered him to take up residence in Monterey. De Neve faced two problems: securing the province against encroachments by other powers and increasing the supply of farm produce for the military garrisons then supplied with wheat and other foodstuffs from Lower California and Mexico. He reasoned that further colonization of the province by Spaniards in civil communities would assist in solving both problems. [. . .]

Evidently Neve hoped to found a number of such communities as part of his general policy of stimulating colonization. In these regulations Neve provided for increased assistance to prospective settlers in the form of agricultural equipment, an annual payment for five years, and sufficient livestock to begin farming operations. In return the settlers were to repay the government for the tools and livestock in horses and mules, and the surplus food produced by the *pueblo* was to be sold to the *presidios*. [. . .]

Under Spanish, and, later, Mexican rule, the *pueblos* remained small and relatively unimportant settlements. A view of Los Angeles in 1853 is reproduced in Figure 5.3 and shows the condition of the town as it must have looked within a few years of its founding. [. . .]

The Spanish colonial *pueblo* was more than a town; it was intended as a self-contained urban–rural unit. The *pueblo* lands generally were square, ten thousand *varas* or five and a quarter miles, on each side. The town proper occupied a site in or near the middle of this tract. Here each of the original settlers, or *pobladores*, had a house lot, or *solar*. Farming plots, or *suertes*, were laid out in rectangular fields beyond the town proper and were allotted to each settler. Apparently settlers did not receive absolute title to their lands, holding them instead in perpetuity subject to prescribed duties of cultivation. Nor could lands be sold. In this respect the *suertes* closely resembled the common fields of the New England communities.

Certain farm tracts were reserved to the king, the *realengas*, and were to be used in

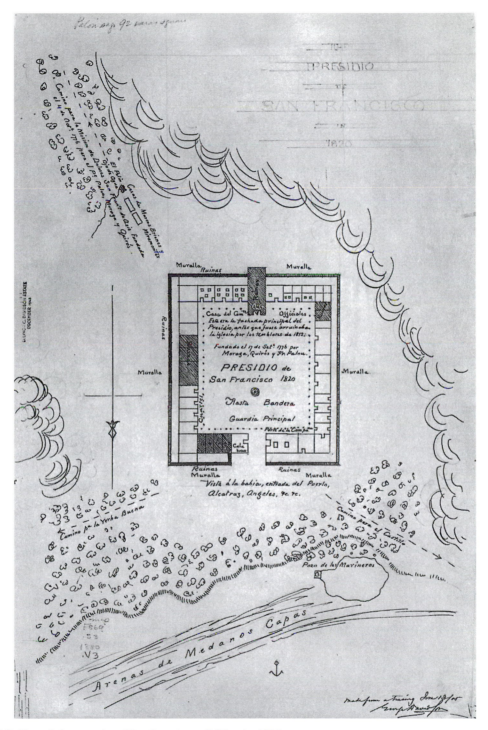

Figure 5.2 Plan of the *presidio* of San Francisco, California, 1820.

Figure 5.3 View of Los Angeles, California, 1853.

making grants to later settlers. Other farm tracts, called *propios*, remained *pueblo* property to be rented, with the income being devoted to community purposes. Common pasture lands and common woodlands were also set aside for general community purposes and not allotted to individuals. Finally, on one or more sides of the town, or completely surrounding it, were the *ejidos*. These also were common lands not under individual jurisdiction and apparently were for the purpose of permitting the enlargement of the town should additional streets and *solares* be needed.

The Spanish *pueblo* pattern of land division closely resembles that of the early New England towns. In the case of the English colonies this system of land distribution clearly stemmed from the conditions of land tenure in rural England at the time of American colonization. This in turn had slowly evolved from feudal land law. It seems a justifiable hypothesis that the Spanish *pueblo* land system similarly owed its origins to the pattern of feudal land holding in Spain. Indeed, throughout most of Western Europe these feudal patterns were common. In America the municipality, Spanish *pueblo* or New England town, replaced the feudal lord. It was, in fact, a fairly sophisticated concept of collective ownership and communal land management that guided the early years of these relatively simple communities. [. . .]

Notes

1 Eugène Duflot de Mofras, *Exploration du Territoire de L'Oregon, des Californies et de la mer Vermeille* (Paris, 1844), translated and edited by Marguerite Eyer Wilbur, *Duflot de Mofras' Travels on the Pacific Coast*, Santa Ana, 1937, 1, 134–35

2 *Ibid.*, 142–43

6

FRENCH LANDSCAPES IN NORTH AMERICA

by Cole Harris

Source: Cole Harris, ''French Landscapes in North America'', in M.P. Conzen (ed.), *The Making of the American Landscape*, London, Routledge, 1990, pp. 63–79

[. . .]

From early in the 16th century some [French] fishermen traded with natives for furs. Late in the century a few ships began to be outfitted expressly for the fur trade. As this happened the focus of the fur trade shifted westward toward the St. Lawrence River, the principal conduit for the furs of the interior. In 1600 fur traders overwintered at Tadoussac at the western end of the Gulf of St. Lawrence; eight years later another group (led by Samuel de Champlain) established a post at Québec, the head of deep sea navigation on the St. Lawrence River. This time the French were on the St. Lawrence to stay. Trois-Rivières was established in 1634. Montréal, founded as a mission in 1642, soon became the most interior outpost of the fur trade. In these years French traders did not venture beyond the St. Lawrence Valley; the fur trade was in the hands of their Indian allies. [. . .]

By the 1650s [. . .] French traders themselves began to venture inland, in the process mastering the birchbark canoe, learning native languages, and, in 1670, building the first trading post west of Montréal – Fort-de-la-Baie-des-Puants on Green Bay. In this interior world of shifting military and trading alliances and declining local supplies of beaver, canoe and fur post facilitated the remarkable territorial expansion of French commerce. Before the end of the 17th century there were French posts on each of the Great Lakes, along the Illinois and upper Mississippi Rivers, on Lake Nepegon north of Lake Superior, and even on James Bay where the French captured posts built by the Hudson's Bay Company (Fig. 6.1). [. . .] [B]y the 1730s there were French trading posts as far west as the lower Missouri and Lake Winnipeg.

The fur post was a palisaded, frequently garrisoned settlement in native territory. The largest – Fort Detroit and Michilimackinac – were entrepôts laid out in a grid of streets and defended by cannon mounted in small angled towers at the corners of curtain walls (Fig. 6.2). The smallest, comprising a few buildings surrounded by a palisade some 12 feet high, could be constructed in a few weeks to provide minimal accommodation for a few traders and soldiers overwintering among potentially hostile natives. White women were absent at such posts, and the traders themselves would leave after a year or two, not necessarily to be replaced. The fur post was, characteristically, an ephemeral outlier of French commerce and the French military, built to house and protect trade goods and personnel, a point of contact between native and European worlds in the wilderness. Wooden palisades and buildings made of squared timbers laid horizontally and

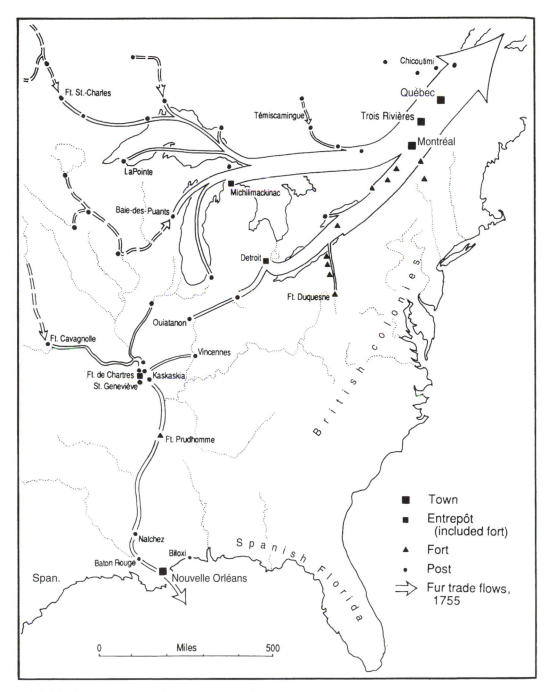

Figure 6.1 The French arc of settlement in North America in about 1755. The fur trade linked the web of settlements together. Trading posts were the most far-flung sites of French presence, guarded by forts in areas contested by the British. The Illinois country served as a breadbasket for many western operations, and the chief towns developed at the outflows of the St. Lawrence and the Mississippi Rivers.

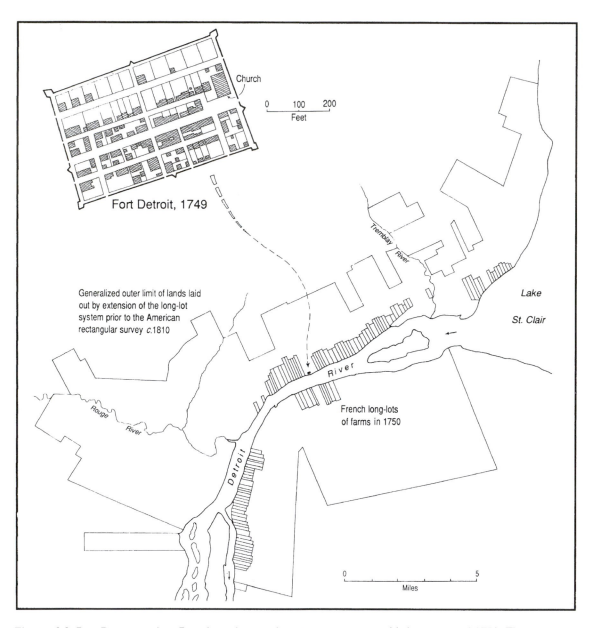

Figure 6.2 Fort Detroit and its French settlement district in easternmost Michigan around 1750. The town developed as a compact unit, but did not survive American takeover, which produced a grandiose new plan for the city of Detroit centered several hundred yards to the east. The rural long lots endured, however, and with their pre-American extensions created a framework that still controls the land parcel pattern of central Detroit and the adjoining city of Windsor, Ontario.

tenoned to posts at the corners and at intervals along the walls had not been used for centuries in military construction in France. [. . .]

The French crescent: St. Lawrence to the Mississippi

All these French settlements in North America developed within the context of the larger military struggle between France and England for control of a continent. In this regard the Treaty of Utrecht (1713), which ended the long Anglo-French hostilities known as the War of the Spanish Succession, was calamitous for France in North America. The Treaty confirmed English title to much of Acadia, ceded Newfoundland (France retained fishing rights in the north), and returned the forts on Hudson Bay. France had bargained for European advantage with North American territory. In the aftermath of the Treaty of Utrecht, France sought to strengthen her diminished North American position by building a massive fortress town, Louisbourg, on Cape Breton Island at the entrance to the Gulf of St. Lawrence; and by encouraging trade and settlement along the Mississippi. It was hoped that a crescent of French power from the Gulf of St. Lawrence to the Gulf of Mexico might contain the British east of the Appalachians. [. . .]

The year in which France decided to fortify Louisbourg (1717) she moved to strengthen her hold on the Mississippi Valley by granting a merchant company title to Louisiana and a trading monopoly for 25 years. The company was to establish 6,000 free settlers and 3,000 slaves. The next year the company founded New Orleans. [. . .]

New Orleans, like Louisbourg, was laid out in a rectangular grid of streets and, like Montréal, was walled on three sides. As local stone was not available, most buildings were of timber frame construction with brick infill. Otherwise, New Orleans looked much like a smaller version of the other French towns in North America. [. . .]

On the eve of the Seven Years' War (1757–1763) the French claim to North America extended from Labrador to Texas, including the Gulf of St. Lawrence and St. Lawrence Valley, the Great Lakes, the whole drainage basin of the Mississippi, and, except for a rim of land acknowledged to be British, most of the territory draining into Hudson Bay. Britain also claimed the Hudson Bay drainage, the eastern Great Lakes, and the Ohio Valley. In fact most of this enormous territory was still controlled by natives. French claims, advanced against British counterclaims, had a cartographic and geopolitical vitality they did not have on the ground. Nevertheless, the French fur trade operated through much of the continent, the French fishery to Newfoundland and Labrador was 250 years old, and there were widely distributed patches of permanent French settlement: some 13,000 people, by the early 1750s, on or near the marshlands around the Bay of Fundy; some 5,000 or more on Cape Breton Island; just over 60,000 along the lower St. Lawrence; some 2,000 (including slaves) in the Illinois country; 1,000 scattered in dozens of fur posts; and perhaps 6,000 (including slaves) along the lower Mississippi.

These were not many settlers to hold the larger portion of a continent. There were several colonial jurisdictions: Cape Breton Island, what remained of Acadia, Canada, and Louisiana. There were several unrelated export economies: the fishery, the fur trade, and the various trades of the Mississippi. There were several isolated regional cultures. Canadians and Acadians, descended from different immigrant stocks, lived in different northern agricultural niches, and after a time were different peoples. Most of the settlers in the Illinois country had come from Canada but, on the edge of the prairie and the plantation economy, were no longer Canadian habitants. The subtropical lower Mississippi was another realm,

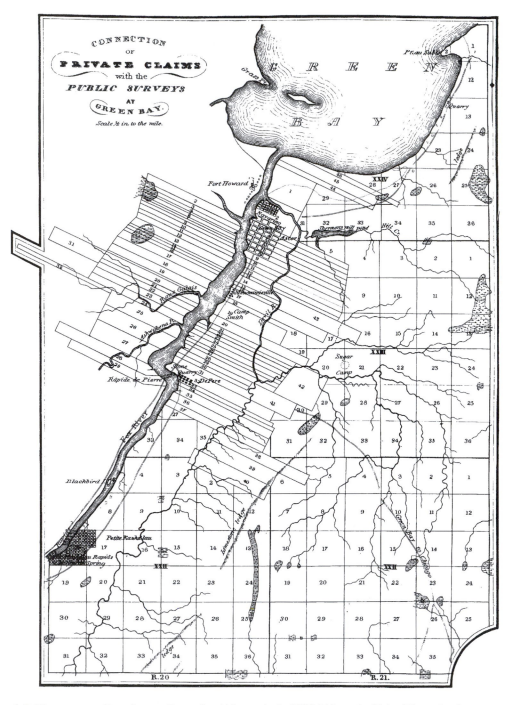

Figure 6.3 The pattern of long lots at Green Bay, Wisconsin, in 1809. When the United States land surveys reached the area, they gridded all land not previously laid out. Authorities honored the long lots as existing "private claims," and their outlines became embedded in the subsequent evolving pattern of land ownership, still very evident today.

differing in settlement history, economy, and local cultures from any other patch of French settlement in North America. A more official France was superimposed on these scattered, varied settlements, but its impact focused on the towns and weakened rapidly away from them. The townscapes of Québec, Montréal, Louisbourg, and New Orleans all reflected the outreach of official France, whereas the rural landscapes of French North America revealed the dynamics of local cultures.

The legacy

During the Seven Years' War France lost almost all her North American territory. [. . .]

Today, the French imprint on the American landscape is most widely discernible in the distribution of French placenames. [. . .] French patterns of land division endure with remarkable clarity in the vicinity of major settlements, such as Green Bay [Figure 6.3], St. Louis, Vincennes, and Prairie du Chien, where later American land survey studiously avoided established claims. French town planning is most evident in the cities of the St. Lawrence Valley, and New Orleans within the United States, partly in street patterns both regular and irregular, and partly in building forms that contrast strongly with standard American styles. The French imprint in the United States is sparse, muted, and mostly blurred, but in a few localities, most notably along the Mississippi River, it stands in bold defiance of patterns of later American dominance that have nevertheless failed to erase it completely.

7

NEW ORLEANS

by John W. Reps

Source: John W. Reps, *The Making of Urban America: a history of city planning in the United States*, Princeton, NJ, Princeton University Press, 1965, pp. 81–5

[. . .]

Of all the French settlements in the United States, [. . .] it is mainly in New Orleans that substantial portions of the French legacy remain today. [. . .]

As early as 1718 Bienville sent a few emigrants from Canada to clear the site and to erect temporary buildings, but not until 1722 did the actual planning and town building begin. [. . .]

An early drawing of the Bienville–Pauger plan and the surrounding area appears in Figure 7.1. Although in its details of the city the drawing is inaccurate in some particulars and in its depiction of the countryside it is highly stylized rather than precise, it gives a good general picture of the new settlement. Charlevoix's comment about its difficulty of execution may well have been directed to the problems of construction in the low-lying, marshy terrain. The swamps and ponds that show on the map long caused severe health hazards, and the high water table still creates problems for building construction.

An accurate plan of the town forty years after its founding is shown in Figure 7.2. The focal point of the city is the *place d'armes*, the modern Jackson Square. As at Mobile, this appears at the water's edge. But this open square was plainly intended as more than a parade ground. On the inland side, facing the square and the river beyond, stood the principal church. To give this building further architectural promi-

nence the planners introduced an extra street dividing the central range of blocks terminating at the square. This strong axial treatment doubtless reflected the intentions of Bienville to create a capital city of beauty as well as utility. Stretching each way from the central square along the river ran the quay, broadening at either end where the river curved away from the town site.

As the capital city, New Orleans became the favored spot for French settlers, and it enjoyed a mild prosperity. Above and below the city on both sides of the Mississippi plantations were developed, laid out on the typical French pattern with relatively narrow river frontages and stretching back from the river for great distances. Yet even by 1797 we are told that not all of the land within the original city boundaries enclosed by its fortifications had been built on. [. . .]

New Orleans at this time was under Spanish rule, a period lasting from 1763 to 1801, and little was done to further the development of the city. France regained possession for a brief period, and then, with the Louisiana Purchase in 1803, American control began. It is from this time that the city began its steady expansion beyond the original boundaries, expansion that in its initial years at least was orderly and controlled.

Figure 7.3 shows the enlarged city in 1817. As in many European walled cities, when the

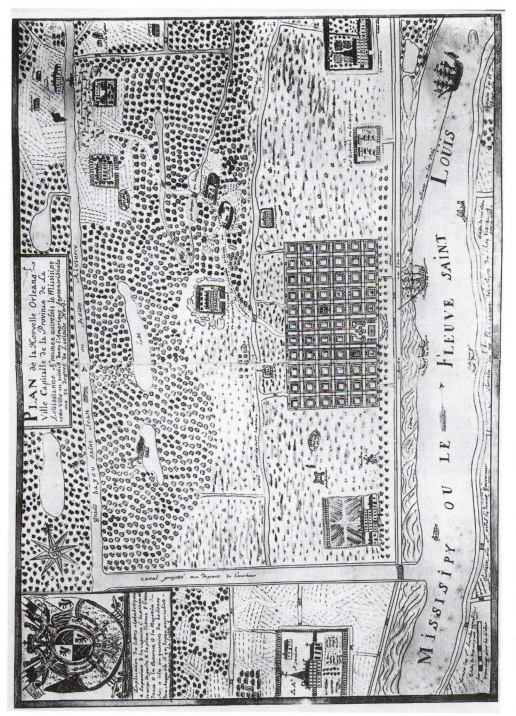

Figure 7.1 Plan of New Orleans, Louisiana, c. 1720.

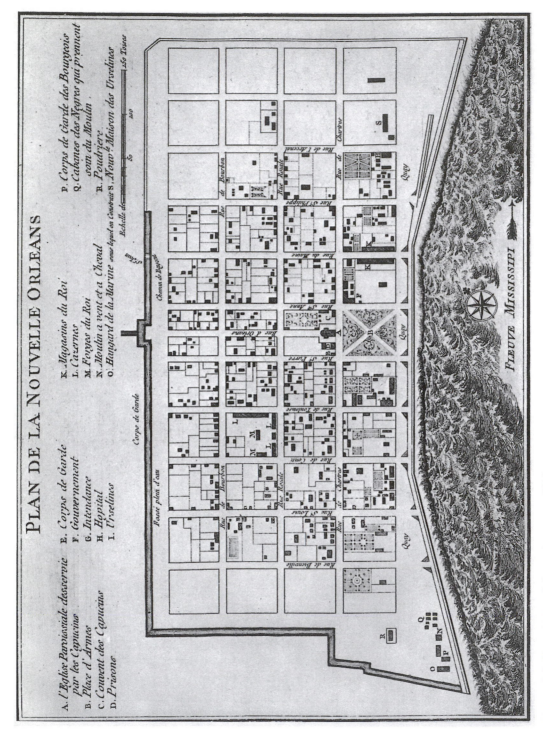

PLAN DE LA NOUVELLE ORLEANS

A. *l'Eglise Parroissiale desservie*
   par les Capucins
B. *Place d'Armes*
C. *Couvent des Capucins*
D. *Prisons*

E. *Corps de Garde*
F. *Gouvernement*
G. *Intendance*
H. *Hopital*
I. *Urselines*

K. *Magazins du Roi*
L. *Caserne*
M. *Forges du Roi*
N. *Moulin a vent et a Cheval*
O. *Bangard de la Marine vous lequel on construit S. Nouv.e Maison des Ursvelines*

P. *Corps de Garde des Bourgeois*
Q. *Cabanes des Negres qui prennent*
   soin du Moulin
R. *Poudriere.*

FLEUVE MISSISSIPI

Figure 7.2 Plan of New Orleans, Louisiana, 1764.

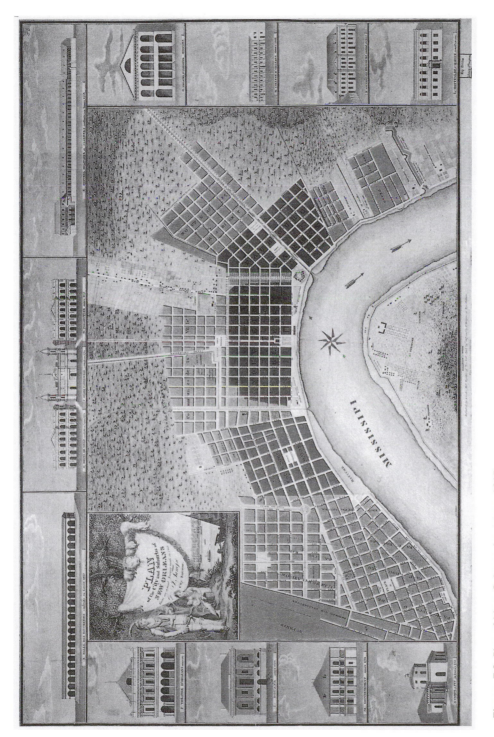

Figure 7.3 Plan of New Orleans, Louisiana, 1815.

old fortifications of New Orleans were pulled down broad boulevards replaced them: the Canal Street, North Rampart Street, and Esplanade Avenue of the present day. This ring established a pattern followed in other sections of the city. Above and below the old city new suburbs or "faubourgs" were laid out. The curving river bank suggested a new orientation for the grid plan of these city extensions, a change in street directions that could have been disastrous except for the skill with which these were connected to the streets of the old city. This was done with particular effectiveness downriver, where the Faubourg Marigny was joined to the city with great care so that only three streets rather than four come together at the angled intersections.

Good planning and the influence of topography combined in another way to fashion a street pattern of beauty and interest as well as one that functions well even with modern traffic loads. As the long, narrow plantations adjoining the expanding community were subdivided, each was laid out in a little gridiron strip. But, because the land back from the river was the least desirable owing to its location and its swampy character, the first subdivision of land occurred along the river. Each grid developed independently and rather slowly. Developers recognized the importance of short cross streets and arranged for connections across property lines.

New Orleans adopted one feature of its original plan, which was repeated at intervals as its boundaries expanded. This was the open square, often combined as a space opening off or terminating one of the numerous boulevards. [. . .] In this respect, although the city was under American rule, it remained European in its use of the urban square as a major element of planning and growth. [. . .]

8

THE NORTHEAST AND THE MAKING OF AMERICAN GEOGRAPHICAL HABITS

by Peirce F. Lewis

Source: Peirce F. Lewis, "The Northeast and the Making of American Geographical Habits," in M.P. Conzen (ed.), *The Making of the American Landscape*, London, Routledge, 1990, pp. 80–103

[. . .]

[. . .] [F]rom the days of earliest European settlement, it was in the Northeast that Americans formed some of their most persistent geographical habits.

Many of those habits had very tangible results, for over the course of time they came to be etched into the face of America's ordinary human landscape. Northeastern ideas would determine where cities would be situated and how their streets would be laid out. They would determine what ordinary houses would look like and how they would be placed in relation to streets and gardens. They would determine where roads would be built, and who would build them; where farmers would live and how they would design barns to house their crops and livestock; and a host of smaller matters. In concert, these ideas and habits would produce a set of ordinary human landscapes highly distinctive in appearance, in turn underlain by a set of ethical, esthetic, and even religious ideas about how humans should treat the land.

If these geographical ideas and habits had been restricted to the northeastern corner of the nation, they would be of little more than local interest today. The Northeast, after all, is only a small part of the United States. But the Northeast was the source from which most of the Midwest and West were eventually settled. In consequence, what began as a congeries of rather peculiar regional quirks was carried westward and ultimately stamped as standard patterns of human geography across an enormous part of the American nation (Fig. 8.1).

Originally, many of the basic precepts of organized society were not American at all, but started out English. America, after all, was English long before it was American – and for most of the 17th and 18th centuries, most transatlantic settlers were content that it should remain so. The name, New England, for example, was not chosen by accident, and it announced clearly that America was not intended to become a *new* world, but instead a new version of an old one. [. . .]

A different sort of place

But America was not merely a duplication of England. From the time of earliest settlement, American geographical behavior diverged sharply from that of England – in ways that often made America seem perverse, uncouth and eccentric – at least in the eyes of European spectators.

Much of this seeming eccentricity was a matter of plain necessity. Ways of managing land that had worked in the old country often did not work in America, and Americans quickly learned (sometimes the hard way) about the virtues of keeping an open mind,

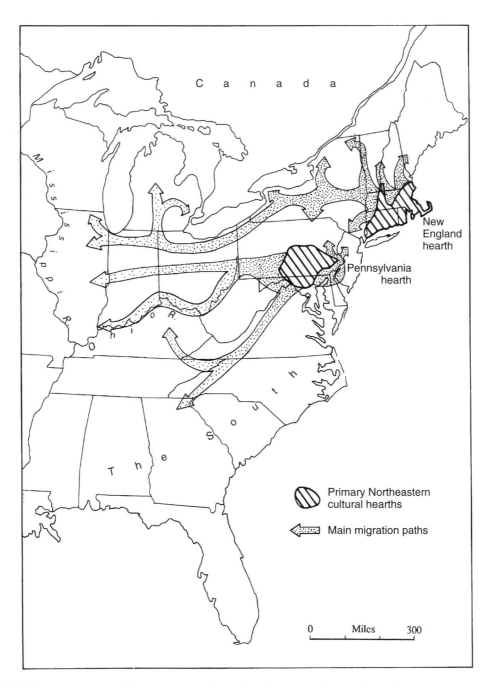

Figure 8.1 Westward spread of Northeastern cultures. New England culture originated from a broad stretch of the Atlantic coast, but as it spread westward was squeezed into a narrow corridor between the Adirondacks and the Catskills – hence via the Erie Canal along the southern edge of the Great Lakes. Pennsylvania, by contrast, started with one small foothold on the Delaware River, but spread westward in a broadening diffuse fan that covered much of the continental interior. Even in the flatlands of the Midwest, however, the two streams of eastern settlement remained quite separate from one another.

and abandoning traditional ways when the new geographical circumstances seemed to call for it.

Such constant experimentation did not always produce attractive results. [. . .] Bringing in a harvest [was] more important than making one's fields look pretty. If the timber was wasted and the land disfigured in the process, no matter. There was always more timber, more soil, more land – or so it seemed. [. . .] [E]conomics commonly took precedence over esthetics, especially in the early days. Unlike England, America was a big country, and it rewarded those who seized its riches quickly. But such ambitions did not make for a tidy landscape, and they did not encourage habits of geographical thrift.

Nor did they make for habits of permanence. For people who had already migrated once, there was always a propensity to migrate again – and yet again. It was all very well for Englishmen to have special attachments to special places – indeed, to take their names from the places where they and their ancestors had lived since the beginning of time. In England (indeed, in the Old World in general), one knew one's place, both socially and geographically. That was never the American way. Mobility – the willingness to abandon places when they had served a particular purpose – was the key to success, whether success was defined in economic or social terms. [. . .]

Two regions of the Northeast

But British North America was not a homogeneous place. Within a short time after initial settlement, major differences had begun to emerge along the northeastern seaboard of what would become the United States. Two quite different culture hearths had begun to evolve, which by the time of the Revolution had expanded to dominate the northern half of colonial America (Fig. 8.1). One was New England, a little theocracy settled by post-Eliza-bethan puritans, who had broken away from the Anglican church at precisely the time when Britain's religious wars were raging hottest. Not surprisingly, these New England puritans took religion seriously, and went to great pains to organize their landscape in a way that would ensure the continuity of their ideas, and the rigorous exclusion of folk who did not agree with them. [. . .]

But philosophy does not bake bread, and for all of New England's high-minded social aspirations, it immediately became obvious that New England was a meager land. The initial arrivals had expected to settle down and become farmers and, in the beginning, most of them did. Indeed, by the mid-19th century they had cleared the forests from all of southern New England and much of the mountainous north as well. But the climate was fierce, and except for the fertile bottomlands of the Connecticut River valley, soils were marginal at best, impossible at worst. [. . .]

Thus, ambitious New Englanders could choose one of several options. They could take to the sea for trading or fishing or whaling, and many of them did so in preference to grubbing stones from sterile fields. By the mid-19th century, New England ships were trading and whaling all over the world and bringing profit to dozens of colorful ports along the rockbound coast. Or, they could learn to manufacture things, and they did that too, considerably before most of America had thought of doing so. As a result. New England got a head start in all kinds of useful industries, and the region became a major center of America's industrial revolution. Industry was densely concentrated in places like Manchester, Lawrence, and Lowell, crowded along the Merrimac River [in New Hampshire and Massachusetts], where waterfalls generated power for spinning thread and weaving cloth. Along the north coast of Long Island Sound, Connecticut Yankees earned a world-wide reputation for manufacturing high-quality machined products, guns and locks and machine tools – useful and highly profitable things in a country like America that was expanding by leaps and bounds. Or, finally, a disgruntled Yankee farmer could simply pack up his family and chattels and go looking for better land west of the Appalachians. By the early 19th century, New Englanders were swarming westward across New York State, first by turnpikes, then by the Erie Canal, later still by way of the New York Central Railroad. Many New Englanders went as far as western New York's fertile Genesee County, liked what they saw, and stayed, ultimately converting upstate New York into an extension of New England. Others, still footloose, headed yet farther west along the southern shores of Lakes Erie and Michigan, and then fanned northward to convert the upper Great Lakes states into a vast Yankee preserve blanketed with Yankee houses, Yankee towns, and Yankee placenames. Even today, rural landscapes of Michigan and Wisconsin still have a very Yankee look to them, as do the northern parts (but not the southern) of Ohio, Indiana, and Illinois. By the end of the 19th century, this "Yankee Exodus," to use Stewart Holbrook's term, had almost depopulated most of rural New England;[1] by the middle of the 1900s, most of New England had reverted to forest. [. . .]

It is hardly surprising that New England was not an attractive place for non-Englishmen, and the region's population remained almost totally British in national origin until well into the 19th century. Only then did a second wave of migrants begin to arrive, chiefly Catholic Irish refugees from the potato famine of the 1840s and, starting in the last third of the 20th century, waves of Italians and Portuguese. Although all of these later migrants originally came from rural places in Europe, when they moved to New England the farmland was gone, and they consequently settled in the only places that jobs were available, cities like Boston, Providence, New Haven, Waterbury,

Fall River, and a host of others. By the end of the 19th century, New England had become an overwhelmingly urban place, an archipelago of hundreds of cities and towns, set down in a vast unbroken ocean of second-growth forest.

The Pennsylvania culture region

But there was another part of the northeastern United States, and it was a very different sort of place from New England. Across the Hudson River to the south and west lay Pennsylvania – or, more accurately, "the Pennsylvania culture region."[2] Like New England and the South, Pennsylvania is seen not as a political state but rather as a multistate region with a distinctive set of cultural traits and has exercised a potent and pervasive influence on the larger national culture, on a par in importance with New England and the South. For just as New England has powerfully flavored the upper Great Lakes region, Pennsylvania's influence spilled westward in a great swath that stretches across much of the nation's midriff (Fig. 8.1).

The character of Pennsylvania was indelibly stamped by the manner of its founding in 1682, when William Penn arrived with a band of English Quakers to create his new colony, and build de novo his city of Philadelphia. It was a lucky time to found a new colony, for England's fiercest religious wars were finally drawing to a close, and northwestern Europe was about to embark on the unknown seas of industrial revolution. The spirit of the times was changing, and there were opportunities for political and social experimentation that would have been unthinkable only a few years before. Penn made good use of these new opportunities, as he set about to prove that one could follow one's religious conscience, tolerate the religious views of others, and prosper economically at the same time. Penn's "Holy Experiment," therefore, started out with very different assumptions than did the early settlers in New England, where religious conformity was the order of the day, and social order was considered a higher virtue than human freedom. Pennsylvania, by contrast, would be a haven of religious diversity, but it would also be a business venture, to make money for Penn and his fellow investors, and for any settlers whom he could persuade to buy land from him.

[. . .] From 1700 onwards, migrants flooded to Pennsylvania through the new port of Philadelphia, soon to become the biggest city in North America and the largest English-speaking city in the world outside England itself. And then, around 1740, for the first time in the American colonies, settlers began arriving from the European continent, speaking languages other than English. Overwhelmingly, these new non-English migrants were German and Swiss pietists from the upper Rhine. By that time, however, the immediate outskirts of Philadelphia had already been occupied by immigrants from England and Wales, so the Germans leapfrogged beyond them to the west, and settled in the rich Piedmont land that stretches from Allentown to Reading to Lancaster to York, a region which today constitutes the heart of the "Pennsylvania Dutch" (*Deutsch*) country. By the time of the American Revolution, those of German immigrant stock came to number more than one third of Pennsylvania's population, and they turned Penn's "Holy Experiment" into the least English of all of Britain's Atlantic colonies. More than was true for any other of those colonies, however, the promise of Pennsylvania was a portent of America's promise – a place where the highest values were freedom, tolerance, and the ability to make money. It was a quite different set of values than motivated the New England puritans; values from a different period in English history applied to a different region of America. [. . .]

There were other major differences between Pennsylvania and New England. At the same time that Massachusetts Yankees were struggling to root boulders from their sterile plots, Penn had stumbled across some of the most productive country in eastern North America, a place with rich soils and a genial climate – at least by American standards. [. . .] Thus, over the years, while most of rural New England has reverted to forest, the bulk of southeastern Pennsylvania remains farmland – and profitable farmland at that. [. . .]

The two landscapes of the Northeast: differences in vernacular architecture

The appearance of domestic houses is a case in point. Until well after the Revolution, important public buildings looked much the same in Boston as they did in Philadelphia or Savannah, and so did the houses built by affluent merchants and landowners. Indeed, on both sides of the North Atlantic, power-brokers and taste-makers were all attached to the same British system of ideas and values and, not surprisingly, they often possessed correspondingly similar tastes in food, drink, clothing, and architecture. In particular, high-style buildings tended to look alike, for the simple reason that all were designed by the same English academic architects, or by a small number of American architects who had learned their craft in England.

When regional differences in architecture began to appear, well before the Revolution, they came not in high-style houses but in the vernacular houses of ordinary people. Furthermore, those differences were exaggerated between the Revolution and the Civil War, a time when settlers were moving away from the coast and its Atlantic connections, into the American interior where information traveled slowly and new environments challenged the utility of traditional ways. In the new western territories of the United States during the half century after the Revolution, regional differences had grown sharper than at any other time in American history. And it was during that same time that the greatest differences emerged between the look of the Pennsylvania landscape and that of New England.

Pennsylvanians stuck to the old architectural ways longer than did New Englanders, a fact that suggests a kind of ingrained conservatism in Pennsylvanian domestic life that was not found in New England. As Pennsylvanians moved inland, they took with them the British habits of domestic building which they had contracted along the coast. The streets of inland Pennsylvania towns like Carlisle and Reading and York were lined with red-brick Georgian row-houses, much as in their English namesakes. [. . .]

New Englanders, however, exhibited much greater independence of mind. Brick row-houses were built in sizeable numbers only in a few large coastal cities, Boston most conspicuously. By the time New Englanders had migrated a few miles inland, however, they had abandoned the use of brick and begun to build in wood. It was not just wood for framing, but exterior wood as well – shingles and clapboards, and a rich variety of wooden embellishments. To colonial Pennsylvanians, to build a wooden house was at best inelegant, at worst an admission of poverty. To New Englanders, it was an opportunity for exuberant experimentation and, by the time of the Revolution, even rich and fashionable people were opting to build their mansions out of wood, even in coastal towns where brick construction had until only recently been the ruling norm.[3]

Why did it happen that way? Differences in environment cannot explain it. Wood was no cheaper nor more abundant in New England than it was in Pennsylvania, and clay for making brick was available almost everywhere. One can only guess that there was some cultural predisposition for New Englanders to experiment, and Pennsylvanians to stick with what was tried and true. The reasons for that, in turn, are less than obvious.

The Yankee inclination to experiment with

their common houses shows up in another very striking way. At the same time that New Englanders were shifting their favor from brick to wood, they were beginning to experiment with new locations for their houses. Only a short distance inland from the coast, New Englanders began to abandon the tradition of building urban row-houses, and instead started to build free-standing houses on spacious lots and set the buildings well back from the street. Thus, by the time of the Revolution, their towns had taken on a very different look (compare Figs 8.2 & 8.3). The Pennsylvanian town still seemed very European, but the New England village had begun to assume an open and rather countrified appearance. On the western frontier, with plenty of wood and plenty of space, it was an obvious way to do things. [. . .] [F]rom the early 1800s onward, Americans everywhere west of the Appalachians took up this New England model – and house construction has followed this pattern in most of the United States ever since. [. . .] [T]he American dream house remains a single-family free-standing house, standing independent of all others on a lot of its own, an ornamental landscaped lawn in front, and a less tidy backyard for gardens and children's play. That now familiar arrangement turned out to be one of New England's most successful inventions.

There were other architectural differences as well. Shortly after the Revolution, Classical Revival architecture had begun to make its way into the United States, a style promoted by Thomas Jefferson, who argued that Greco-Roman classical architecture was more fitting in a republican democracy than traditional Georgian styles, which symbolized, after all, the most detested of British monarchs. From 1790 onward, indeed until the middle of the 19th century, important public buildings throughout the United States came increasingly to be modeled after the Parthenon or the Roman Forum. [. . .]

Figure 8.2 Village in southwestern Vermont, c. 1975. The countrified landscape of the classic New England village has become the apotheosis for suburban America: single-family houses, separated from each other, and set back from the street, with large front lawns under a canopy of shady trees. Note the extensive use of wooden construction, as reflected in the white clapboard exteriors of buildings, a sharp contrast with red-brick Pennsylvania.

Urban forms

It was in cities, however, where the American landscape began to deviate most extremely from old European forms. The most radical departure of all was in Pennsylvania, where Penn laid out the city of Philadelphia in advance of settlement, using a grid plan that called for wide streets laid out at right angles to each other – north–south streets given numbers, east–west streets named after trees (Fig. 8.4). The grid plan itself was nothing new; it had been used across the world since time immemorial – in ancient China, throughout the Roman Empire, and throughout Spanish America – to name but a few places. But it was Penn who introduced the idea in British America, and once implemented, the system spread across the Appalachians all over the United States. From Ohio, everywhere west-ward, it is the rare town where streets do not cross each other at right-angles. [. . .]

There has been endless speculation about the reasons why Penn's Philadelphia grid plan was so enthusiastically adopted by people who were laying out towns for the new American republic. Some have suggested that Americans liked the plan because it was democratic, but that idea does not stand up under scrutiny – despite the practice of designating streets by names and numbers instead of naming them after military heroes. There was nothing in the plan to prevent rich people from buying up big blocks of land, nor were those blocks democratically uniform in slope or drainage. (More than a few unwary buyers were sold city blocks that turned out to be swampland, or even worse, located completely under water in the middle of a river or bay.) But the grid plan had several important virtues in an

Figure 8.3 Elfreth's Alley, Philadelphia, a well-preserved remnant of 18th-century Philadelphia, is a standard bit of British urban morphology. Such brick row-houses continued to be built in Pennsylvania cities and villages until the mid-19th century, long after New Englanders had abandoned the idea.

expanding entrepreneurial republic. Most important, perhaps, it was flexible, with plenty of room for variety within and between the presumably anonymous blocks. There was plenty of room for planning, and it was not uncommon for those plans to go awry. Penn himself had expected that his big Philadelphia blocks would permit farmers to live in town and plant large gardens around capacious houses, each block a kind of mini-farm which would in combination produce a park-like "greene towne." But land in Philadelphia soon became too valuable to fritter away on mere gardens, and land speculators divided the rectangles into narrow slices, and sold them to other speculators who promptly

chopped down the trees to make room for row-houses. And in Washington, D.C., when Major L'Enfant planned the street pattern for the new capital city (a grid overlain by circles and spokes), he had expected the central business district to grow eastward toward the Anacostia River. Thus, the national Capitol was built with its formal face in that direction. In fact, things turned out exactly the opposite. The Anacostia bottoms became a noisome industrial slum, while commercial and ceremonial Washington expanded toward the Potomac and Georgetown to the west. One curious result of L'Enfant's mistake is that for two centuries presidential inaugurations have taken place on the "wrong" side of the building.

Figure 8.4 Plan of Philadelphia, 1682. Penn of course did not invent the grid plan, but Philadelphia's success was largely responsible for the later adoption of the grid by town planners all over the United States.

No matter. If mistakes were made, the grid would accommodate them.

Most alluring of all, perhaps, the grid made it very easy to lay out new towns in advance of settlement, and that was a huge virtue in a booming country where population was pressing rapidly into new and townless territories. The grid also made it easy to describe rectangular parcels of land on a map, so that speculators could buy and sell those parcels sight unseen. At the same time, its mathematical regularity greatly reduced the room for surveyors' errors and consequent legal disputes over the location of boundaries. All in all, the urban grid plan was a perfect godsend for real estate speculators, not only in Philadelphia, but in all of the American towns, real and imaginary, that were strewn across the land to become new Philadelphias.

The grid was occasionally tried out in New England cities, but the effort was half-hearted. The core of New Haven, Connecticut, for example, was laid out in a grid, but New Haven is an exception. Most New England cities grew in the old-fashioned European way, with main streets following old paths, and new streets and alleys added in haphazard bits and pieces as the need arose. The street plan of Boston is typical – a tangled skein of crooked streets that looks more European than American (Fig. 8.5). And, when those crooked streets are lined with red-brick Georgian row-houses, as on Beacon Hill in Boston, the effect is very British indeed.

Despite the unplanned street pattern of many New England cities and villages the geographical arrangement of *towns* was very much a planned affair – and that planning reflects the way that Yankees thought about themselves and about their communities. The New England town was conceived not as a geographical thing, as most Americans think of towns, but as a religious and civic community of people. When set down in a particular geographical place, a town's natural territory turned out to be a bounded chunk of land that was large enough to support a church and its congregation, but small enough to permit all its inhabitants to attend services at the same church on a regular basis.

The geographical result was predictable. New England was divided into a mosaic of politically bounded "towns," 40 or 50 square miles in area. Near the center was a church, spaced 5 to 10 miles from its nearest neighbor. More often than not, villages grew up around the church, first by the building of a tavern or general store, and subsequently other commercial buildings and, usually, a town(ship) hall.

The New England village center was not designed as a marketplace, although commerce usually tended to accumulate there. Visually, its most conspicuous feature was its open "green" of common land, fringed by a church or two, a town hall, and perhaps a grange or fraternal building – mostly demurely classical in design and, of course, painted white. This assemblage of white buildings around a village green has become a powerful image for many Americans, the quintessence of Yankee New England, the visual symbol of small-town simplicity and virtue. One can debate whether that is true or not, but the New England village was clearly a very different sort of thing than the version that developed in Pennsylvania, where the center of town was a busy intersection or market square, suitably laid out at right-angles, with shops crowding to be near the center. Today, many Americans view New England villages through a haze of nostalgic imagery, and see them as quaint vestiges of a bygone age. In one respect, they are quite correct. West of the Appalachians, when westward-moving Americans got down to the serious business of creating towns, there was no room for greens and churches in the middle of town. In most parts of the American West, the Pennsylvania model held sway. As in Pennsylvania, the business of an American town was business – only incidentally the creation of social community.

It is ironic today that the tight-packed Penn-

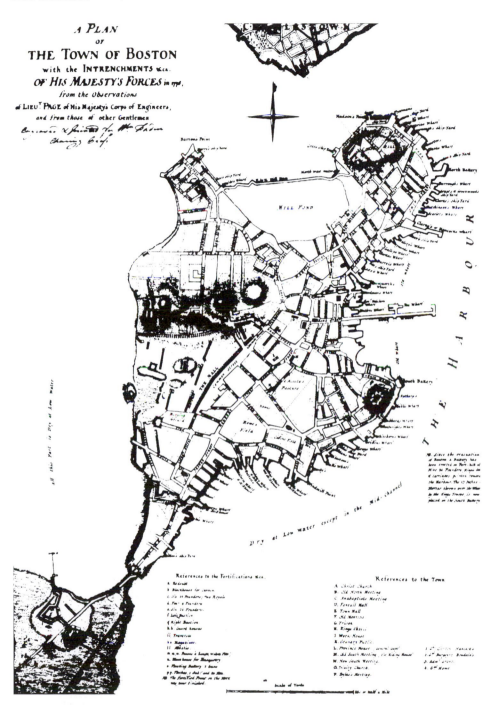

Figure 8.5 Street map of Boston, 1776. The streets of New England cities and villages were laid out *ad hoc*, as they had been laid out for millennia in the Old World. To Americans, accustomed to grid plans that imitated Philadelphia, Boston still looks rather foreign.

sylvania model of the American town, originally thought to be so practical and businesslike,
has been routinely and unsentimentally abandoned by the practical businessmen for whom
it was designed. It worked very well as a commercial center during the 19th century, when
people and goods were delivered to town at a
central railroad station, and proximity to the
station was a requisite for prosperity. But that
was before the advent of the automobile.
Ironically, it was commercial success that was
the undoing of that businesslike town. Commerce causes traffic jams, anathema to red-
blooded American motorists. To avoid that congestion in the early part of the 20th century,
bypasses were built around town centers, and
the traffic that supported downtown prosperity
was siphoned off elsewhere. More recently,
when suburban shopping centers were built
to suit the convenience of motorists, Pennsylva-
nia-model downtown commercial districts
began to decay all over the country. It is an
additional irony that a good many New England
villages, so long believed to be quaintly obsolete, have recently discovered that quaintness is
a marketable commodity. In picturesque village
after picturesque village along the northern
fringes of megalopolis, prosperity has arrived,
brought first by tourists, then by affluent refugees from urban congestion – stockbrokers and
3-day-a-week corporate executives – who were
hotly pursued by purveyors of expensive real
estate, expensive foreign automobiles, and exo-
tic up-scale groceries. In sum, both the Pennsylvania town and the New England village have,
to put it kindly, taken on new functions, while
at the same time they have abandoned the original purposes for which they were so carefully
designed. [. . .]

The cultural–geographical baggage goes west

So it was that when Americans crossed the
Appalachians into the interior of the continent,
they carried two geographical traditions with
them – and borrowed from both in highly
selective ways. The New England tradition
and the Pennsylvania tradition, however,
were geographically separated from each
other, not only along the eastern seaboard,
but west of the mountains as well. The reason
had to do with topography and transportation
routes, for Pennsylvanians went west by a very
different set of routes than did the New Englanders, and those routes led respectively in
quite different directions. New England's avenue to the West was a narrow lowland that
followed the Mohawk River between the
mountain bulwarks of the Adirondacks and
the Catskills and led to the great open plains
along the shores of the lower Great Lakes –
thence, as we have seen, into the northern
part of the old Northwest Territory: northern
Ohio, Indiana, Illinois, and the better parts of
southern Michigan, and Wisconsin. Pennsylvanians, by contrast, had a wider range of
choices. They could head west, by way of
what became the National Road, via Wheeling,
Columbus, Indianapolis, and on to Saint Louis.
Alternatively, they could move down the Ohio
River from Pittsburgh, toward the Kentucky
Bluegrass and the middle Mississippi Valley.
Or they could avoid the mountains altogether,
and drift southwestward down the Shenandoah
Valley into western Virginia, North Carolina,
and the whole upland South. The New England
stream, in short, was narrow and confined until
it reached the lower Lakes. The Pennsylvania
stream spread out in a great fan that eventually
covered much of the interior. But both streams
retained a kind of cultural purity as they poured
westward – and they remained separate for a
considerable distance west of the mountains.
[. . .]

But the migrants were selective about the
geographical ideas they carried with them,
and the ordinary landscapes of middle America include elements from both New England
and Pennsylvania, both in turn much altered

from ancient English models. The mixture is eclectic. The interchangeable American grid-pattern town is pure Pennsylvania, of course, and one can argue that the widespread use of the Philadelphia city plan paved the way for acceptance of Jefferson's idea of a gridded land division system for the rural lands of the whole Northwest Territory. But even that system is a combination of the two regional traditions. The basic unit of land division is a square township, 6 miles on a side, and rigidly oriented to the cardinal directions of the compass. The rectangular geometry springs from Jeffersonian rationalism, but the 6-mile dimensions are those of the ancestral New England town. Towns, too, are mixtures. The middle of midwestern and western towns was consigned to business, and that was the Pennsylvania way of doing things. But the residential areas, with their widely spaced houses, big yards, and tree-shaded streets are quintessentially New England. [. . .]

Large parts of this old landscape seem obsolete today, overlaid by new technologies, new people, and new canons of taste. But despite all efforts, old patterns which were etched into the landscape are not easily erased, even though Americans have a seemingly infinite capacity to redesign and find new uses for things that have apparently outlived their usefulness. [. . .] Meantime, a huge part of the United States continues to bear the imprint of geographical ideas that were imported from England three centuries and more ago, and subsequently reworked by colonial Americans in a small corner of the Northeast. That imprint is still visible today, and its patterns continue to shape our lives.

Notes

1 Holbrook 1950
2 Glass 1986
3 Paint, however, was another matter. In colonial and early national times, paint was very expensive and only wealthy people could afford to use it. As a result, the bulk of New England's wooden houses simply went unpainted and turned a weathered gray. The tendency to paint all New England houses white is a fashion of fairly recent date. Meantime, Pennsylvanians often painted their red bricks red – presumably to help preserve them from the weather

References

GLASS, J.W. (1986) *The Pennsylvania Culture Region: A View from the Barn*, Ann Arbor, UMI Research Press
HOLBROOK, S.H. (1950) *The Yankee Exodus: An Account of Migration from New England*, Seattle, University of Washington Press

9

TIMBER FRAMING IN COLONIAL AMERICA

by Carl W. Condit

Source: Carl W. Condit, *American Building: materials and techniques from the first colonial settlements to the present*, 2nd edn, Chicago, IL, University of Chicago Press, 1982, pp. 3–20

[Colonial building materials]

[. . .] The non-existence or high cost of transportation and the expenses and difficulties of quarrying meant that the builder had to use locally available materials which could be readily worked by hand-and-tool techniques. Wood was immediately available in large quantities everywhere except the Southwest, with the result that timber construction was dominant throughout most of the colonial area. Local supplies of suitable clays were extensive enough to warrant the use of brick or adobe masonry for more durable structures, but only after the economy had advanced to the point of supporting their greater cost. Dressed stone masonry laid up in mortar was rare in the colonial period, the usual form being loose rubble, which could be laid in irregular courses if cobbles or fragments of stratified rock lay close to the ground surface. [. . .]

[English techniques]

[. . .] By 1630 the English settlements included a good proportion of artisans either trained in the techniques of medieval carpentry or able to revive them through necessity and the recollections of their homeland experience. Tools began to be imported in quantity from England, and the establishment of the power-driven saw-mill at Jamestown in 1625 greatly increased the output of squared framing timbers. The carpenter's tools, like the first timber building forms, were entirely of medieval origin. Implements for cutting and shaping included the felling ax, the frow ax for splitting logs, the single-handed and the two-handed pit saw, the adz, plane, chisel, and gouge. Punching and driving tools were the hammer, drill, awl, and punch or drift pin. For accurate squaring and fitting there were the square, rule, compass, plumb, and prick or marking iron. The rough dried skins of various fish served in lieu of sandpaper. [. . .]

A number of structural developments lay behind the large timber-framed house of medieval England, which had reached its maturity by the fifteenth century (Fig. 9.1). It was more elaborate than anything the colonists were to build, but it served as the prototype for all the forms in the English colonies. Several elements of the medieval frame either did not appear or were used in greatly simplified form in the American work. Chief of these were the numerous bracing members and the embryonic trusses of the gable frames. [. . .]

The colonial framed house, also called a *fair house* or *English house*, appeared early in the seventeenth century. [. . .] The structural frame was covered with clapboard siding, which is also of medieval English origin. The

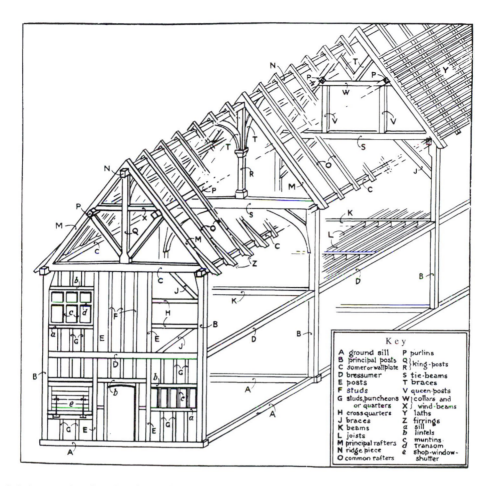

Key

A	ground sill	P	purlins
B	principal posts	Q	king-posts
C	somer or wallplate	R	
D	bressumer	S	tie-beams
E	posts	T	braces
F	studs	V	queen-posts
G	studs, puncheons	W	collars and
	or quarters	X	wind-beams
H	cross quarters	Y	laths
J	braces	Z	firrings
K	beams	a	sill
L	joists	b	lintels
M	principal rafters	c	muntins
N	ridge piece	d	transom
O	common rafters	e	shop-window-shutter

Figure 9.1 Late medieval timber frame for a two-storey house. Framing systems such as this constituted the basis for all timber construction in the English colonies up to the mid-eighteenth century.

open spaces between framing members in the wall were filled with clay or mud reinforced with straw or sticks. Later, in the more expensive homes, this infilling (called nogging) was brick laid up in mortar. By the eighteenth century the builders adopted the practice of using wood sheathing as the nailing base for the clapboards, with the result that the clay or brick nogging could be abandoned.

The frame itself was originally an unbraced and simplified version of its late medieval antecedent (Fig. 9.2). Although this system of framing was most common in New England, where

timber construction enjoyed the most vigorous life, it was widely used in all the English colonies. The primary members were the sills, corner posts, and a variety of girts (girders). The posts were fixed to the sills, which in turn rested on a stone foundation of loose rubble. Closely spaced joists spanning between the girts carried the floor boards. The roof sheathing was nailed directly to the rafters, which sloped upward from the plates on top of the posts to the ridge line. By the end of the seventeenth century the developed frame included diagonal braces, usually set in the corners

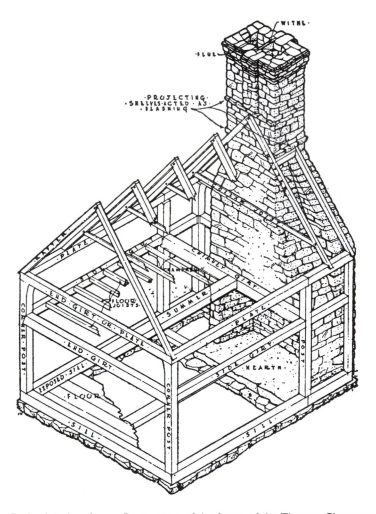

Figure 9.2 A New England timber frame. Restoration of the frame of the Thomas Clemence house, Manton, R. I., c. 1680.

between posts and sills, a ridge beam running the length of the roof and carrying the upper ends of the rafters, and purlins extending along the long dimension of the roof and serving as a nailing base for the roof sheathing.

The abandonment of clay and brick nogging between the posts required the addition of studs to the wall frame. These are light boards set vertically and framed into the sills and girts to provide support for the wall sheathing. The outer wall covering was nearly always clap-board siding in the English colonies. After the Dutch introduced the shingle covering for walls about 1650, on Long Island, the shingle house migrated to New England. The shingle roof was used in all the colonies before the mid-century. The more durable but more expensive slate roof was introduced in Boston in 1654, and tiles were used before the end of the century in the more elegant and costly houses.

The structural members were tightly joined

by means of a variety of devices for framing one member into another. The simplest joint was a notch cut into the side of the post to receive the end of the girt, or into the girt to take the end of the joist. The notch was carefully cut to the size of the member it was designed to receive. The dovetailed joint was used to make a tight connection between larger, more heavily loaded members, such as the girts. In this type, a keystone notch was cut in one girt to match the dovetail end of the other, the diagonal cuts making it impossible for the load to pull one member away from the other. The most secure joint, however, was the mortise-and-tenon, which was commonly used to join any members subject to tensile and shearing stresses, such as the connection between the girt and the post. In the customary practice a slot or mortise was cut into the post with chisel and mallet, and the end of the girt was reduced in section to a tongue or tenon shaped to fit snugly in the mortise. The assembly was secured by passing various kinds of tapering wooden pins known as treenails through the mortised piece and the inserted tenon.

As the framing members grew larger for churches and other public buildings, the mortise-and-tenon joint became more elaborate. By the mid-eighteenth century the carpenters were using double tenons for very heavy girders, one tenon above the other with the face between the two projecting beyond the end plane of the girder in either a vertical or an inclined plane. Both devices served to reduce the shearing stress on the tenon. To reduce both shearing and bending forces under the heaviest loads knee braces were introduced into the angle at the connection of the beam and the post. These had a variety of shapes, the larger forms having been adapted from the so-called ship's knees used to join the deck beams to the ribs in the hull of a ship.

In addition to wooden fasteners, colonial builders began early in the seventeenth century to use connecting devices of iron. The oldest and simplest of these was the hand-wrought nail, which was used in the English colonies throughout the colonial period. The great variety of nails in medieval timber construction were manufactured in quantity by smiths who sold them at a certain price per hundred nails, the cost depending on the size. This practice led to the designation of nails by price, such as a threepenny nail, or a fivepenny nail. They were imported into the colonies up to the Revolution, although local production was carried on in New England and the Philadelphia area from the mid-seventeenth century on. During the period from 1790 to 1830 the hand-wrought nail gave way to the machine-cut variety, the manufacturing process being drastically altered by the introduction of power-driven cutting machinery.

In the eighteenth century other wrought-iron connections appeared, chiefly the dog, the stirrup, and the wedge. The first was a bar with the ends bent down and hammered to points so that they could be driven into the wood. It was ordinarily used as a tie set diagonally across the corners of a right-angle joint to prevent the separation of the two members. The stirrup was a heavy U-shaped strap ordinarily used at a connection involving a high concentrated load imposed by a vertical member on a horizontal. It commonly appeared in the latter part of the century to bind a post to the bottom chord of a truss, the whole assembly being drawn tightly together by means of narrow iron wedges inserted through aligned holes in the post and the stirrup.

Roofs were developed into a variety of shapes during the seventeenth century, but the framing system remained essentially the same. The only structural difference was that between the purlin and the rafter roof. In the purlin type, which was the earlier form, the longitudinal purlins lay above the rafters. The roof sheathing ran from the ridge to the plate, at right angles to the ridge line, and was nailed to the purlins. In the rafter roof, heavy purlins were framed into the principal rafters, while a secondary group of light rafters rested on the purlins. The sheathing, running parallel to the ridge line, was nailed to the light rafters. [. . .]

The straightforward column-and-beam framing of the colonial house, whether of English, Dutch, or German origin, could be easily enlarged and altered for any kind of building – barns, storehouses, churches, forts, governmental and commercial structures. The carpenter-builder followed the pragmatic tradition of the craftsman, relying on his own experience, his ingenuity, and the accumulated knowledge of his medieval forbears. For larger buildings carrying heavier loads, his solution was to increase the size and number of structural members and to introduce bracing for stability against lateral forces. [. . .]

As the Dutch at New Amsterdam and at New Amstel in the Delaware valley grew wealthy in the fur and tobacco trade, they began, shortly after the completion of the Manhattan fort in 1626, to build substantial houses and commercial structures. Their timber construction differed from the English system only in the flaring, steeply pitched roof and the extremely massive floor beams, almost as heavy as the girts of the English house. Heavy floor beams of this kind were sometimes so long that it was necessary to reduce the clear span by inserting brackets or knee braces at the corners between posts and beams. [. . .]

The colonists who followed William Penn to Philadelphia had the advantage of a comprehensive building guide prepared by Penn himself and included in his *Information and Direction to Such Persons as Are Inclined to America* (1684). The work contains a complete description of all framing members, such as posts, girts, sills, and the like, together with "several other small pieces, as *Windbeams, Braces, Studs*, etc., which are made of the waste Timber." Penn also gives information on the making of iron nails and on the cost of carpentry work.

The German settlers of the Philadelphia region brought other variations on the fundamental techniques of timber framing, following the medieval antecedents of their homeland. They divided the wall frame of houses and barns into a series of bays by setting the posts close together and omitting the studs of the English practice. Their chief innovation was an extensive system of diagonal bracing in which the individual brace extended across two or more bays of the wall frame. [. . .]

Philadelphia by this time was rapidly approaching the position of the richest and most cosmopolitan city in the American colonies and the second largest in the British Empire. It was here, accordingly, that the colonial building craft passed from the vernacular to what we might call the protoprofessional stage. The event that signalled this transition was the establishment in 1724 of the Carpenters' Company of the City and County of Philadelphia. Based on the Worshipful Company of Carpenters of London, which had been founded in the early fourteenth century, the Philadelphia company was thus much like the medieval guild, and the master carpenters who constituted its membership were comparable to the master builders of the guild. The organization was somewhat analogous to the modern architects' and contractors' associations, since the master carpenter of the eighteenth century, like his medieval forebear, acted as both architect and contractor.

In 1734 the Carpenters' Company began to assemble a library of English treatises on the building arts. The systematic study of these works and the practical application of the knowledge gained from them eventually transformed the members into trained professionals who knew arithmetic, geometry, surveying, and drafting and who were responsible not only for the design of a building but also for estimating its cost and supervising its construction. As a consequence, Philadelphia soon became the center of the most advanced building arts in the English colonies. By the mid-eighteenth century associations of brick-layers, stonecutters, plasterers, and various mechanical artisans were founded in Philadel-phia and other large cities like New York and Boston, but none was as wealthy and as author-itative as the Carpenters' Company. [. . .]

With sophisticated techniques such as these available to them, the builders were prepared by the mid-century to face the challenge of constructing large public buildings like churches and the assembly houses of the var-ious political units. Containing relatively large open areas within the walls, these buildings posed peculiar problems in the framing of roofs and ceilings over the interior enclosures. This was especially true in the case of churches, where it was desirable to keep the nave free of columns for unobstructed sight lines. The simplest solution was to use long, heavy rafters spanning between the ridge beam and the side walls or wall columns, and to introduce tie beams extending transversely between each pair of rafters on either side of the ridge. If the span was so great as to produce excessive flexibility in the frame, a king-post could be inserted between the junction of the rafters and the center of the tie (Fig. 9.1 shows king-posts in the frame). Rigidity and strength could be further increased by adding a second set of ties that would lie between the main tie beams and the ridge beam. Within this general form a number of additions and variations were possible. [. . .]

The building that sums up the state of timber construction in the most highly developed form it was to achieve in the eighteenth cen-tury is also the most famous and historically most important work of colonial architecture. The national landmark known as Independence Hall in Philadelphia was built between 1732 and 1748 as the original State House of the Colony of Pennsylvania. [. . .] The main block of the Hall is a brick structure divided into three parts by two internal brick walls, the central area being a lobby and the two enclo-sures on the east and west, respectively, the Assembly Room and the Supreme Court Room.

These rooms are very nearly 40 feet square on the inside and are constructed without intermediate supports, so that the framing of the second floor and attic over the open areas called for heroic ingenuity on the part of the builders to keep the space free of structural members. The ceiling and floor of any one level are supported chiefly by a two-way system of massive girders framed into each other in such a way as to form an interlocking pattern some-what like the fireman's basket carry (Fig. 9.3). The girders do not span from wall to wall but nevertheless form a rigid framework that sup-ports the joists to which the floor planking is fixed. The two-way system, composed of gir-ders less than half the full span in length, makes possible a great reduction in the bending force and hence the tensile stress to which a single 40-foot girder would be subject. [. . .] Supple-menting the girders in the work of carrying the floor frame of the Hall was a vertical trusslike frame as deep as the full height from the sec-ond floor to the attic and extending across the full width of each of the two rooms along the center line. [. . .]

[French techniques]

The major part of French building in the Mis-sissippi valley was heavy framed construction at first comparatively primitive in character. The earliest structural technique and probably the antecedent of most subsequent forms was known as *poteaux-en-terre*. In this system the building wall consisted of a row of stout cedar or cypress posts, hewn to flat surfaces on oppo-site sides, set upright in a trench, and held in place by backfilling tightly around the lengths below grade level. The posts were spaced no more than a few inches apart, the interstices being filled with clay and grass or Spanish moss. There were no footings: the posts con-

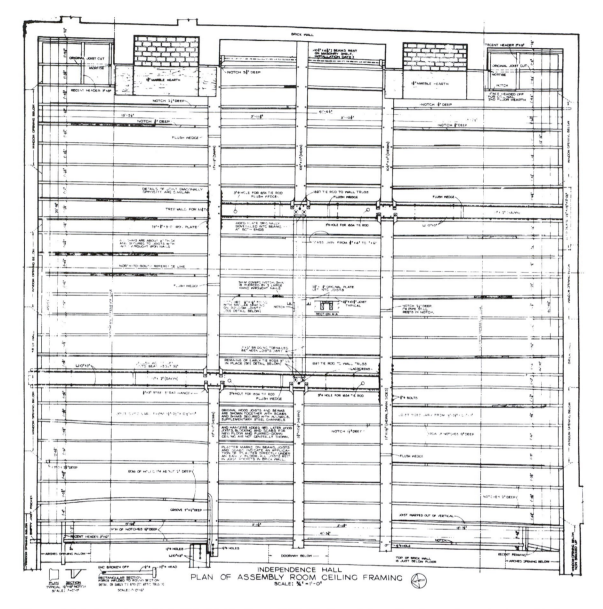

Figure 9.3 Independence Hall, Philadelphia, Pa., c. 1732–48. Plan of the ceiling frame above the Assembly Room. Renaissance precedents lay behind this ingenious system of girders designed to support a floor over a wide enclosure.

stituted both a foundation and a wall. Rafters or horizontal beams spanning between two parallel rows of posts carried the roof planking.

Posts set directly in the ground quickly rotted, especially at their exposed lower ends, where destructive microorganisms could easily penetrate the wood. To protect them from deterioration the French builders developed the technique known as *poteaux-sur-sole*, a method of obscure origin but possibly derived from techniques used by Huron and Iroquois Indians to construct log buildings. In this system the posts rested on a timber sill (*sole* appears to be the French cognate), in turn supported by a row of flat stone blocks set in a trench. [. . .]

The most advanced structural system used by the French in Louisiana was a kind of half-timber construction known as *briqueté-entre-poteaux*. It was introduced into New Orleans by the military engineers who had charge of civic building in the colony, and was based partly on medieval precedents, partly on techniques described in later handbooks of military engineering, especially *La Science des Ingenieurs dans la conduite des Travaux de Fortification et Architecture Civile* of Bernard de Belidor (1693–1761). In this technique a brick foundation set into a trench carried a timber sill into which the heavy squared posts were framed. The posts were spaced about three feet on centers, and heavy single or double diagonal braces were set between successive pairs of posts. A plate running horizontally over the tops of the posts carried the lower ends of the roof rafters. The triangular spaces enframed by sills, posts, and braces were filled with soft brick nogging. The timbers were originally left exposed, as in Tudor English half-timber framing, but the humid climate of Louisiana caused them to rot in a few years. To prevent this the whole surface of wood and brick was covered with lime plaster and later with brick. [. . .]

[Wooden bearing-wall construction]

The structural systems we have so far described have been examples of framed construction, in which the external covering, whether wood siding or brick, is a protective curtain rather than a bearing element. To combine the two functions requires that the covering be built up as a solid bearing wall, a common and very old technique in masonry which was adapted to timber construction by medieval builders. The simplest and possibly the earliest form in the colonies was the blockhouse or the garrison house, built to function as a fortified place as well as a dwelling. In the garrison house the walls were built up of logs hewn to square or rectangular section and laid up longitudinally one above the other. At the corners the logs were joined by any one of a variety of halved, mortised, and dovetailed joints, in which matching notches cut in the ends of the timbers made a tight fit. In some cases the timbers were mortised into corner posts. Garrison houses date from the early seventeenth century, although the term blockhouse (or palisadoed house) does not appear to have been used before King Philip's War, from 1675 to 1677. The war posed a serious threat to the New England colonies, with the consequence that the carpenters in some of the towns were called upon to build garrison houses large enough to shelter all the inhabitants during an Indian attack. These large structures probably embodied a combination of bearing-wall construction and interior braced framing for greater strength and support for military equipment on an upper floor.

A derivation from the garrison suitable for the domestic scale is the so-called plank house, which appeared at Plymouth as early as 1627, spreading from there to the Connecticut valley and thence to New Hampshire. The simplicity and durability of the form made it popular in north-central New England, where it continued in use up to 1860. In the fully developed house of the eighteenth century, the exterior walls and interior partitions were solid bearing structures of stout vertical planks with their ends reduced to tenons and set in slots cut in the sills and plates, the joints secured by wooden pins or treenails. The butted planks, trimmed to fit tightly together, were usually covered on the exterior with weatherboarding of clapboard siding. The openings for doors and windows were cut directly into the planks. In the case of two-story houses a bearing plate was introduced at the second-floor level to carry both the planks of the upper story and the joists of the second floor; the similar plate for the attic floor, however, was often simply pinned by means of dowels to the inner face of the planks. The thickness of the planks ranged from 1¼ to 3 inches, but the smaller size was rare because it required a supplementary framework to provide the necessary rigidity. Planking with a minimum thickness of two inches provided a rigid and durable structure while at the same time offering considerable insulation against heat loss. Its great defect was that it was wasteful of wood, since an open framework would have performed the necessary structural role with far less material.

The best known type of bearing-wall construction in wood is the log house, which became a symbol of American frontier life. Houses with walls built up of logs laid horizontally were a medieval invention of various North European peoples and by the end of the seventeenth century were common in the Scandinavian countries, Finland, and the Baltic provinces of Russia. The form was brought to America by the Swedes who established the colony of New Sweden on Delaware Bay in 1638. [. . .]

10

MASONRY CONSTRUCTION IN COLONIAL AMERICA

by Carl W. Condit

Source: Carl W. Condit, *American Building: materials and techniques from the first colonial settlements to the present*, 2nd edn, Chicago, IL, University of Chicago Press, 1982, pp. 26–36

The proudest achievements of colonial architecture were constructed with brick or stone bearing walls carrying the beams, joists, and rafters of the floor and roof frames. Brick has always been the more useful building material for a number of reasons that are still important today but were compelling in the colonial period: the clays from which it is made are widely distributed and lie close to the surface of the ground; the process of manufacture before mechanization was a simple handicraft technique, far less expensive than the tedious and exacting techniques of quarrying and dressing stone; finally, brick walls are strong, durable, highly resistant to weathering, and more resistant than stone to the corrosive effect of smoke-laden air. Colonial brick-building, like the techniques of timber framing, was derived from medieval and Renaissance precedents. [. . .]

Stone masonry is equally old and was carried to levels of complexity and refinement in the late Middle Ages that have never been surpassed. But to people with the modest economic resources of the English-speaking colonists, building in stone was likely to be very rare. Cobbles and loose fragments, available on the ground or in the beds of small streams, could be easily collected and placed in the wall by hand, but more refined forms involve the costly processes of quarrying, dressing, and transporting the stone and of laying the heavy blocks in carefully leveled beds of mortar. Domed and vaulted construction requires elaborate systems of scaffolding and centering which would have taxed the ingenuity of the carpenter as well as that of the mason, with the consequence that it was simply prohibitive in cost. Yet the American colonies were blessed with extensive resources of building stone; and where the money and skill were available, stone masonry became fairly common in certain regions by the mid-eighteenth century.

As in the case of the carpenter, the tools of the quarryman and the mason were inherited from the artisans of the Middle Ages. They included the iron-shod wooden shovel, the pick, wedge, crowbar, and maul for breaking the stone, and assorted poles, hooks, and beaks for disengaging the pieces from the natural mass. The hammer ax, broaching ax, chisel, saw, drill, and mallet were used for shaping and dressing the rough blocks. The lime required for mortar to lay up brick and stone walls was often derived by the colonists from the shells of molluscs, but they also followed the medieval practice of burning limestone or chalk. None of the limes so produced contains the alumina and silica necessary for the formation of hydraulic cement, which will set in water as well as in air. Limited to the non-hydraulic varieties, the colonial builders were barred from rigid masonry construction under water.

[. . .] Building in brick began shortly after the initial essays in timber framing. The town of Henrico, Virginia, near Jamestown, had several two-story framed houses with first stories of brick as early as 1611. The second church at Henrico rested on brick foundations, which had been laid in 1615. Brick kilns were built at New Amsterdam in 1628, and a brickyard, possibly with associated kilns, was established at Salem in the following year.

The most rapid development of brick construction came in Virginia, where aristocratic taste demanded and wealth made possible the elegant plantation manor. Moreover, the settlers of Virginia came from all parts of England, including regions where brick construction was common, whereas the New Englanders came chiefly from the eastern and southeastern counties, where it was rare. [. . .]

The Dutch, who were the most skillful brick masons of Europe, brought their art to New Amsterdam shortly after the colony was founded. They were the first to build with glazed brick and tile and the first to make brick in a variety of colors, such as red, salmon pink, orange, yellow, and purple, achieved by varying the temperature and the time of baking in the kilns. The presence of various salts in the clay also affected the color and was the source of the glazed surface. The Dutch bricks were usually narrower than the English, about 1½ inches thick as against 2½ inches, the latter being standard today. By using a variety of bonds – common, English, Flemish, and Dutch-cross [Figure 10.1] – along with contrasting colors and glazes, they built walls in elaborate patterns, the lively geometric effect enriched by colored roof tiles and stepped gables. Interior timber construction, however, was usually simpler than the New England system: plank floors rested directly on a closely ranked series of heavy beams spanning from wall to wall.

Many of the masterpieces of colonial Georgian architecture were derived without essential structural change in the masonry work from the traditions of the seventeenth century. Brick became dominant for the exterior walls of all major buildings; for the interior structure, variations and extensions of heavy braced framing continued to be used to support floors, roofs, towers, and spires. The only new materials added to the eighteenth-century building were paint and lime plaster. [. . .]

Nearly all masonry work in the English colonies belonged to the categories of straight bearing-wall or pier construction, with arches largely confined to those parts of the wall surmounting arched openings for windows

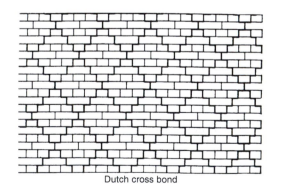

Dutch cross bond

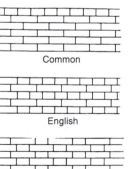

Common

English

Flemish

Figure 10.1 Styles of brick bonding.

and entranceways. Full domes and vaults were never used, chiefly because of the expense of building the necessary centering and false-work. [. . .]

Special forms of structural arches were employed in Independence Hall in connection with the heating and smoke dissipation systems. Since the fireplaces on the second floor could not have hearths resting on the timber frame, each was carried on a half-arch (or more strictly, a half-vault) springing from the brick end wall and supported at its free end by wedge-shaped pieces fixed to blocks extending on either side of one of the floor girders. A larger and more daring form of arch construction is associated with the flue and chimney structure. The four-chimney group at each end of the hall rests on a massive brick arch set against and parallel to the end wall, thus bringing the load of the four chimneys down to two pierlike masses projecting from the wall on either side of the two fire-places located at each end of the building. The chimney arch in the east wall is semicircular, whereas that in the west is a flattened segmental form made up of three circular segments drawn about separate centers.

Stone masonry, as we have seen, was rare in the colonies beside brick and timber construction. It was concentrated mainly in the Hudson valley of New York, northern New Jersey, and eastern Pennsylvania. In New England the extensive granite outcroppings lie in enormous masses, discouragingly expensive to quarry with the colonist's limited means, and in the Piedmont and along the coast the soil is composed chiefly of marine and alluvial sediments. The original form of stone masonry, first confined to chimneys and later used for walls, was an irregular rubble of stone fragments found lying naturally on the ground, but sometimes broken from ledges or split off from weathered masses. Up to the mid-seventeenth century the practice was to lay the pieces in clay bound with straw, but mortar made from sand and oyster-shell lime soon became standard for dur-

able building. There was a steady improvement in masonry techniques throughout the century: stones were carefully selected for uniformity of size and shape and laid roughly to course; eventually the stones were dressed and laid with even joints to form what is called ashlar masonry. In some areas, a special kind of stone construction known as cobble masonry was used, in which the wall is made of small rounded boulders set in a thick matrix of mortar. Cobble houses became common in central New York, chiefly in the Mohawk valley, where there was a plentiful supply of these glacially distributed stones. The walls of all stone buildings tended to be very thick running from 18 to 36 inches even for ordinary residences.

Stone houses were common in the Dutch building of the New York area, and a few still survive in the city itself. [. . .]

The great works of masonry construction were built in the Spanish colonies that once stretched without break from Florida to California. The Spaniards, who came to find precious metals and to convert the Indians, established their permanent settlements on feudal and hierarchical lines, with the consequence that the wealth of church and state was available for the construction of missions, fortifications, and government buildings. The richness and complexity of Spanish Baroque architecture could thus be duplicated in the New World in forms and on a scale impossible to the bourgeois economy of the English colonies and antipathetic to their Protestant spirit. Whereas the building inheritance of the English colonists [was] initially medieval, that of the Spanish domains was derived from neo-Renaissance precedents at the start.

The first permanent settlement of the Spanish in the area of the United States was founded in 1565 at St. Augustine, Florida. Here a century later the colonial governor built the first great work of military architecture in North America and one of the early masterpieces of masonry construction. The Castillo de San Marcos, standing today as a national monument, was begun in 1656, largely completed between 1672 and 1675, and extended in 1756. The Castillo is typical of the bastioned fortification developed for artillery combat in the period from the fourteenth to the seventeenth century and given its ultimate form by the French military engineer Sebastien de Vauban (1633–1707). The walls of the bastion, or the curtains, are built up of dressed blocks of coquina limestone laid in mortar made from sand and oyster-shell lime. The stone is a soft white limestone composed of broken shells and coral cemented under pressure into durable strata. The outer curtains of the fort are 25 feet high, 12 feet thick at the base, and 7 feet at the top. The coursed masonry constitutes the outer covering of a rubble core, a composite masonry form that was to survive in the construction of large dams until about 1910. The most remarkable feature of the Castillo is the presence in one of the cells of a concrete floor made from a mixture of sand, oyster-lime cement, and a crude aggregate of broken shells, the surface of which was rubbed smooth and finished with linseed oil. This undoubtedly represents the first use of concrete in America.

The fully developed Baroque architecture of Spain appeared in the large mission churches that were built from Texas to California in the eighteenth century. The earliest of the major works is the Mission San Jose y San Miguel de Aguayo, constructed at San Antonio, Texas, 1720–31. The solid bearing walls are built up of rubble masonry laid in lime mortar. The bearing stone is tufa limestone, a porous, calcareous stone somewhat like travertine in its appearance and physical properties, and the ornamental details are carved in brown sandstone. The exterior surface is covered with stucco, which hides the irregular jointing under a smooth coating. [. . .]

The leading structural archievement among Spanish ecclesiastical buildings is the Church of San Xavier del Bac, constructed at Phoenix, Arizona, between 1784 and 1797. The walls throughout are built up of kiln-baked brick laid in lime mortar and covered with white stucco. The most remarkable features of the church are the buttresses that serve to maintain the stability of the relatively narrow octagonal towers above the flanking wings. A substantial part of the tower load is carried to the corner piers of the wings by flying buttresses whose surfaces are arched below and developed into scrolls above. The nave, transept wings, and apse are covered by low domes of brick; a high dome on an octagonal drum roofs the crossing. This drum is carried to the four corner piers at the junctions of the nave and transept walls by means of inverted pyramidal blocks corbeled out from the pier faces (these may be regarded as pendentives with flat triangular faces). The extremely thick walls of the church sustain the outward horizontal thrust of the domes, and the opposing thrusts along the inner segments of the peripheries of contiguous domes cancel each other. These technical details indicate how nicely the builder calculated the distribution of material for structural purposes in the San Xavier Church.

The less pretentious ecclesiastical buildings were combinations of timber and masonry construction. [. . .]

For the smaller churches and the secular buildings of New Mexico the Spanish colonists often used adobe construction, the only form of permanent building derived from native traditions of the New World. [. . .]

The secular buildings and small churches were constructed by the Indians under Spanish direction but largely according to native traditions of workmanship. The chief innovation introduced by the Spaniards for high-walled churches was to build one wall thicker than the other to serve as a working platform and as a fulcrum for lifting the heavy log beams into position. [. . .]

II

CHICAGO: NATURE'S METROPOLIS

by William Cronon

Source: William Cronon, *Nature's Metropolis: Chicago and the Great West*, New York, W.W. Norton, 1991, pp. 46-65, 91-3, 278-84, 307-8

Chicago and its Hinterland [pp. 46–54]

Reading Turner backwards

Few people read the boosters[1] anymore. Their unabashed optimism about progress and civilization has long since gone out of fashion, and their prose is alternately too dry and too baroque for modern tastes. [. . .] The boosters expressed what many Americans believed – or wanted to believe – about the expansion and progress of the United States and its Great West. They offered seemingly rational arguments to reinforce the visionary faith that sustained many who lived and invested in the region. As a group, they present a strikingly consistent picture of how the western landscape would be absorbed into a commercial system revolving around a small number of urban centers. Natural advantage and the movement of human populations together determined how individual cities, towns, and villages would fit into that system. Many such places would prosper, said the boosters, but only one would emerge as the central city of the Great West. [. . .]

The West of the great emporium and its satellites bore little outward resemblance to the West of Frederick Jackson Turner's frontier.[2] In contrast to the boosters, Turner consistently chose to see the frontier as a rural place, the very isolation of which created its special role in the history of American democ-

racy. [. . .] For Turner and his followers, frontier development had been slow and evolutionary, with cities appearing only after a long period of rural agricultural growth. Cities marked the *end* of the frontier. For the boosters, on the other hand, western cities could and did appear much more suddenly. They grew in tandem with the countryside and played crucial roles in encouraging settlement from a very early time. City and country formed a single commercial system, a single process of rural settlement and metropolitan economic growth. To speak of one without the other made little sense. [. . .]

The chief difference between Turner and the boosters hinges on a seemingly minor point: Turner's Chicago rose to power only as the frontier drew to a close, whereas the boosters' Chicago had been an intimate part of frontier settlement almost from the beginning. In this, the boosters saw more clearly than the historian. When they argued that the city grew by drawing to itself the resources of an emerging region they also implied that urban markets made rural development possible. "Chicago, the inevitable metropolis of the vigorous northwestern third of the prairie world," wrote James Parton in 1867, "has taken the lead in rendering the whole of it accessible."[3] Making a landscape "accessible" meant linking it to a market, which meant fostering regular exchange between city and country.

Urban–rural commerce was the motor of frontier change, a fact which the boosters understood better than Turner.

In the twentieth century, the body of theory which analyzes urban–rural systems of the sort that both Turner and the boosters were trying to understand goes by the name of central place theory. Curiously, it traces its roots back to a contemporary of the boosters, writing in Germany at about the same time. Johann Heinrich von Thünen, an educated gentleman farmer in Mecklenburg, published the first edition of his book *The Isolated State* in 1826.[4] In it, he tried to produce a rigorous mathematical description of the spatial relationships and economic linkages between city and country. Neither Turner nor the boosters appear to have read it, and yet it may offer a way to resolve the apparent differences between them. [. . .]

Von Thünen's abstract principles had strikingly concrete geographical consequences [see Figure 11.1]. A series of concentric agricultural zones would form around the town, each of which would support radically different farming activities. Nearest the town would be a zone producing crops so heavy, bulky, or perishable that no farmer living farther away could afford to ship them to market. Orchards, vegetable gardens, and dairies would dominate this first zone and raise the price of land – its "rent" – so high that less valuable crops would not be profitable there. Farther out, landowners in the second zone would devote themselves to intensive forestry, supplying the town with lumber and fuel. Beyond the forest, farmers would practice ever more extensive forms of agriculture, raising grain crops on lands where rents fell – along with labor and capital investment – the farther out from town one went. This was the zone of wheat farming. Finally, distance from the city would raise transport costs so high that no grain crop could pay for its movement to market. Beyond that point, landowners would use their property for raising cattle and other livestock, thereby creating a zone of even more extensive land use, with still lower inputs of labor and capital. Land rents would steadily fall as one moved out from the urban market until they theoretically reached zero, where no one would buy land for any price, because nothing it might produce could pay the prohibitive cost of getting to market.

Von Thünen acknowledged that his abstract thought experiment departed from reality in several important ways. No city was as isolated as this one. All were surrounded by a variety of smaller towns and villages which complicated the hinterland picture. No region was as homogeneous as the hypothetical plain. The natural resources of any real landscape clustered in random patterns that inevitably distorted the abstract zones. Moreover, towns almost always appeared along rivers or canals, which drastically reduced transportation costs for lands along their banks, thereby introducing still more distortions to von Thünen's ideal geography. But none of these "distortions" undermines von Thünen's underlying principles. Each in fact suggests how those principles express themselves in the more complicated geography of the real world. Von Thünen radically simplified his landscape to demonstrate what the nineteenth-century boosters knew intuitively, and what modern central place theorists have confirmed with formal mathematics. Where human beings organize their economy around market exchange, trade between city and country will be among the most powerful forces influencing cultural geography and environmental change. The ways people value the products of the soil, and decide how much it costs to get those products to market, together shape the landscape we inhabit.

Von Thünen's idealized geography suggests how the boosters' urban theories might combine with Turner's rural history to produce a new way of understanding the history of colonization in the Great West and elsewhere.

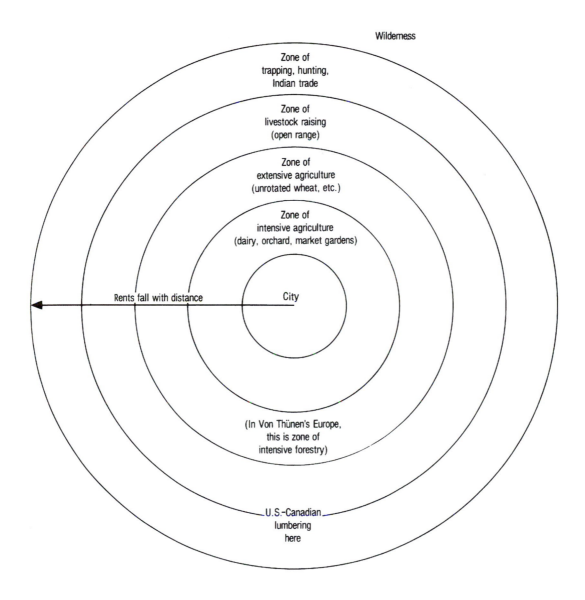

Figure 11.1 Von Thünen's Isolated State.

One has only to imagine his central city in a nineteenth-century American setting – Chicago in 1870, for instance – and then travel outward through the surrounding rural countryside, to experience an odd sense of déjà vu. Leaving the city and its factories behind, one first passes through a zone containing densely populated farm settlements practicing intensive forms of agriculture. Truck [market] gardens, dairies, and orchards dominate the landscape, with many signs that farmers are investing their profits in outbuildings, fences, fertilizers, and other technologies for "improving" agriculture. As one travels farther west, these intensive farms gradually give way to newer and more sparsely settled communities. They practice more extensive agriculture, exploiting the prairie soil by raising unrotated crops of corn and wheat. Farther west still, these give way to the open range, where ranchers and cowboys raise animals rather than crops, on vast stretches of land with very few people and low capital investment. (This is also, in nineteenth-century North America, the zone of the forest, which was lumbered much more extensively – and wastefully – than in von Thünen's Germany, and so was located on lands of low value at greater distance from the city.)

Von Thünen's idealized landscape ended in the livestock-raising zone. But in America, to borrow Turner's admittedly problematic language, an additional zone beyond the pastoral still belonged to "the Indian and the hunter," both of whom had long since welcomed, like the Potawatomis, "the entrance of the trader, the pathfinder of civilization" – a pathfinder whom we can now recognize as an emissary from the metropolitan marketplace.[5] To read von Thünen in this way is suddenly to realize that one is reading Turner backwards, and that Turner's frontier, far from being an isolated rural society, was in fact the expanding edge of the boosters' urban empire. Seen from the midst of the city, Turner's "frontier" stages – hunters, traders, cattle raisers, extensive grain farmers, intensive truck gardeners, and urban manufacturers – look like nothing so much as the zones of von Thünen's Isolated State. Frontier and metropolis turn out to be two sides of the same coin.[6]

One can read von Thünen's map too literally and fall victim to the same sorts of distortions and simplifications for which Turner's thesis has been rightly criticized. The Great West of the nineteenth century was a much more diverse and complicated landscape than these broad zones suggest, and the sweeping abstractions of an idealized geography do little justice to the different historical experiences of the real people who lived within it. Even on its own terms, this application of von Thünen's model raises technical questions about modern central place theory. Where precisely, for instance, should we locate the metropolitan core of the nineteenth-century American city system? It presumably lay somewhere off to the east – whether on the American or the European side of the Atlantic – but it clearly contained many more cities than one. What does this imply about von Thünen's zones, and where do cities like Chicago or St. Louis or San Francisco fall within those zones?

I will return to such questions in the closing chapters of this book, but for now I would offer just two observations about von Thünen's geography. First, he reminds us that city and country are inextricably connected and that market relations profoundly mediate between them. A rural landscape which omits the city and an urban landscape which omits the country are radically incomplete as portraits of their shared world. The zoned hinterland of the Isolated State may oversimplify the diverse realities of the Great West, but it nonetheless suggests the sorts of underlying market principles that have linked city with country to turn a natural landscape into a spatial economy. Chicago's story remains incomprehensible without some knowledge of von Thünen's principles. But – and this is the second point – von Thünen, like many modern central place theorists, made no effort to place his city–country system in *time*. The lone city in the midst of the featureless plain *had no history*, and so poses real problems when one tries to apply it to the extremely dynamic processes that reshaped city and country in the nineteenth-century West. As a historical explanation, the Isolated State is as wanting – and as teleological – as Turner's frontier.

The concrete example nearest at hand is the

best case in point: von Thünen's zones, for all they may suggest about Chicago's later hinterland, shed little light on the city's explosive growth during the 1830s. In just three or four years, a tiny village suddenly increased its population twenty fold, the value of its land grew by a factor of three thousand, and boosters began to speak of it as a future metropolis. Capitalists from the largest cities in Europe and America – London, New York, Boston, Philadelphia, and others – raced to invest in the would-be city. To understand these events, we have to combine von Thünen's abstract geography with the booster theories that persuaded New Yorkers like Charles Butler and William B. Ogden to invest substantial sums to help make Chicago's urban dream come true. Booster models of urban growth were nothing if not dynamic, for the simple reason that they sought to prevail on wealthy investors to turn predictions of urban greatness into self-fulfilling prophecies. "If our National Wheel of Commerce have its Hub immovably pivoted by Nature and by Art," wrote John Wright at the beginning of a 475-page book promoting Chicago, "should not every Business Man know it?"[7] Much as they might claim that the city's growth was "natural" and "inevitable," boosters knew that they were whistling in the dark if they failed to attract outside capital to make their prophecies come true.

Capital held one of the most important keys to metropolitan empire, which was why boosters wrote so many tracts making their case to potential investors. Repeatedly in the nineteenth century, western cities came into being when eastern capital created remote colonies in landscapes that as yet contained relatively few people. Movements of capital helped explain why large cities developed so much more quickly in the West than Turner's evolutionary frontier stages suggest. By linking frontier areas to an international system of cities, these centers of capital investment emerged as

urban markets which drove the region's growth.[8] [. . .]

Chicago had been a marketplace long before boosters proclaimed it a metropolis. Its Potawatomi and French inhabitants made it the focus of a thriving fur trade many decades before it reached the watershed of the 1830s. The Potawatomis had been shrewd traders who bargained well for wealth and power as they understood those things, linking themselves to distant urban markets to pursue their own cultural sense of the good life. These were no "savages," noble or otherwise. They understood the market as it applied to such things as animal skins and alcohol and blankets and guns, and they had at least an equal hand with Euroamerican traders in dominating local trade before 1833. If the frontier was an expanding "wave" of market activity, they were well within its leading edge. [. . .]

For many, if not most, Americans, "the discovery, cultivation, and capitalization" of land meant bringing it into the marketplace and attaching it to the metropolis. They might articulate their visions in terms quite different from those of the boosters' urban empire – speaking of freedom, or community, or family, or getting ahead in the world – but even these noneconomic dreams generally presupposed a growing commercial intercourse between city and country. Frontier and metropolis, and the ideas that lay behind them, would reshape the Great West together. [. . .]

The land was not free but taken. Moreover, even if it became free in the moment that it passed from Indian control, it soon ceased to be free again as it entered the market place. Never again would it be without a price. Tallgrass prairies, oak openings, white pine forests, herds of bison, and the people who might choose to live amid these things: none would ever be the same again. As village became metropolis, so frontier became hinterland. The history of the Great West is a long dialogue between the place we call city and the place we call country. So perhaps the best vantage point from which to view that history is not with Turner, in the outermost of von Thünen's zones, but in the place where Turner himself said "all the forces of the nation intersect."[9] Viewed from the banks of the Chicago River, the Great West is both an urban empire and a countryside transformed. [. . .]

Rails and Water [pp. 55–65, 91–3]

Market in the mud

The boosters spoke much about "natural advantages." Resources, waterways, and climatic zones loom so large in their writings that one can almost forget that *people* have something to do with the building of cities. A river, a lake, and a fertile plain present many opportunities that intimately influence those who live nearby. And yet people make such different choices about nature's opportunities: one could hardly confuse the French–Anglo–Indian fur-trading village that was Chicago in 1830 with the speculative American boomtown that had replaced it half a decade later. Geographical arguments do not explain how the one became the other; only culture and history can do that. [. . .]

The harbor was just one example of the many "improvements" that Chicagoans and other American settlers would bring to the landscape of the Great West. In addition to advantages and opportunities for growth, nature threw up obstacles which those who dreamed of human progress had to overcome at every turn. Each new improvement meant a shift in regional geography – a dredged harbor here, a canal or a road there – so the advantages sustaining the city came to have an ever larger human component. A kind of "second nature," designed by people and "improved" toward human ends, gradually emerged atop the original landscape that nature – "first nature" – had created as such an inconvenient jumble.[10] Despite the subtly differing logic that lay behind each, the geography of second nature was in its own way as compelling as the geography of first nature, so boosters and others often forgot the distinction between them. Both seemed quite "natural." Nowhere was this more true than in the new artificial transportation technologies that changed the ways people and commodities moved back and forth between city and country. Although early boosters had believed that rivers and lakes would carry an irresistible flow of resources to the city they favored, canals and especially railroads finally proved more important in building Chicago and other cities of the Great West. Second nature defined the corridors of commerce at least as much as first nature.

By the 1840s, before any railroad had yet reached Chicago, its merchants were doing a good business with the expanding farm communities of northern Illinois and Indiana. But the difficulty of moving agricultural produce across the landscape discouraged a wider trade and limited the city's growth. [. . .]

[M]erchants and farmers of the 1840s [. . .] had to live with the seasonal transportation challenges that afflicted city and country alike. The practical difficulties of the mud season were but the flip side of the very advantages that had led speculators to identify Chicago as a prime center for water-based trade in the first place. If people wanted a town that would benefit from a natural harbor (however bad) and a natural canal corridor (however undug), they would have to live with a little water and mud. Trade and transportation therefore waxed and waned with the seasons. Just as von Thünen had predicted, the regional economy was shaped primarily by distances between city and country, expressed not in miles but in the time and expense devoted to transportation. Periods of slow trade and difficult travel became part of the cost of doing business, a kind of natural excise tax paid on virtually all movement and trade. The more time people devoted to waiting for customers or traveling to market, the less time they had for more productive activities.

Even when roads were in decent condition, the only vehicles that could use them were horse-drawn wagons, which had limited capacity and became uneconomical for moving agricultural produce over any great distance. Farmers drove such wagons to Chicago from surprisingly remote places, bringing to market their most valuable commodities – apples, ham, butter, feathers, chickens, wheat – from as far away as the Rock River in northern Illinois and the Wabash River in southern Indiana, well over a hundred miles distant. But they could not make such journeys often, and that limited the entire economy. Furthermore, because farmers could carry only small loads in such vehicles, the costs of wagon, horses, and driver consumed a sizable portion of any money they earned. Wagons offered few economies of scale, and so set well-defined limits to how far one could afford to travel in them. As one Chicago businessman observed, it took a nearby farmer on the Rock River five days just to bring an average-sized wagonload of thirty bushels of wheat to market, so the cost of the journey "took off nearly all the profit."[11] [. . .]

The farmers chose Chicago as their destination because they received more cash for their crops there, and because they could buy more and better supplies at lower prices. River towns in the interior – Peoria, Springfield, Vincennes, even St. Louis – did not have the cheap lake transportation to the east that gave Chicago its price advantage. Wheat, for instance, often brought anywhere from ten to sixty cents more per bushel in Chicago than in downstate communities.[12] [. . .]

What the farmers found in Chicago was the western outpost of a metropolitan economy centered on the great cities of Europe and the American Northeast. Chicago, located in one of von Thünen's outer zones was able to buy and sell so successfully because the lakes, the Erie Canal, and the Hudson River gave it better access to eastern markets – especially those of New York – than any other city in Illinois. Other lake cities had comparable advantages: Cleveland and Toledo offered their hinterlands the best markets in Ohio, and Milwaukee played a similar role in Wisconsin. As in these other places, most of what farmers bought in Chicago during the early years came not from the city itself but from the Northeast. Chicago's advantage in selling such merchandise derived from its favorable price structure. Its merchants could buy goods at eastern wholesale prices in ship-sized quantities with no markup for expensive land transport. For the same reasons, they could also offer the best prices in the region for farm produce moving east. Low prices for eastern goods, and high prices for western ones: the combination was a sure recipe for success. [. . .]

Cities with the greatest access to the East would become the new metropolises of their region; towns with less direct eastern ties would rely on western wholesaling centers for the bulk of their merchandise and develop only a local retail trade of their own.[13]

Here was the hidden foundation of the boosters' geographical determinism: natural avenues of transportation might play important roles in shaping a city's future, but the preexisting structures of the *human* economy – second nature, not first nature – determined which routes and which cities developed most quickly. Chicago enjoyed its favorable price structure because New York merchants and bankers had already consolidated for their city the role of national metropolis. By midcentury New York had developed the most direct access to European markets, the most extensive trading hinterland, and the most powerful financial institutions of any city in North America. Without New York, the natural advantages of Great Lakes shipping would have meant little. Had New Orleans, and not New York, been the chief entrepôt between Europe and North

America, the evolution of western trade would surely have followed a different course. [. . .]

Artificial corridors

The muddy roads and shallow harbor gave Chicago its early hinterland, attracting farmers and other customers from a hundred or more miles away during the 1840s. But the considerable disadvantages of these early transportation routes also limited the city's business. [. . .]

Boosters had initially expected that Chicago would float to greatness on the proposed canal between Lake Michigan and the Illinois River, following the old glacial channel from the days when the lake drained south toward the Mississippi. Surveying the canal route had helped trigger the city's real estate boom in the 1830s, though the canal itself took much longer to get going. [. . .]

[T]he Illinois and Michigan Canal finally opened for traffic in April 1848.

Just as its early promoters had predicted, the canal brought striking changes to the regional economy. During its first season of operation eastern corn shipments from Chicago multiplied eightfold as farmers in the Illinois River Valley suddenly discovered an alternative to St. Louis as an outlet for their produce. The explosion of corn sales furnished convincing proof of the boosters' arguments in favor of water transport. By avoiding the risks and frustrations of the muddy roads leading to Chicago, farmers could bring much more of their produce to market, and purchase greater quantities of urban manufactured goods as well. Over 90 percent of the new corn shipments came to Chicago via the canal, which was henceforth the city's chief source of corn until after the Civil War. Lumber receipts at Chicago from the forests of Michigan and Wisconsin nearly doubled in 1848, and one-fourth of this wood moved south down the canal, to be used for houses, fences, and farm buildings on the Illinois prairies. By decreasing the difficulty and cost of transportation, the canal enabled larger quantities of heavier and bulkier goods to extend their geographical reach both to and from Chicago. It was as if a corridor of relatively cheap transport had suddenly appeared like a fault across the various zones of von Thünen's isolated city, displacing them and necessitating a complex series of adjustments in the region's spatial economy. The canal almost instantly expanded Chicago's hinterland southward to the Mississippi River just above St. Louis.

Before people had fully adjusted to trading via the canal, however, a second artificial corridor, which would bring even more dramatic changes, augmented Chicago's access to its surrounding region. In 1836, the year canal construction began, a new company was chartered to construct a railroad. [. . .]

By fulfilling the role that the railroads had assigned it – serving as the gateway between East and West – Chicago became the principal wholesale market for the entire midcontinent. Whether breaking up bulk shipments from the East or assembling bulk shipments from the West, it served as the entrepôt – the place in between – connecting eastern markets with vast western resource regions. In this role, it became a key participant in a series of economic revolutions that left few aspects of nineteenth-century life untouched. Henceforth, Chicago would be a metropolis – not the central city of the continent, as the boosters had hoped, but the gateway city to the Great West, with a vast reach and dominance that flowed from its control over that region's trade with the rest of the world.

[. . .] The changes that the railroad system initiated would proliferate from Chicago and fundamentally alter much of the American landscape. As the city began to funnel the flow of western trade, the rural West became more and more a part of its hinterland, mimicking the zones von Thünen had described even before Chicago existed as a town. The isolation that had constrained the trade and production of frontier areas would disappear. [. . .] Wherever the network of rails extended, frontier became hinterland to the cities where rural products entered the marketplace. Areas with limited experience of capitalist exchange suddenly found themselves much more palpably within an economic and social hierarchy created by the geography of capital. [. . .]

The railroads had made Chicago the most important meeting place between East and West. But they also continued the process begun long before with the harbor and the canal, and before that in the trading village and the booster dreams that transformed it. In Chicago and its hinterland, first and second nature mingled to form a single world. The boosters had been indulging their rhetorical mysticism when they likened the railroads to a force of nature, but there can be no question that the railroads acted as a powerful force *upon* nature, so much so that the logic they expressed in so many intricate ways itself finally came to seem natural. In the second half of the nineteenth century, city and country, linked by "the wild scream of the locomotive," would together work profound transformations on the western landscape. On the farms of Illinois and Iowa, the great tallgrass prairies would give way to cornstalks and wheat fields. The white pines of the north woods would become lumber, and the forests of the Great Lakes would turn to stumps. The vast herds of bison on which the Plains Indians had depended for much of their livelihood would die violent deaths and make room for more manageable livestock. Like von Thünen's isolated city, Chicago was remote from all of these events. And yet no place is more central to understanding why they occurred. [. . .]

Chicago, a Gateway City [pp. 278–84, 307–8]

[. . .] In economic and environmental terms, we should think of a city and its hinterland not as two clearly defined and easily recogniz-

able places but as a multitude of overlapping market and resource regions. This suggests in turn that we should revise von Thünen's suggestive but simplistic map of concentric agricultural zones surrounding an isolated metropolis. His core insight remains sound: goods do travel to market according to their value, weight, bulk, and ability to pay their cost of transportation. But von Thünen's model becomes much more complicated as soon as we recognize that no real city exists in such grand theoretical isolation. Precisely because a city's markets create so many different regions of supply and demand neighboring cities and towns inevitably share hinterlands. "Thus each commercial city," wrote a nineteenth-century economist, becomes "very sharply the rival of every other commercial city."[14] Merchants in different places compete with each other to sell goods in surrounding areas wherever they can offer attractive prices. Economic geographers have struggled in the century and a half since von Thünen to understand the spatial implications of this competition among urban merchants, and the result has been the arcane body of mathematical models known as central place theory. [. . .]

The urban hierarchy

Central place theorists seek to explain the geographic phenomenon I noted at the start of this chapter: the tendency of human settlements to organize themselves into hierarchies.[15] All cities in the modern capitalist world – not just Chicago during the nineteenth century – exist within *systems of cities*. A few large metropolises link with a larger number of big cities, and each of those links in turn with a still larger number of small towns.[16] Urban populations arrange themselves into rank order by size: population increases exponentially with rank, so the higher a city's rank, the more people it contains. By 1890, the year when Chicago finally surpassed Philadelphia to become the second-largest metropolis in the United States, there were only 3 cities in the nation with populations greater than 1,000,000. [. . .]

The difference between a high-order metropolis like Chicago and a lower-order town like Peoria or Burlington was not merely Chicago's much larger population. Chicago's high rank meant that its market attracted customers for many more goods and services from a much wider region. No less important, it attracted demand for much more specialized goods and services. Just as one can rank human settlements according to the number of people who live in them, so can one rank all economic goods according to the number of people and concentrations of wealth needed to create a market for them. The hierarchy of urban settlements is also a hierarchy of markets. [. . .]

A city like Chicago was first and foremost a center of *wholesale* trade.[17] Merchants in small towns and medium-sized cities sold principally to the retail customers in their immediate hinterlands. To do so, they bought their own supplies from wholesale merchants in Chicago, New York, and other metropolitan markets. Chicago earned its high rank partly by being a retailer itself, offering its customers a greater variety and number of retail establishments than any other city in the Great West. These ranged from large department stores selling every conceivable product to small firms specializing in exceptionally narrow lines. But the city's metropolitan status derived above all from the ability of its wholesalers – many of them just as specialized as its retailers – to supply discounted goods to virtually any retailer in the country. The city was a shopkeeper for shopkeepers, a market for other markets. [. . .]

A metropolis like Chicago contained within its hinterland hundreds and even thousands of smaller places. Hinterland villages and towns sold food and clothing to their immediate retail customers, and medium-sized cities sold more specialized retail products to the towns and farms that surrounded them. But all bought their supplies from the wholesale markets of

Chicago, just as local farmers, ranchers, and lumbermen sold their output to the city's grain elevators, packing plants, and lumberyards. The map of towns and settlements reflected this hidden network of markets within markets, low-ranked places within the fields of high-ranked ones. Modern central place theorists have offered elaborate formal geometries to describe these nested urban hinterlands, with intricate layers of large and small hexagons describing like so many honeycombs the markets for high- and low-ranked goods in high- and low-ranked places.

Central place theory has an elegant mathematical simplicity as it confronts the complex hierarchies of human settlement and trade, but it shares with von Thünen's agricultural zones one great flaw: it is profoundly static and ahistorical. Reading the treatises of the German theorists who originally developed it, one is struck by the abstract neatness of this geography. Its nested hexagons have none of the messiness one expects of real historical places and landscapes. In its original, most undiluted form, central place theory offers a purely formal explanation of how market hierarchies evolve. In the fantasy of a flat, featureless plain which the central place theorists share with von Thünen, population grows until small village centers begin to appear with the expansion of local market demand; they in turn eventually create a market for medium sized towns; they in turn create larger cities; they in turn create a great metropolis. Like the economic logic of capitalism itself, the entire process easily comes to seem second nature, as organic and evolutionary as Darwin's model of biological change.

But the growing city system in the region west of nineteenth-century Chicago followed a more precipitous course. Far from being a gradual, bottom-up process in which villages called forth towns, towns called forth cities, and cities at last called forth the metropolis of Chicago, nearly the opposite was true. The highest-ranking regional metropolis consolidated its role at a very early date, and promoted the communities in its hinterland as much as they promoted it. The region underwent its greatest growth during a period when urban–industrial capitalism had already established itself on the eastern seaboard, tying the American economy to an international trade system that stretched across the Atlantic to European ports and markets. Because Chicago enjoyed unique transportation advantages by virtue of its position on the divide between the Great Lakes and Mississippi watersheds, and because the profound centralizing tendencies of the railroads amplified those advantages, the city emerged as a metropolis strongly linked to eastern markets long before the villages and towns of its expanding hinterland had filled out their eventual hierarchy of settlements. In so doing, Chicago disrupted the trade patterns that had already been developing in the region to its west. By the terms of central place theory, Chicago grew too large, too high-ranked, too quickly. This in turn suggests that something other than gradual market evolution was responsible for its metropolitan status.

The hierarchy of city, town, and country that appeared so quickly in the Great West during the second half of the nineteenth century represented a new phase of American frontier expansion, far more rapid than anything Frederick Jackson Turner described. Its accelerated pace was driven by the new rail technologies, but the growth of Chicago's metropolitan hinterland was an extension of the urban hierarchy that had already emerged in the East, particularly in relation to New York City. Chicago's high-ranking urban functions as a wholesaler and financial metropolis flowed directly from its special relationship to the city on the Hudson. By choosing Chicago to be the greatest concentration of railroad capital on the continent, and by giving Chicago merchants special access to the credit and discounts that made wholesaling possible, New Yorkers and other eastern capitalists placed it atop the western city system at the very moment that settlement in the region began its most explosive growth. [. . .] The Canadian geographer A.F. Burghardt [. . .] described this process. In his phrase, Chicago became a "gateway city" by serving as the chief intermediary between newly occupied farms and towns in the West and the maturing capitalist economy of the Northeast and Europe.[18] Although the city's gateway status lasted only a few decades, while the capitalist economy completed its colonization of the Great West, that was long enough to make Chicago second only to New York in the reach and power of its markets. [. . .]

Gateway cities were a peculiar feature of North American frontier settlement. To return to the argument of the Canadian geographer A.F. Burghardt, they were not central places, and did not conform to the expectations of central place theory that a metropolis should sit like von Thünen's isolated city at the center of a symmetrical network of medium- and low-ranked cities, towns, and farms. Instead, the gateway served as the entrance and exit linking some large region with the rest of the world, and it therefore stood at one end – usually the eastern end – of a large tributary hinterland that had no other means of communication with the outside. Often, as in the case of St. Louis and then Chicago, the gateway city's hinterland was extremely elongated, stretching hundreds and even thousands of miles to the west but a much shorter distance to the east. It was undoubtedly a metropolis, but there was nothing *central* about it.

Compared with that of a typical central place sharing its high rank, the gateway city's economy was much more committed to long-distance transportation and wholesaling, and for a simple reason: it was the principal colonizing agent of the western landscape. The gateway city served as the go-between linking the settlements and natural resources of the Great West with the cities, factories, and commercial networks of the Northeast. On one side, westerners used it as the most effective way to gain access to the eastern markets that could transmute their land and labor into cash. On the other, eastern capitalists used it to design a system of transport and commerce that would concentrate western supply and demand – and profit – at their doorsteps. The two groups met at the gateway to do business, and so joined east and west in a single market system. The gateway metropolis represented a revolution in political economy, a complex transformation of culture, and an ecological watershed all at the same time. [. . .]

Notes

1 [Optimistic nineteenth-century city-promotors]

2 [See Turner, *Frontier in American History*]

3 Parton, "Chicago," 327

4 Johann Heinrich von Thünen, *Von Thünen's Isolated State* (1826, 1842), trans. Carla M. Wartenberg, ed. Peter Hall (1966)

5 Turner, *Frontier in American History*, 11

6 Articles that apply von Thünen's model to the expansion of agriculture in the nineteenth century include John T. Schlebecker, "The World Metropolis and the History of American Agriculture," *Journal of Economic History* 20 (1960): 187–208; J. Richard Peet, "The Spatial Expansion of Commercial Agriculture in the Nineteenth Century: A Von Thünen Interpretation," *Economic Geography* 45 (1969): 283–301; and Peet, "Von Thünen Theory and the Dynamics of Agricultural Expansion," *Explorations in Economic History* 8 (1970–71): 181–201

7 Wright, *Chicago*, frontispiece

8 A metropolitan perspective has traditionally come more easily to historians in Canada than to those in the United States. Despite the roughly similar settlement histories of the two countries, historians of the American West, with a few notable exceptions, have continued to follow Turner's emphasis on the region's rural aspects; even where they have focused on cities, they have rarely emphasized the city–country relationship. Canadian historians, on the other hand, have more or less rejected the Turnerian frontier and have pursued what has come to be called the metropolitan thesis. Those familiar with the nineteenth-century boosters will instantly recognize its main features. As J.M.S. Careless, a leading Canadian historian, describes it, metropolitanism "implies the emergence of a city of outstanding size to dominate not only its surrounding countryside but other cities and their countrysides, the whole area being organized by the metropolis, through control of communications, trade, and finance, into one economic and social unit

that is focused on the metropolitan 'centre of dominance' and through it trades with the world." In Canada, this city was Montreal; in the United States, it was New York. Beneath the central metropolis were smaller cities playing similar but less extensive roles, since "the metropolitan relationship is a chain, almost a feudal chain of vassalage, wherein one city may stand tributary to a bigger centre and yet be the metropolis of a sizable region of its own." The examples here were Toronto in Canada, and Chicago in the United States. J.M.S. Careless, "Frontierism, Metropolitanism, and Canadian History," *Canadian Historical Review* 35 (1954): 17

9 Turner, *Frontier in American History*, 150

10 On the Hegelian concept of "second nature" as the cultural transformation of "first nature," see Alfred Schmidt, *The Concept of Nature in Marx* (1971), 42–43; for a critique and elaboration of the concept, see Neil Smith, *Uneven Development: Nature, Capital and the Production of Space* (1984), 19–20

11 Cleaver, *History of Chicago*, 111

12 Yetter, "Chicago Commercial Growth," 32; Pierce, *History of Chicago*, 1:127

13 [. . .] Lewis E. Atherton, *The Frontier Merchant in Mid-America* (1939; reprint, 1971), 59–162; [. . .] James E. Vance, Jr., *The Merchant's World: The Geography of Wholesaling* (1970)

14 Nimmo, *Rept. Int. Commerce* (1881), 103

15 The early classics of central place theory were almost entirely by German geographers trying to elaborate the insights of von Thünen, *Isolated State*. The most important works are Alfred Weber, *Theory of the Location of Industries* (1909), trans. Carl J. Friedrich (1929); Walter Christaller, *Central Places in Southern Germany* (1933), trans. Carlisle W. Baskin (1966); August Lösch, "The Nature of Economic Regions," *Southern Economic Journal* 5 (1938): 71–78; Lösch, *The Economics of Location* (1939), trans. William H. Woglom (1954); Edward Ullman, "A Theory of Location for Cities," *American Journal of Sociology* 46 (1940–41): 853–64; Edgar M. Hoover, *The Location of Economic Activity* (1948); and Walter Isard, *Location and Space-Economy: A General Theory Relating to Industrial Location, Market Areas, Land Use, Trade, and Urban Structure* (1956)

16 The classic essay on city systems is Brian J.L. Berry, "Cities as Systems within Systems of Cities," *Papers and Proceedings of the Regional Science Association* 13 (1964): 147–63. [. . .] See also L.S. Bourne and J.W. Simmons, eds., *Systems of Cities: Readings on Structure, Growth, and Policy* (1968)

17 Vance, *The Merchant's World: The Geography of Wholesaling* remains the classic on this subject, and offers an important historical critique of the ahistorical models that characterize other central place theorists

18 Burghardt, "Hypothesis about Gateway Cities."

References

ATHERTON, L. (1930) *The Frontier Merchant in Mid-America*, reprint 1971, Columbia, University of Missouri Press

BOURNE, L.S. and SIMMONS, J.W. (eds) (1968) *Systems of Cities: Readings on Structure, Growth, and Policy*, New York, Oxford University Press

BURGHARDT, A.F. (1971) "A Hypothesis about Gateway Cities", *Annals of the Association of American Geographers*, 61, pp. 269-85

CHRISTALLER, W. (1933) *Central Places in Southern Germany*, trans. BASKIN, C.W. (1966) Englewood Cliffs, NJ, Prentice-Hall

CLEAVER, C. (1892) *History of Chicago from 1833 to 1892*, Chicago, privately printed

HOOVER, E.M. (1948) *The Location of Economic Activity*, New York, McGraw-Hill

ISARD, W. (1956) *Location and Space-Economy: A General Theory Relating to Industrial Location, Market Areas, Land Use, Trade and Urban Structure*, Cambridg, MA, MIT Press, New York, John Wiley

LÖSCH, A. (1939) *The Economics of Location*, Trans. WOGLOM, W.H., New Haven, Yale University Press

NIMMO, J., Jr. (1877-81) *Report on the Internal Commerce of the United States*, US Treasury Department, Bureau of Statistics, Washington, DC, US Government Printing office

[PARTON, J.] (1867) "Chicago", *Atlantic Monthly* 19, pp. 325-45

PIERCE, B.L. (1937-57) *A History of Chicago*, 3 vols, New York, Alfred A. Knopf

SCHMIDT, A. (1971) *The Concept of Nature in Marx*, London, New Left Books

SMITH, N. (1984) *Uneven Development: Nature, Capital and the Production of Space*, New York, Basil Blackwell

TURNER, F.J. (1920) *The Frontier in American History*, reprint (1962), New York, Holt, Rinehart and Winston

VANCE, J.E., Jr. (1970) *The Merchant's World: The Geography of Wholesaling*, Englewood Cliffs, NJ, Prentice-Hall

VON THÜNEN, J.H. (1826, 1842) *Von Thünen's Isolated State*, trans. WARTENBERG, C.M., ed. HALL, P. (1966), New York, Pergamon Press

WEBER, A. (1909) *Theory of the Location of Industries*, trans. FRIEDRICH, C.J. (1929), Chicago, University of Chicago Press

WRIGHT, J.S. (1870) *Chicago: Past, Present, Future Relations to the Great Interior and to the Continent*, Chicago

YETTER, R. (1937) "Some Aspects of the Commercial Growth of Chicago, 1835-1850", MA thesis, University of Chicago

12

COMPARATIVE PERSPECTIVES ON TRANSIT IN EUROPE AND THE UNITED STATES, 1850–1914

by John P. McKay

Source: John P. McKay, ''Comparative Perspectives on Transit in Europe and the United States, 1850–1914'', in Joel A. Tarr and Gabriel Dupuy (eds), *Technology and the Rise of the Networked City in Europe and America*, Philadelphia, PA, Temple University Press, 1988, pp. 3–21

When one considers the city in a broad historical perspective, it is clear that public transit is a recent phenomenon and that sophisticated transit technologies are even more recent. [. . .]

In fact, the old transportation technology of animate power – the horse – proved more adaptable to emerging transit needs than the new one of steam and iron. Thus horse-drawn omnibuses appeared as the first significant form of urban public transportation. [. . .] [O]mnibus service played a rather significant role in the mid-century growth of new middle-class residential areas located 1 to 2 miles from the historic city center in both Europe and the United States.[1] [. . .]

By 1860, horsecars were replacing omnibuses in many American cities, and the trend accelerated during the Civil War. Once established, per capita usage grew rapidly. In New York, for example, horse railway traffic almost tripled in the 1860s, reaching 115 million passengers by 1870 – or about a hundred rides per year per capita.[2] In Europe, in contrast, only a few large cities had established the new form of transit by 1869, and the 1870s marked only the beginning of serious street railway development in most British, French, and German cities. Moreover, whereas streetcars quickly replaced omnibuses in most American cities, their European counterparts often used streetcars to complement omnibuses, which remained in service into the twentieth century in some instances.

Although Europe's lag and America's lead in adopting and developing the new transit technology may seem a minor or perhaps accidental matter, the early lack of synchronization in increasing the supply of urban public transportation reflected significant differences between the two urban environments. Indeed, a comparative study of urban transportation enables us to isolate forces and institutions that have shaped purely technical developments – so often the primary interest of transportation historians – and to analyze the consequences of alternative technologies for economy and society.[3]

First, the rate of urban growth was generally more rapid in the United States, and this growth apparently created a more intense demand for urban transit at an earlier date. More certainly, the difference in timing reflected contrasting physical environments. Fast-growing American cities, with their easily extended grid pattern, had fewer paved streets than their Western and Central European counterparts, and the gain in efficiency and profit from replacing omnibuses with street-cars was correspondingly greater. Moreover, although the larger "walking cities" in the United States were perhaps as congested and over-crowded as their European counterparts – New York's population density surpassed London's in 1850 – the desire to move toward the fringes, toward the suburbs, was also stronger than in continental Europe (though perhaps not in Great Britain). On the Continent, centuries of city walls and intensive utilization of a carefully defined built-up area limited the appeal of an emerging suburban ideal directly dependent on improved transit.

Transit technology was also embedded in different institutional structures. Cities on both sides of the Atlantic relied on privately owned street railway companies, but these private transit firms were more strictly regulated by national and municipal governments in Europe than in the United States. Franchises granted to private entrepreneurs in the United States had few specific provisions – paving between and beside the rails was often the most onerous obligation – and they were often perpetual in their duration. Even when municipalities stipulated a fixed term, what was to occur when the franchise expired was "a subject upon which almost all of the limited franchises granted in the United States are silent," according to an authoritative 1898 report.[4] Instead, with both the public and real estate speculators avid for more urban transit, and politicians susceptible to corruption, American cities generally relied on a multiplicity of competing franchises and market forces to regulate companies and provide good service. Around 1880, New York had seventeen urban transit companies and Boston had seven; Philadelphia had more than twenty by 1865.

In Europe, however, street railway companies, like their mainline railroad predecessors, were regulated by public authority from the beginning. This European regulatory framework placed restraints on transport entrepreneurs, scrutinizing plans to replace omnibuses with streetcars, for example, and often checking the spread of the American innovation. Generally speaking, European states established a comprehensive juridical framework and well-defined procedures for granting "concessions" to petitioning entrepreneurs, who worked out specific terms with local governments and technical experts from the central administration. Concessions were normally for long periods – forty to fifty years – but on expiration, all the immovable property – mainly the tracks – reverted to the city free of charge. In addition to contractual provisions on fares, routes, paving obligations, and frequency of service, national and local authorities had extensive powers allowing them to regulate the speed and capacity of cars, or require or reject modifications as a matter of course. As a result, European cities were more disposed to accept one or a very few transit companies and generally avoided a proliferation of competing franchises.

Here, as elsewhere, there were variations on European themes. Supposedly laissez-faire Great Britain laid down rigorous control in the Tramways Act of 1870. Municipal governments were permitted to build tracks and own lines, although they were required to lease them to a private operating company for not more than twenty-one years. France, a model for much of Europe in street railway legislation, tightened up initial regulation in its tramways law of June 11, 1880. This law called on state engineers to verify the expenditures of tramway constructors and check the speculative inflation of the capital account noted in the tramway boom of the 1870s. German regulation was strict, but at the state rather than the imperial (that is, national) level.

All these factors continued to operate in the 1880s. And there can be little doubt that in 1890, America continued to lead in the development of urban transit. In the first place, the urban population in the United States simply used urban transit to a much greater degree than its European counterpart. In 1890, when rapid electrification had barely begun, per capita transit usage was three to four times as high in the United States as in Europe. For example, in large German cities, the arithmetical average of annual rides per capita was 39; in large American cities, it was 143. In 1890, the inhabitants of the four largest American cities averaged 195 rides annually on all modes of urban public transportation, while the inhabitants of Great Britain's four largest cities averaged but 56.

This quantitative advantage in usage was probably the critical difference for most Americans who thought about the matter. Thus the 1890 American transportation census admitted that "comparisons much to our disadvantage are frequently made between street railway systems of our own and foreign cities, the grounds of conclusions favorable to the foreign service being that it is managed with more regard for public convenience and made to contribute more to municipal expense." But with evident pride in the American achievement, the writer concluded that in their "most important function" – that of carrying passengers – "our lines have attained a development unapproached in Europe. . . . The street railway system of Berlin, justly regarded as the most perfect in Europe, carried in proportion to the inhabitants of the city less than one-third as many passengers as the city of New York."[5]

The 1890 difference reflected substantial growth in per capita usage in the United States in the 1880s. In Europe, in contrast, the rapid growth of the 1870s was followed by stagnation in the 1880s in such cities as Paris, Leipzig, and Vienna. Once again, there was apparently less demand for more transit in Europe than in the United States in these years. Part of the reason was macroeconomic – less robust economic growth in Europe. But more important was urbanizing America's greater areal dispersion, most strikingly evidenced by its "streetcar suburbs" – new, relatively low-density residential areas linked by daily commuting and a symbiotic relationship to a sharply defined central business district. To cite but one example, the well-studied growth of such suburbs in Boston coincided with a sharp rise in rides per capita, from 118 in 1880 to 175 in 1890.[6] In Europe, horse-drawn tramways had succeeded in tapping an expanded clientele with lower fares, but before 1890 they clearly contributed much less to spatial diffusion than their American counterpart.[7]

Finally, as is well known, the United States took the lead in applying mechanical traction to urban public transportation. What is less well known is that innumerable experiments first proceeded on both sides of the Atlantic and that, in addition to the rather obvious greater demand in the United States, the different institutional patterns we have noted and the contrasting aesthetic perceptions we have hinted at played significant roles in the differential development and diffusion of new transit technologies.

Steam technology was one line of attack. Steam engines suitable for urban transit purposes seemed to promise faster, more flexible, and potentially cheaper service than horse traction. Steam traction would also be more dependable – horses were subject to illness and disease – and would reduce the hygiene problem inherent in horse droppings. The challenge, of course, was to meet environmental objections to noisy, dangerous engines in busy public streets, either by building specially designed dummy locomotives for street railways or special, self-propelled steam streetcars. Yet, in spite of many efforts, such as ingeniously enclosed coke-fired boilers capable of reducing noxious smoke and cinders, steam-powered tramways never succeeded in meeting environmental requirements and simultaneously achieving a clear economic advantage over horse traction. Only on a few purely suburban lines did they attain real urban significance in Europe, and apparently in the United States as well.

Yet there were important differences in steam-traction development. Americans appear to have experimented less and Europeans more with pollution-reducing steam engines and self-propelled cars. Moreover, steam street railways carried 287 million passengers in the United States in 1890, almost one-fourth as many as were carried by horse traction and a far larger portion than in European countries.[8] Above all, in a single but extremely significant case, New York City allowed massive smoke-belching, mainline railroad locomotives to run through the streets on elevated tracks, despite the obvious drawbacks of these "environmental monstrosities."[9] To be sure, as Charles Cheape has shown, New York's use of elevated railroads for mass transit grew incrementally out of an unsuccessful cable-powered franchise, and he has argued further that environmental damage was thus accepted because it was not immediately apparent. Yet the fact that major elevated extensions were granted in 1875, after the initial elevated system was fully functioning, suggests that transit needs took precedence over cultural–aesthetic–environmental considerations in New York's acceptance, however grudging, of new technology. By contrast, in the largest European cities, steam-powered elevated railways in city streets were never accepted. Although the similarity is not perfect, it seems clear that the same factors accounted for the Americans' greater use of steam for urban transit as for its earlier, more wholehearted acceptance of horse-railway tracks in city streets.

The extensive development of the cable car was another aspect of the decisive US lead in mechanization. Developed in 1873 in San Francisco, where extremely steep grades combined with straight streets to provide strong incen-

tives and ideal conditions for a continuously moving cable laid between the tracks, cable systems were installed in twenty-six American cities by 1891. Requiring heavy investment and therefore suitable only in high-density areas, cable cars ran twice as fast as horsecars and were nonpolluting.[10] By 1890, they were carrying 373 million passengers in the United States, as opposed to 1.23 billion on all horse-drawn street railways (and 287 million on all forms of steam-powered street railways).[11] In contrast, cable cars were used on only a few lines in Europe, mainly in London, Edinburgh, Birmingham, and Paris. Differences in physical environments were probably the most critical factor. Moving well on straight, grid-set thoroughfares and turning poorly on curves, cable cars were ill suited for the sinuosities of Europe's historic urban landscape.

Efforts to apply electric traction to urban transit, which were eventually to prove highly successful, are particularly interesting from a comparative point of view. A whole series of pioneers and inventors in many countries made important technical contributions before the decisive breakthrough, a familiar pattern in the process of strategic innovation. [. . .]

Until 1882, largely because of Siemens' work, Europe led in the development of electric traction.

Yet Europeans failed to capitalize on their initial advance, in part because Siemens did not believe electricity would be permitted in public streets, and Americans achieved the decisive breakthrough between 1883 and 1888. In those years, following Thomas Edison's momentary interest in 1880, several inventors and businessmen competed to create what became the standard American technology for urban street transit. That technology – first successfully synthesized in Richmond, Virginia, by Frank Sprague in February 1888 – combined improved electric motors with an overhead conductor method. An under-running trolley on the streetcar picked up current for the motor from a single copper-wire conductor attached to poles installed for that purpose. The rails were used for the return circuit.

Overhead electric traction promised – and then delivered – substantially greater profits to street railway companies by cutting per unit operating costs on horsecar operations. Although there were severe difficulties in calculating exactly how much savings were involved, costs per rider declined by roughly 50 percent even as fixed investment per unit of line doubled. Still more important than these cost reductions, electric traction broke the long-standing bottleneck of inadequate supply of urban transit. It permitted companies to increase enormously their capacity to carry passengers on "saturated" lines and networks. Increased supply capacity, so critical for economic growth in general and not merely urban transport development, resulted principally from increased car speed, larger cars (and/or double-deckers and trailers), and much better performance on hills and in snow. Moreover, the increased quantity of transport was of better quality because electric trams rode more smoothly, stopped more easily, and did not lurch from side to side. In short, the gains from technology were very substantial, however they might be distributed between passengers, operating companies, equipment producers, and taxing authorities.

As we might expect at this point, the initial responses of American and European cities to practicable electric traction were quite different. Electrification shot through the American street railway industry like current through a copper wire. Within two short years after Sprague's Richmond installation, one-sixth of all street railway track in the United States was electrified. By the end of 1893, total street railway track had doubled from the 1890 level to 12,000 miles; ten years later, it had reached 30,000 miles. Of this greatly increased total, fully 60 percent was electric by the end of 1893, and 98 percent was electric by the end

of 1903. In the words of a leading transportation historian, the electric streetcar in the United States "was one of the most rapidly accepted innovations in the history of technology."[12]

In Europe, however, electric traction proceeded slowly amid great discussion for several years. As one of many European observers put it in 1892, the development of electric street railways had been prodigious in the United States, since it became a "practical commercial success in 1888, while Europe has shown scarcely any improvement."[13] Then, from the end of 1893, Europe increased its electric line about tenfold, to almost 1,900 miles at the end of 1898, when an electric tramway boom was in full swing. By the time the depression of 1901 arrived, electric systems were either in operation or under construction in almost all European cities. As might be suspected, there were national differences. Germany clearly led the way in European street railway electrification, and Great Britain lagged behind Germany by about a decade. France and indeed most continental countries occupied the middle ground, leaning more toward the German than the British extreme. More surprisingly, although the standard trolley-wire conductor used almost exclusively in the United States clearly predominated in Europe, European cities also used a complex variety of electric traction systems well into the twentieth century. The extent to which these different electric systems were used in different countries may be determined with some precision through 1898, thanks to the annual tramway censuses of the French journal *L'Industrie Electrique*.

In explaining contrasting patterns of European and American street railway electrification, factors evident in the horsecar era continued to exert a strong influence. Quite clearly, the greater preference for spatial diffusion and central-business-district concentration in the United States combined with more rapid urban growth to create a stronger demand for urban transit. Yet continuing greater pressure from demand and a more vigorous supply response in the form of more rapid mechanization and technical change – the kind of argument preferred in economic analysis of technological improvement – is only part of the story. Equally significant was the way cultural–aesthetic judgments regarding the new technology continued to combine with the previously noted institutional differences to determine the course of diffusion and modification of American innovation.

Judging by a wealth of reports in the technical press, municipal debates, and scattered monographic studies, there is no doubt that large numbers of influential Europeans thought that overhead-wire conductors and their support poles – the very essence of the American innovation in surface transit – were extremely ugly and aesthetically unacceptable. Better, they argued, to send inventors and engineers back to the drawing boards than accept such "visual pollution." This cultural reaction was most pronounced in connection with putting poles and wires along broad boulevards and through historic squares, and was less pronounced regarding many suburban lines and lines through working-class and industrial districts. And, unlike early objections to electric traction that were related to things like safety for horses or telephone reception, the aesthetic critique rested on subjective values that could not be refuted by technical experts.

An 1897 report by Birmingham's Subcommittee on Tramways of its special tour of various continental systems fairly suggests the scope of antitrolley feeling.

> Your subcommittee would point out that the unsightliness and otherwise objectionable features of the overhead wire are recognized in many of the places they have visited. The municipal authorities at Paris, Vienna, and Berlin stated that in no case would they permit overhead wires in the central portion of those cities. At Brussels several miles of conduit are being laid to avoid

them; at Dresden and Budapest they are not allowed in the principal streets.

And the list of cities went on and on, leading the subcommittee to "recommend strongly that no consent be given for the erection of overhead wires in any part of the city."[14]

Interestingly, there was also initial hostility toward overhead conductors on aesthetic and safety grounds in the United States. While installing his pioneering Richmond operation, Sprague himself admitted to street railway operators that the overhead system "may be unsightly" and "somewhat in the way," so it "perhaps would not be tolerated in twenty large cities in the country."[15] Yet only in New York and Washington did such considerations delay the installation of the trolley for any considerable length of time.

Although aesthetic opposition to overhead electric traction was by every indication less in the United States than in Europe, differing institutional arrangements were the decisive factor. American cities inherited from the horsecar era a lack of effective control, and street railway operators could ignore or easily outflank antitrolley opposition. Thus most companies in the United States were able to electrify their lines almost as quickly as they saw that it paid, and without even having to modify their existing franchises.

Technical freedom in the United States had consequences that went well beyond more rapid diffusion and aesthetic controversies. First, companies in the United States were usually able to maintain a flat 5-cent fare, which had already emerged as the standard in most franchises during the horsecar era, and they were not required to share the cost-reducing gains from technology through fare reduction. Or, more accurately, companies reduced fares by extending lines and providing longer rides for the same price, as electric traction subsidized suburbanization on a massive scale. Second, with their perpetual or quasi-perpetual franchises intact, American companies did not need to worry much about amortization in a given time period of the very large capital increases that electrification necessarily required and that were swollen further by manipulation and watered stock.

In European cities, where technological departures in urban transit customarily required regulatory permission, opponents of poles and wires could call on existing institutional arrangements to assess the "environmental impact." Thus one result of time-consuming opposition to overhead conductors was a strong stimulation of European efforts to develop technological alternatives to meet aesthetic–environmental objections all through the 1890s, long after the technical issue was a closed book in the United States.

While continuing earlier efforts with battery-powered cars, Europeans in the 1890s focused primarily on developing various surface-conduit systems, as well as highly imaginative "mixed" systems – systems that "mixed" an overhead conductor and an alternative such as the conduit conductor on the same car and same line. One variant, perfected by Siemens and Halske, was used in Berlin, Prague, Budapest, and Vienna, for example. Another, developed by General Electric's French affiliate, Société Thomson-Houston, was used in Bordeaux, Marseilles, and especially Paris. In both cases, cars used a more expensive, less dependable surface conduit – an electric wire laid in a special tube between the tracks in the street – to traverse historic or very fashionable areas. The same car then switched to an overhead-wire conductor in parts of the city where aesthetic objections could be overcome. In addition, European constructors modified early American practice to improve the visual effect, reducing especially the unsightliness of overhead wires and ornamenting their poles.

These experiments and modifications allowed European street railway constructors, regulators, and the general public to explore the possibilities and tradeoffs of different modes of electric traction, the installation of which was then negotiated between the parties and spelled out between them in a new concession. In spite of important variations, a fairly common pattern may be discerned for most continental cities. Basically, the street railway companies and the electrical equipment producers that financed them promised lower fares, more frequent service, extension of existing horsecar lines, somewhat higher municipal taxes, and some improvement in working conditions. In return, they required permission to introduce electric traction in general, a delineation of areas where trolley wires were unacceptable, and a substantial prolongation of the concession. The extra expense of antitrolley limitations to preserve urban beauty had to compete therefore with other desirable goals – lower fares, expanded service, and the rest.

This pattern of careful regulation and renegotiation, so different from that in the United States, produced some significantly different results. First, in Europe, aesthetic–environmental concerns – probably part of a broader cultural commitment to a harmonious urban landscape – were incorporated into street railway electrification to a larger extent than in the United States. Many a European street railway had to make do grudgingly with more expensive alternatives, like conduit traction, on at least some of their lines for many years, although the number of such lines declined with time and continued reevaluation stemming from company complaints and proposals.

Use of more expensive technologies declined to permit maximum fare reduction in general, and low-cost workman's tickets for morning and evening commuting in particular. Fare structures varied between individual cities, but a dramatic reduction in European fares actually paid was unmistakable and unmatched by a similar fare reduction in the United States – the second notable difference. In France, for example, where the carefully renegotiated contract with private industry reigned supreme, the average fare per tramway rider fell from 15.4 centimes in 1896 to 8.8 centimes in 1910, or from about 3 cents to about 1.7 US cents. As a confidential internal report by financial experts at France's largest deposit bank concluded in late 1900, "It is above all the public which has benefited from the application of electric traction. The companies have not lost by it, but the financial benefits have rarely been as high as could have been expected."[16]

Sharply reduced fares and reasonable profits in Europe are related to the third difference: the general lack of outrageous inflation of capital stock in Europe compared to the United States. Although financial manipulation and speculation were not unknown in Europe, the verification of actual expenditures by state engineers and municipal authorities limited unhealthy abuses. Capital accounts reflecting genuine cash outlays also laid a solid foundation for the systematic (and very un-American) amortization of fixed investment in the long period before the prolonged concession expired and all immovable property reverted to the city. Finally, and less certainly, it would appear that Europe's regulated, renegotiated framework of tramway electrification short-circuited the hostility and sense of injustice that Americans seemed to develop toward their transit companies. This greater hostility in the United States may have encouraged the growth of a more general American dissatisfaction with public transportation per se, and combined with other well-known factors to contribute to a more rapid decline of transit systems in the United States than in Europe after 1920.[17]

This study has concentrated on the street railways found in every city and slighted the rapid transit of elevateds and undergrounds found only in a very few of the largest agglomerations before 1914; thus it has stressed American–European differences in transit development and downplayed the obvious similarities, which have seemed less significant in a comparative perspective. Nevertheless, two important and growing similarities dating from the electrification era should be noted, if only briefly.

First, whereas electric traction evolved out of ongoing mechanization in the United States, it produced a sharp, revolutionary break in Europe's public transportation history. And as a result of Europe's more rapid rate of change, European cities closed a very large part of the previously existing transit gap, as measured by per capita usage and similar indicators. Thus per capita usage in the four largest British and German cities was almost four times as great in 1910 as in 1890. In 1910, European per capita usage had come to represent about 70 percent of the level in comparable American cities, as opposed to less than 30 percent in 1890. Moreover (and more work needs to be done here), electric traction facilitated areal expansion and suburbanization in Europe as it did in America, allowing a reduction in overcrowding at the center. The familiar American argument that electric street railways would encourage generally desired suburban development found many echoes in Europe. It was important in overcoming opposition to overhead electric traction, and it played a part in the partial or total abandonment of the long-standing zone system in favor of American-style flat fares in many French and German cities.

Second, there were growing similarities in terms of industrial structure. In the United States, electrification allowed nimble entrepreneurs and street railway magnates to combine competing horsecar franchises into unified, monopolistic transit systems in many cities. Thus the American pattern moved toward that in Europe, and in both areas a unified transit operation faced the riding public and regulators. A related similarity was on the equipment producer side. In both the United States and leading European countries, the electrical industry quickly became highly concentrated, so electric installations on street railways and subsequent expansions were planned and executed by a small number of firms locked in intense but clearly monopolistic competition. The desire of such firms to sell their equipment diffused electric traction around the world. For example, Belgian firms led in carrying tramway electrification throughout the Russian empire.[18]

Our focus on European and American comparisons has mentioned differences between European countries only in passing. A few of these, like the contrasting patterns of electric traction adoption noted earlier, were important. Above all, Great Britain's slow acceptance of electric traction was tied up with the impending municipalization of private operating companies, as the twenty-one-year leases came due. Securing finally the authority to operate as well as own their systems, British cities then made rapid progress and emerged as the new leaders in public transportation in some eyes, particularly American eyes.

Yet while some American reformers, such as Frederic Howe, urged their countrymen to follow the municipal ownership model of almost all British (and some German) cities, at least one famous British expert – James Dalrymple, general manager of Glasgow Tramways – disagreed. In 1906, called to study Chicago's street railway system by a reformist mayor, Dalrymple agreed that the system was "altogether out of date" and that "there is no wonder that the inhabitants are intensely dissatisfied with their transit facilities." But Dalrymple counseled against municipalization, accepting implicitly the American streetcar industry's contention that in the United States, politicians and cities were too dishonest and corrupt to establish efficient, successful municipal enterprises on the European model. It was another indication of how, by 1914, Europeans had closed the earlier large gap in urban transit and perhaps even moved ahead of their American cousins in some important ways.

Notes

1 For further discussion of omnibus development and citation of sources, see John P. McKay, *Tramways and Trolleys: The Rise of Urban Mass Transport in Europe* (Princeton, N.J., 1976), pp. 9–13

2 Charles W. Cheape, *Moving the Masses: Urban Public Transit in New York, Boston, and Philadelphia, 1880–1912* (Cambridge, Mass., 1980), p. 25

3 For some discussion of sources, see McKay, *Tramways and Trolleys*, pp. vii–ix, 247–255; and Glen E. Holt, "The Main Line and Side Tracks: Urban Transportation History," *Journal of Urban History* 5 (1979): 397–400

4 State of Massachusetts, *Report on the Relations between Cities and Towns and Street Railway Companies* (Boston, 1898), p. 74

5 US Bureau of the Census, *Report on Transportation Business in the United States, 1890*, part 1, *Transportation by Land* (Washington, D.C., 1895), p. 685

6 Sam B. Warner, Jr., *Streetcar Suburbs: The Process of Growth in Boston, 1870–1900* (Cambridge, Mass., 1962); Cheape, *Moving the Masses*, p. 214. Joel A. Tarr, *Transportation Innovation and Changing Spatial Patterns in Pittsburgh, 1850–1934* (Chicago, 1978), concludes that street railways (and commuter railroads) had gone far toward creating the shape of a modern metropolis in Pittsburgh, complete with considerable suburban expansion, by 1888

7 The impact of horse tramways on European living patterns has been little studied, but see McKay, *Tramways and Trolleys*, pp. 22–25, 205ff.; and see the important article by David Ward, "A Comparative Historical Geography of Streetcar Suburbs in Boston, Massachusetts and Leeds, England: 1850–1920," *Annals of the Association of American Geographers* 54 (1964): 477–489

8 US Bureau of the Census, *Transportation, 1890*, 1: 682

9 Clay McShane, *Technology and Reform: Street Railways and the Growth of Milwaukee, 1887–1900* (Madison, Wis., 1974), p.7

10 Edgar Kahn, *Cable Car Days in San Francisco* (Stanford, Calif., 1940), pp. 27–42; and George W. Hilton, *The Cable Car in America* (Berkeley, 1971), pp. 21–50. There is a good study on Melbourne, Australia, which had the largest cable system outside the United States: John D. Keating, *Mind the Curve! A History of the Cable Trams* (Melbourne, 1970)

11 US Bureau of the Census, *Transportation, 1890*, 1: 682

12 George W. Hilton, "Transport Technology and the Urban Pattern," *Journal of Contemporary History* 4 (1969): 126

13 *Electrical Engineer* (London) 9 (1892): 196

14 Report of Birmingham Subcommittee on Tramways, as quoted in *Electrical Engineer* (London) 19 (1897):

569-570. See McKay, *Tramways and Trolleys*, pp. 84-88, for similar reactions

15 Remarks of Mr. F.J. Sprague on Electricity as a Motive Power,'' American Street Railway Association, *Proceedings* 6 (1887-88): 60-68. Also see Harold C. Passer, *The Electrical Manufacturers, 1875-1900: A Study in Competition, Entrepreneurship, Technical Change, and Economic Growth* (Cambridge, Mass., 1953), p. 216

16 Archives du Crédit Lyonnais, Paris. Etudes Financières. ''Etude sur les résultats de la traction électrique . . . en France,'' December 1900

17 See the suggestive comments in Edward S. Mason, *The Street Railway in Massachusetts: The Rise and Decline of an Industry* (Cambridge, Mass., 1932), pp. 185-186

18 John P. McKay, *Pioneers for Profit: Foreign Entrepreneurship and Russian Industrialization, 1885-1913* (Chicago, 1970), pp. 100-102

13

THE TROLLEY AND SUBURBANIZATION

by Kenneth T. Jackson

Source: Kenneth T. Jackson, ''The Trolley and Suburbanization'', in *Crabgrass Frontier: the suburbanization of the United States*, Oxford, Oxford University Press, 1985, pp. 118–37

The electric streetcar was vital in opening up the suburbs for the common man. In a pioneering study of Boston, Sam Bass Warner, Jr., has shown that, beginning in the 1870s, the introduction of improved street railway lines made possible a continuing outward expansion of the city by ½ to 1½ miles per decade. In practical terms, this meant that the outer limits of convenient commuting (by public transit as opposed to steam railroad) stood at about six miles from City Hall in 1900 as compared to two miles in 1850. That was the distance that one could reasonably be expected to traverse in one hour or less.[1] [. . .]

Two policies of the streetcar entrepreneurs were especially important in facilitating the outward movement of population. The first was the practice of extending the lines beyond the built-up portion of the city and into open country. As will be discussed later, this had the practical effect of enabling heads of households to see that a convenient transportation mode would be available from their homesite.

The second essential policy of the trolley companies was the five-cent fare. Unlike the European streetcar systems, which depended upon high prices instead of high passenger volume and adhered to the practice of the zone fare – or payment according to distance – American firms usually adopted a flat fee with free transfers, thus encouraging families to move toward the cheaper land on the periph-

ery. The cheap fare thus served the social purpose of preventing congestion and of reducing the necessity for tenement dwellings.

The pattern was as follows. First, streetcar lines were built out to existing villages, like Tenally Town in the District of Columbia, Hyde Park in Chicago, Idlewild in Memphis, and Weequahic in Newark. These areas subsequently developed into large communities. Second, the tracks actually created residential neighborhoods where none had existed before. In 1886, when President Grover Cleveland bought a house northwest of Georgetown between Wisconsin and Connecticut Avenues in Washington, the area was undeveloped countryside. Ten years later, the trolley had helped turn the section into thriving Cleveland Park.

The best evidence on the relationship between mass transit and urban growth comes from the work of Sam Bass Warner. Examining Boston, Warner found that Henry Whitney and his associates regarded the endless expansion of ridership as the key to profit. To this end, they laid 238 miles of track in Boston, Lynn, and Cambridge. Theoretically, as more areas had access to the city, more people would have a valid reason to ride the cars.

This is not to say that a trolley line could itself determine the pace of change; as a matter of fact transportation is and was only one of a number of variables influencing development. [. . .]

Trolley tracks and suburban developers

The close American relationship between land speculation and the construction and location of streetcar tracks can be demonstrated through an examination of three cities in which transit tycoons were less interested in the nickels in the fare box than they were in their personal land development schemes. They learned quite early that transit access would make undeveloped farmland attractive to potential commuters and thus raise its value. Their goal was simply to reap the land-speculation profit by dictating in their corporate chambers the direction and extent of transportation lines.

Oakland

F.M. "Borax" Smith was such an investor. After making a multimillion-dollar fortune by marketing borax as a cleanser, Smith bought a controlling interest in several electric railway companies in the East Bay area [near San Francisco] in 1893. Over the next ten years he added other companies to his portfolio, and by 1903 Smith's Oakland Transit Company was operating seventy-five miles of track and carrying thirty million riders per year.

Smith cut costs with a vengeance. Previously, the various companies had rendered service on tracks of different gauges which could not be adapted to each other. On one street alone there were six sets of tracks, not to mention duplicating power plants. Smith eliminated unprofitable lines and franchises, especially those that replicated existing services, and he achieved economies of scale by consolidating eight previously separate managerial and accounting operations.

Smith's land-development plans were closely tied to his trolley operations. In 1895 he associated himself with several other local businessmen to form the Realty Syndicate. This consortium promptly went about the task of purchasing thirteen thousand acres of undeveloped land extending from Mills College to Berkeley, as well as farm properties inaccessible to existing towns. These tracts were then held until the growth of the towns necessitated new subdivisions. The Realty Syndicate would then divide the larger parcels into small building lots and arrange for a trolley line to extend only to that section while bypassing the land of other real estate concerns. The Realty Syndicate not only installed the streets, sidewalks, and sewers, but also provided financing for home builders, an unusual practice available only to heavily capitalized firms. Land that was accessible to an electric railway, namely that owned by the Realty Syndicate, was usually the first choice of middle-class families who depended on public transportation to Oakland to get the breadwinner to work.

So important was real estate to Smith's overall money-making plans that in 1903 he created a commuter line called the Key System that was never intended to make money. Directly competitive with the Southern Pacific Railroad in the Bay Area, the Key System was instead designed to sell real estate in the manner of Smith's other ventures. Smith expected that the profits that the Realty Syndicate would earn on land would more than offset the operating losses of the transit system. In the words of one of Smith's senior executives: "The matter of operation or relations between the two companies (the Key System and the Realty Syndicate) has been very similar to the relation between two pockets in the same man's trousers." [. . .]

Los Angeles

Los Angeles provides the premier example of the confluence of street railway entrepreneurs and real-estate development. Land speculation was part of the tradition of the City of Angels from the northern conquest in 1847. Properties inflated 200 percent between 1865 and 1866 and another 500 percent in 1868. During the railroad boom of the mid-1880s the process repeated, fueled in part by Robert Maclay Widney, Francis P.F. Temple, and their associates, who organized the first streetcar line in 1874. Other powerful Los Angeles real-estate moguls soon organized other lines. Although land prices collapsed in 1888–1889, and although over $14 million in property values disappeared in Los Angeles County, transit moguls, especially Moses H. Sherman and Eli P. Clark, bought more properties in the 1890s.

The greatest of Los Angeles transit barons was Henry E. Huntington. At the turn of the century, even as the Southern Pacific was battling "Borax" Smith in Oakland, one of its former owners was himself engaged in almost exactly the same land-transit enterprise six hundred miles to the South. Between 1890 and 1910 Huntington amalgamated several shoe-string transit lines into the potent Pacific Electric Railway Company, and he put down tracks throughout the Los Angeles Basin, from Santa Monica to San Bernardino and from Pasadena to Balboa. Technically called "interurbans" because the cars were larger and faster than those which operated solely within city limits, Huntington's "big red cars" soon became as familiar as the orange groves through which they swayed and clattered.

Like "Borax" Smith, Huntington concentrated on short-haul passenger traffic, and like Smith, he was more interested in selling land than streetcar tickets. In any case, Pacific Electric had too many branch lines to be a big money-maker. Deciding that the best financial opportunities were in the manipulation of property values, Huntington formed a land company in 1901 (which was composed of Moses Sherman and Otis and Henry Chandler [. . .] among others) to select and promote residential sites. He studied weekend transit patronage to determine which areas appealed to riders, and he focused his real-estate promotions there. Laying out rectangular blocks upon which row upon row of tidy houses could be built, Huntington even operated water companies, usually at a loss, in order to encourage development in the desert-like region. Most importantly, Huntington connected the widely separated towns of the Los Angeles area with one of the nation's best transit systems, and as much as any single person, he initiated the southern California sprawl that still baffles visitors.[2]

Washington, D.C.

On the opposite side of the continent, a more exclusive type of suburb was growing up near the nation's capital as the brainchild of Senator Francis G. Newlands of Nevada. A longtime Washington resident who married the daughter of wealthy Senator William Sharon (Nevada, 1875–1881), Newlands was among the first investors to perceive the speculative implications of Frank Sprague's invention. Newlands dreamed of extending Connecticut Avenue into Maryland as a corridor for high-class suburban expansion. In 1888 his Chevy Chase Land Company received a charter to run a trolley line along the Avenue that would connect with the regular District of Columbia transit system. With the Congressional assistance of Senator William Morris Stewart, who himself bought $300,000 worth of the first issue of the Chevy Chase Land Company stock, Newlands not only received a favorable charter for the street railway, but he also arranged for the creation of Rock Creek Park. As Roderick S. French has noted: "Not only did the presence of the park raise the value of nearby properties owned so largely by the Land Company, but at the same time, as Stewart so candidly expressed it, the action took '2,000 acres out of the market.'"[3]

Meanwhile, Senator Newlands and his associates, most notably Colonel George Augustus Armes and realtor Edward J. Stellwag, had been systematically, and at first secretly, purchasing as much farmland as they could along the proposed route. By 1890 they had put together 1,712 acres; landowners who held out for what Newlands regarded as excessive profit were bypassed by a shift in the direction of the road and the trolley. This is why Connecticut Avenue today changes direction somewhat at Chevy Chase Circle.

After completing his purchase, Senator Newlands determined the route of Connecticut Avenue; built and graded a broad, 150-foot-wide right-of-way; and then deeded the thoroughfare to Maryland and the District of Columbia. The action was hardly philanthropy, for Newlands well knew the impact the glamorous road would have on the attractiveness and accessibility of Chevy Chase.

Newlands's aim was to build a totally planned residential area that would serve as a "home suburb" for the national capital. Chevy Chase would be a model of elegance and planning for the entire nation. Alleys were proscribed, and the wide new streets were given appropriate English and Scottish names. Even the trees and shrubs were carefully selected to represent the best in contemporary style and taste. The first subdivision opened in 1893 – a depression year – on 250 acres just beyond the District of Columbia line. Newlands said he wanted "a community where every residence would bear a touch of the individuality of the owner," but that individuality had to be expressed within very specific limits. No commercial ventures and no apartments were to mar the serene landscape. Lots were sold with the understanding that no home on Connecticut Avenue could cost less than $5,000, and no dwelling on a side street for less than $3,000. Individual properties had to be at least sixty feet wide, and houses had to be set back at least twenty-five feet from the street. And to give prospective purchasers some idea of the pretentiousness that was envisaged, the land company built the first four houses itself.

Although most streetcar suburbs had a proletarian or middle-class image, Senator Newlands was successful in creating a different tone for Chevy Chase. More interested in quality than in rapid growth, he did not create an instant suburb; in 1900, when the streetcar line made the trip to the White House in thirty-five minutes, only fifty families were living in the town. And Chevy Chase was to remain a tightly controlled enclave of upper-income Americans long after its great growth in the first half of the twentieth century.

"Borax" Smith, Henry Huntington, and Francis Newlands were linked by their common perception of the impact of mass transit on land values. All used the electric streetcar to generate customers for their building lots. All pretended to be operating as independent entrepreneurs in the best traditions of a democratic society. All in fact manipulated government agencies and employed political favoritism in order to use public streets and to gain public franchises for their private ends.

Although few eyebrows were raised over the way politics and business were mixed in the development of American suburbs, such tactics were unique to the United States. In Great Britain and on the European continent, transit owners were not allowed to speculate in real estate served by their lines, and landowners were not given streetcar franchises. In this country, by contrast, no such prohibitions existed, and many entrepreneurs recognized that the opportunities for fast profit in real estate multiplied with the development of mass transit. Even in Boston, where small operators dominated the residential development of the suburbs, the process was helped along by the self-interest of the transit leadership. The great consolidator of the Boston streetcar system, Henry Whitney, was a big speculator in Brookline real estate. With advance information on the planning and layout of Beacon Street and of the trolley line, his West End Land Company bought farms along the right of way at a fraction of their true worth. After title had passed and the lots were ready for sale, the West End Railway brought willing customers to the property. [. . .]

Cheap land/high wages

The rapid suburbanization of the United States cannot be viewed in isolation from the material prosperity of its people and the sheer abundance of its land. By the middle of the nineteenth century, Americans were already a "people of plenty." The wages of the working man, no matter how meager, were almost invariably higher than those of his counterparts elsewhere in the world. Geographically, the amount of space potentially available to each citizen was also staggering. Other countries – Canada, Australia, and Russia – were as large as the American republic, but much of their space consisted of treeless desert or frozen tundra. The United States, by contrast, was composed substantially of heavily forested or grass-covered ground, most of it habitable by human beings. In comparison with other countries of the world, the real estate of North America was almost literally endless.

In urban areas, and especially in developing suburban sections, this translated into land that was cheap by international standards. The average price of a lot suitable for building ranged widely from city to city, but a figure of more than five hundred dollars would have been high before 1900, and prices of $150 were common. The sales target was often the "little man," the working-class clerks, mechanics, and struggling businessmen who aspired to more security and space for their families and who were sensitive to slight variations in price. [. . .]

Affordable property was in part a function of the continental size of the nation and in part a function of the speed of mass transit. This occurred because the streetcar took advantage of a law of geometry to the effect that the area of land increases with the square of the radius from the center of the city. Thus, by simply doubling the radius, the amount of land available for development would be quadrupled. In economic terms, the marginal cost of transportation did not rise as fast as the price of land fell with distance from the city center. In other words, the least expensive housing option for middle-class families that could afford a commute was to move outward.

The relative affordability of land was greatly aided, especially in the three decades after Appommatox, by a sustained agricultural depression. Between 1865 and 1896, the prices of most commodities fell in real terms, as the American farmer paid more and more and received less and less. This meant that the value of land was falling for agricultural purposes, even as acreage within commuting range of a city was becoming desirable for residential use. On the edges of large communities, therefore, farming could be economically justified only by ignoring the potential value of the property for building lots. Most farmers responded rationally and eventually sold their land; some to speculators and others to individuals. In either case, the amount of suburban-style real estate increased to meet the demand.

The process of land conversion was also speeded by the pre-Civil War introduction of building and loan associations. Even before 1860, realtors claimed "Long Credit and Low Terms." The long credit meant the note was payable in six years, and the low terms were interest due semi-annually at 6 percent. Building and loan associations smoothed the lending process. First organized in Philadelphia in 1831, they enabled an individual of modest means to invest his savings in shares of an association and ultimately to borrow against the value of those shares at low interest. An essential part of the loan contract was that the associations provided for a method of repayment. [. . .]

The role of the savings and loan associations was especially crucial in the nineteenth century because before 1916 most commercial banks were prevented by national legislation from providing long-term credit for real-estate loans. Beginning in the 1890s, such institutions began cautiously to issue a few short-term mortgages for non-income producing residential property, and by 1915, most important banks had set up real-estate departments. Although they acted conservatively and rarely extended large loans to builders, their change in policy during World War I stimulated suburbanization and eased the transition of many families to home-ownership.

The provision of urban services

New balloon-frame communities usually developed in conjunction with the establishment of essential services. Except for immigrant neighborhoods, where improvements were often resisted in order to hold down taxes, massive public investments in roads, storm sewers, street lighting, curbs and gutters, playgrounds, and schools were necessary for growth, and real-estate advertisements usually emphasized the availability of upgraded city facilities. [. . .] And not only were such improvements financed at general taxpayer expense, but their low-density sprawl involved numerous inefficiencies.[4]

Streets are instructive in this regard. Before the Civil War, streets were paved or widened when owners of a certain percentage (usually three-fourths) of the property facing the right-of-way petitioned the city to do so. To finance such improvements, property-owners "abutting and directly affected" paid special assessments. The municipal government played a limited role: the basic decisions as to when and how to pave were made by private individuals. Because the owners would presumably benefit from the increased value of their land after the street was opened, the system had a certain logic and justification. Where the cost of new services would be a financial problem for working-class homeowners, the residents simply delayed the paving.

In the final decades of the nineteenth century, however, a second method of financing became more common, one that passed the cost of peripheral street improvements on to the municipality as a whole. As horsecar lines enabled upper- and middle-income families to move away from downtown, outlying residents placed increasing pressure on the city to build smooth pavements at public expense. By the 1890s engineering publications were hypothesizing that well-paved roads, paid for by the city, would reduce the cost of freight-handling, thus encouraging new businesses and reducing the tax rate. Reformers added their voices to the chorus in the belief that suburbanization and the improved housing it promised would alleviate the evils of tenement districts. By 1900 the changeover was effected. The centralization of street administration meant that all city dwellers subsidized those who moved to the edges. [. . .]

Individual effort and homeownership

Cheap land, inexpensive construction methods, favorable peripheral taxing policies, and the rapid expansion of public utilities only partially explain high homeownership rates in American urban areas. In the last quarter of the nineteenth century, working-class people were offered free horsecar rides, lunch, and entertainment at sales of land through auction. Under the banners of "Get a Slice of the Earth" and "Own Your Own Home," ethnic arrivals from the Piedmont, Lombardy, Holland, Germany, and Ireland built modest homes along the streetcar lines. They aspired to and achieved the detached house and garden that had been so far beyond their reach in the Old World.

As Olivier Zunz, Roger Simon, and Stephan Thernstrom have noted, immigrant groups managed to purchase homes at rates equal to or above those of native Americans because of extraordinary personal and family sacrifice. As Simon has demonstrated for the 14th Ward of Milwaukee, Polish newcomers purposefully delayed the expense of paved streets and city sewers, eagerly rented first floors or basements to tenants, and did without leisure and material comforts in order to buy modest bungalows. In Simon's judgement, they adapted the physical environment to their needs, for as he says, "the new housing stock and public policy permitted an ordering of priorities in which financial security won out over convenience and public health."[5] [. . .]

The process of suburban land conversion

At the heart of all suburban growth is land development – the conversion of rural or vacant land to some sort of residential use. The process involves property owners, speculators, banks, private lenders, builders, and buyers. As land values at the center of the metropolis rise, individual parcels either produce the higher yields to hold their place, or, in the course of a few years, more profitable businesses move in on the site. By the same token, if much the same yield can be earned at a peripheral site of lower value, there is little incentive to remain in and around the central business district. The pattern of urban land investment affects the value of outlying farmlands, which either increase yields by more profitable crops, or, as is usually the case, they give way to more lucrative subdivision and real-estate developments.

A single model cannot possibly describe the entire suburban experience. Before World War I, however, rarely did a single individual or firm buy land, lay out streets, build houses, and finance sales to the ultimate occupants. Instead, a landowner typically hired a civil engineer to determine streets and lots, and then, depending upon local circumstances, either pressured the municipal government to extend pavement at public expense or brought in private crews to construct roads. The land was subsequently sold, often at auction in the nineteenth century, to numerous buyers who would either build houses for their own occupancy for sale, or would retain vacant lots for

speculation. The subdivider often retained some of the land for his own use and built his own home there. Some real-estate syndicates operated in the larger markets, but the predominant force was the small developer.

In contrast to the heavy governmental involvement in the use of land in Europe, residential development in the United States has largely been the work of private interests. Indeed, urban real estate was the single most important source of leisured wealth in the nineteenth century. [. . .]

The subdivision and the real-estate specialist

The basic unit of development in the nineteenth-century suburb was the subdivision. In 1843, for example, lots were auctioned in the twenty-acre Linden Place subdivision in Brookline, where a few years later the 300-acre Longwood development was opened, this time without an auction so that the residents could more easily select desirable neighbors. In the Bronx (then part of Westchester County), subdivisions date from 1850, when Morrisania first experienced a suburban boom. Within a decade, nearby Fordham and Tremont were also laid out, and by the end of the Civil War their combined population approached 20,000. They were typical of early subdivisions in that they were adjacent to railroads which established commuter stations at their centers.

Just as the auction system was not universal, so also was there no single pattern for subdivision development. In cases where a large residential section emerged from a single rural property and a single investor, the developer usually had full responsibility for the street system. Where contractors lacked the capital to construct more than a few houses at a time, they were in no position to question the city engineer's imposition of the grid street plan. Thus, in many peripheral areas within cities between 1875 and 1945, uniform, narrow rectangular lots defined the houses before they were built.

Whether their subdivisions were large or small, real-estate specialists were more active in the city building process than anyone else. The theory that early suburbs just grew, with owners "turning cowpaths and natural avenues of traffic into streets," is erroneous. Subdividers lobbied with municipal governments to extend city services, they pressured streetcar companies to send tracks into developing sections, and they set the property lines for the individual homes. Each city and most suburbs were created from many small real-estate developments that reflected changing market conditions and local peculiarities. And even when the area was developed by large operators, ultimately the land filtered down to the private buyer. [. . .]

For the first time in the history of the world, middle-class families in the late nineteenth century could reasonably expect to buy a detached home on an accessible lot in a safe and sanitary environment. Because streetcars were quick and inexpensive, because land was cheaper in suburbs than in cities, and because houses were typically put up using the balloon-frame method, the real price of shelter in the United States was lower than in the Old World. [. . .]

A [. . .] common criticism of the burgeoning peripheral subdivisions was that their very popularity undermined the results incoming families hoped to achieve. Privacy and solitude disappeared as the populace streamed outward, blighting the rural charm that had lured them in the first place. The environment they found was never quite as open or as isolated as theorists had wished. The typical streetcar suburb featured one- or two-family homes on lots of about three thousand to six thousand square feet (about one-tenth of an acre). Such dimensions were generous by the standards of the walking city, but much more cramped than Andrew Jackson Downing had envisaged. This compactness occurred because the streetcar,

the common man's mode of transit par excellence, required a certain volume of traffic for profitable operation, and because the very availability of a transit line tended to raise the price of land. Thus the trolley, which was supposed to provide space for the middle class, was not feasible unless the residential neighborhoods were closely packed.

Nevertheless, the electric streetcar, and the land developers who were so quick to take advantage of its possibilities, had created a new kind of metropolis by 1900, one that was very different from the walking city of a century earlier. By 1900 the center of the city had become an area of office and commerical uses that was almost devoid of residences. Nearby were the grimy factories, and just beyond them the first tenement districts of the poor, the recent immigrants, and the unskilled, persons unable to afford even the streetcar fare and forced to compete for housing space where real estate was the most expensive and housing the least desirable. Along these same streets, the well-to-do had lived only two generations earlier.

Beyond the compact confines of the walking city lay the new streetcar suburbs, the essence of the American achievement at the turn of the century. The residential structures that filled them were not elegant, but they were spacious and affordable by European standards and they represented an attainable goal.

Farther out, the railroad commuters lived in houses that sprawled in ample yards, thick with trees and shrubbery behind iron or wooden fences. These residences represented a new American ideal. Unlike the intown dwellings of the Old World wealthy and unlike the country houses of the English gentry, these structures were uniquely American and, with their pseudo-Gothic towers and cupolas and mansard roofs, they set a suburban rather than an urban standard for achievement-oriented Americans.

A number of models have been developed to explain residential patterns in relationship to the two most important factors: the quality and cost of housing and the convenience, speed, and cost of transportation. The well-known model of urban growth by Burgess assumed that concentric zones of residential areas develop around the central business district in an ascending hierarchy. William Alonso has added to Burgess's conception the idea that in general the price of land decreases with increasing distance from the center. In order to earn a return comparable to that from other uses, residential development near the center has to be dense and compact. The lower-income groups live near the center in the United States because the factors of centrality and cost are more important than the quality of housing. Among the middle and upper classes, the importance of centrality decreases because of the sinking importance of transportation costs. Thus, affordable housing means that the middle class could set as a priority the quality of the dwelling unit, preferring to live in suburbs with a low population density far outside the city.

Notes

1 Sam Bass Warner, Jr., *Streetcar Suburbs: The Process of Growth in Boston, 1870-1900* (Cambridge, Mass., 1962), *passim.*

2 The best study of suburbanization in Los Angeles is Robert M. Fogelson, *The Fragmented Metropolis: Los Angeles, 1850-1930* (Cambridge, 1967)

3 Easily the best source on this subject is Roderick S. French, "Chevy Chase Village in the Context of the National Suburban Movement, 1870-1900," *Records of the Columbia Historical Society*, XLIX (1973-1974), 300-29

4 Jon A. Peterson, "The Impact of Sanitary Reform upon American Urban Planning, 1840-1890," *Journal of Social History*, XIII (Fall 1979), 83-103; and Harold L. Platt, *City Building in the New South: The Growth of Public Services in Houston, Texas, 1830-1915* (Philadelphia, 1983)

5 Roger Simon, *The City-building Process: Housing and Services in New Milwaukee Neighborhoods, 1880-1910* (Philadelphia, 1978)

14

THE REVOLUTION IN STREET PAVEMENTS, 1880–1924

by Clay McShane

Source: Clay McShane, ''Transforming the Use of Urban Space: A Look at the Revolution in Street Pavements, 1880-1924'', *Journal of Urban History* 5 (1979), pp. 279-307

Introduction

In 1880 George Waring, the noted civil engineer, made the first comprehensive, statistical study of urban street pavements as part of the Census Bureau's pioneer attempt to collect urban data (see Table 14.1). Forty-four years later, in 1924, *American City Magazine* compiled the last complete set of statistics on urban street pavement types. Interest in the subject reached a higher level during this period than at any time before or after as American cities completely transformed the design, materials, and administration of paving. Over half of

urban streets were unpaved in 1880. By 1924 municipalities had paved almost all urban streets. The *American City Magazine* survey did not even bother to ask respondents to specify what proportion of their streets remained without pavement.

Moreover, during these years, cities also greatly changed the types of pavement they used. In 1880 gravel or macadam (a technically superior method of using gravel) pavements accounted for just over half of the streets paved, usually on lightly traveled residential streets. Engineers had developed such gravel pavements over the previous 60 years. The

Table 14.1 The evolution of street paving in eleven large American cities, 1880–1924

	1880	1890	1902	1909	1924
Miles of street	5,001	7,678	12,797	12,782	n.d.
% paved	49.2%	37.3%	51.7%	57.2%	n.d.
% unpaved	50.8%	62.7%	48.3%	42.8%	n.d.
% of paved-asphalt streets	2.5%	n.d.	16.9%	30.2%	47.4%

Sources: U.S. Bureau of the Census, *Social Statistics of Cities, The Tenth Census, 1880*, vols. 18 and 19 (Washington, DC, 1886), *passim*; U.S. Census Office, Bulletin 100, *Social Statistics of Cities, 1890* (Washington, DC, 1891), 5–16; United States Bureau of the Census, Bulletin 20, *Statistics of Cities Having a Population of Over 25,000 (1902–1903)* (Washington, DC, 1905), 114–119; U.S. Bureau of the Census, *General Statistics of Cities, 1909* (Washington, DC, 1913), 73, 149–159; The Asphalt Association, "Street Construction," in *Municipal Index, 1924* (New York, 1924), 109–229. The eleven cities are New York, Chicago, St. Louis, Cleveland, Cincinnati, Buffalo, San Francisco, New Orleans, Detroit, Milwaukee, and Washington, DC, the eleven largest cities in 1900 for which complete data sets were available. The data are maddeningly erratic. For example, Chicago reported 4,222 miles of streets in 1902 and only 2,880 in 1909 because it did not include largely unpaved alleys with the latter year. Some cities seem to have counted gravel streets as paved in 1880 and unpaved in 1890.

most popular form of heavy duty pavement, accounting for about a quarter of all pavements, was cobblestone, which had been used by cities since the Middle Ages. These pavements consisted of small, water-rounded stones, usually set on end in a bed of sand. Advances in stone quarrying, however, cut the price of square granite blocks enormously after the Civil War, and these cheaper, smoother, easier to maintain blocks were replacing cobblestones in urban areas where heavy traffic demanded strong pavements. Paving contractors set both types on fragile sand foundations. Some cities had adopted experimental pavements of wood or asphalt on a few streets, but these two accounted for less than 7% of the total. Experiments with and increased adoption of all five types dated from the 1850s in the United States, but the pace of adoption was slow. [. . .]

The picture in 1924 presented an almost complete contrast. The lightweight gravel and macadam pavements were completely gone, unable to withstand the suction of automobile tires, which tore them apart. While the traditional block pavements of cobblestone and granite block remained on a few downtown streets, most urban public works departments had covered the streets with smooth asphalt and concrete surfaces, the materials which predominate today. Perhaps more important than these surface changes, engineers had enormously improved pavement design in other ways, especially through the use of concrete foundations to replace the traditional sand foundations. This change was the key to better street drainage and more durable surfaces.

The reasons for these enormous changes in the technology of pavements are, as with all historical changes, complex. Pavements evolved, in part, because of a series of technological and scientific changes in related areas. Concrete as a building material, for instance, dates back to Ancient Rome, and American engineers had used natural cement deposits

as early as the 1820s during the construction of the Erie Canal. It was not until the new industrial chemistry of the 1880s thoroughly explained the composition of the material, however, that cheap, manufactured concrete of predictable quality became available. Similarly, natural asphalt deposits provided surfaces with unpredictable life spans until the pioneering chemical research of Allen Dow, Frederick Warren, and Clifford Richardson in the 1890s led to better refining and testing techniques.

Another reason for change in pavements is the shift from the horse to the automobile. Nineteenth-century engineers held that block pavements were suited to the needs of horses. Each horseshoe had a protruding calk which caught in the cracks of the pavements as the horse's hooves scraped along the surface. Teamsters and other horsemen believed that horses obtained traction from catching these cracks rather than from the surface of the street. In lightly traveled residential areas, fragile gravel or macadam surfaces sufficed. As the automobile came into use early in the twentieth century, such paving types did not provide a suitable surface. Clearly, the automobile did not require block pavements since their wheels had no calks. In the case of the lighter pavements, automobile tires were completely destructive, tearing up gravel and raising great clouds of dust even on residential streets with very little traffic.

These scientific/technical explanations alone, however, do not explain the changes in street pavements. The adoption of asphalt cannot be attributed entirely to the auto, since the shift antedated heavy automobile use. By 1902 cities had paved 16% of urban street surfaces with asphalt, by 1909 nearly one-third. The engineering literature rarely discussed the automobile as a factor in street use before 1909, and the auto's major impact on cities seems to date from the great jitney craze of 1914. Early internal combustion automobile innovation and manufacturing took place primarily in Detroit,

Cleveland, and Buffalo, already among the heaviest users of asphalt in the nation.

The advances in chemistry provide an equally weak explanation. In the cases of both asphalt and concrete, entrepreneurial and governmental interests subsidized the scientific research of the 1890s after they had perceived that a market already existed. A process of trial and error led to the use of these materials before industrial chemists had reached a very precise understanding of their nature. European cities had adopted natural asphalt as early as the 1850s. E.J. DeSmedt, a Belgian chemist, laid the first successful American asphalt pavements in Newark in 1871, using natural asphalt from the Island of Trinidad, and, the Army Corps of Engineers installed a highly publicized, durable asphalt pavement in Washington, DC, at the time of the Centennial. As early as 1876, the first comprehensive American paving manual, Q.A. Gilmore's *Treatise on Roads, Streets and Pavements* recommended the use of both concrete and asphalt.

Bricks, grouted with cement, also evolved purely through a trial and error process. Brick manufacturers discovered, without completely understanding the process, that heating bricks to the point of vitrification hardened them enough to bear vehicles. Scientific research in industrial laboratories played no role in its development.

The auto and advances in industrial chemistry did play a role in increasing the use of modern pavement types after 1900, but they merely advanced a trend already underway. Most major cities had stopped installing any new gravel pavements long before the auto rendered them complete nuisances. The work of chemists, such as Dow and Richardson, meant that any asphalt-based oil, such as the oil from the major fields opened after 1900 in California and Texas, could be refined to serve as the binder in an asphalt pavement. This reduced costs enormously, especially since unscrupulous monopolists controlled the Trinidad asphalt deposits, the most accessible natural source of high quality. Much the same thing happened within the cement industry, so that cheaper, more reliable concrete became available. To reiterate, the research in these materials came only because entrepreneurs believed that reducing costs and improving durability would accelerate the shift to their products.[1]

These pavement changes emerged more for social and cultural reasons, growing out of the great nineteenth-century shift in housing tastes from densely packed row housing to detached suburban homes. This shift resulted in two major changes in cities, which led to the revolution in street paving. First of all, there were enormous changes in the perception and uses of urban street space. For most urban Americans at the close of the Civil War, streets served vital neighborhood and family social uses. Pushcart vendors brought their wares to urban housewives, whose mobility was limited by the slow, expensive transportation system of the era. Surviving lithographs and photos show great herds of children playing in the streets, generally the only available open spaces. Such social uses demanded local control, and abutters normally determined how and when their valuable open spaces would be paved. As Lewis Mumford has noted, such urbanites thought that pavements were only for the rich and their horses.[2] By 1900 many, possibly most, urban residents saw streets as arteries for transportation since house yards and porches in the new streetcar suburbs assumed the traditional social and recreational functions of streets.

Second, important changes in municipal administration led to new paving policies. The urban bureaucrats, who by 1900 controlled paving in most cities, had completely lost sight of the traditional functions of streets. The newly powerful municipal engineers planned pavements more to cope with the enormous upsurge in late nineteenth-century street traffic or to facilitate the removal of the huge amount of wastes left in the streets by city horses than to provide a safe, healthy gathering space. [. . .]

The process of change was slow, incomplete, and uneven. It is useful to examine its complexity in some detail, looking initially at the post-Civil War decade of booming urban growth, and then examining the urban scene in 1900, a date by which the changes in paving technology were manifest.

The rejection of change

Early in the post-Civil War years control of urban street pavements rested largely in the hands of abutters. Decentralized, Jacksonian styles of municipal administration prevailed. Who was better able to judge the desired qualities in a street pavement than the people who lived on the street? Abutters decided when and how streets would be paved. Whenever residents holding two-thirds (in some cities a slightly higher proportion, in others, a slightly lower one) of the front footage on a block petitioned to have the street paved, the municipal government would draw up specifications, put the contract up for bids, and collect special assessments from all abutters (whether they signed the petition or not) to pay the contractor. The municipal government played a limited supervisory role in the whole process. The basic decisions, when to pave and the type and quality of the pavement, remained in the hands of abutters.

What decisions were made under such neighborhood control? In residential areas, abutters usually chose gravel, a quiet pavement unsuited to heavy use, if they chose to have their streets paved at all. Cobblestones were a second choice. They were noisier and more expensive, but more durable. Both functioned to allow surface drainage, which would render the street healthier and more valuable for social purposes while impeding traffic. Both cost little to lay although they required frequent repairs.

This rarely bothered the residents since ward or municipal governments were responsible for street maintenance, not abutters. Occasionally, abutters chose granite blocks, a noisy, expensive pavement type, whose smoothness and frequent cracks made it ideal for heavy teaming. The holders of commercial or business property were likely to choose this heavy duty pavement for use on their frequently traveled streets.

But as the 1870s dawned, some city residents began to change their attitudes. That decade saw much experimentation in pavements and their administration. New horsecar lines allowed some residents to move away from downtown, and a few of the new suburbanites sought smooth pavements in their new neighborhoods. As urban economies boomed, some commercial groups began to attack the slow pace of horse-drawn urban freight as a major bottleneck. They believed that smooth pavements would speed up traffic. The new public health advocates wanted well-paved streets to facilitate drainage and the gravity flow of the recently planned subsurface utilities, water and sewage, which they deemed vital to health.

However, a significant shift in urban housing tastes was perhaps the most important factor in the new attitude toward streets. [. . .]

In the absence of decent streets to provide access to the peripheral areas of cities where the new suburban life style was possible, the middle class turned to horse-operated street railways in the years just before and after the Civil War. [. . .] The consequent reduction of fares from 25 to five cents allowed some middle-class urbanites to secure the suburban housing which they so avidly desired.[3]

The horsecar encountered little resistance from urban residents, who feared high-speed traffic on their streets. It initially seemed like only a slight change from existing means of transportation. The horsecars generally ran on streets that were primarily commercial in nature. Moreover, the financial burden for street improvements rested on commuters since the transit company paid for the rails. Commercial interests welcomed the horsecar. In addition to the benefits of increased market area that radial transit service offered downtown merchants, the horsecar companies also provided surfaces for freight traffic to move on. Cities usually offered them franchises which required that they pave the street within their tracks and for several feet outside. Wagons were able to use this heavy duty pavement and tried to keep their wheels on the smooth tracks as well. [. . .]

Much of the improvement in pavements during this decade came for public health reasons. The new public health movement of the post-Civil War years had provided what seemed to most Americans to be scientific, authoritative proof that environmental conditions, especially polluted water and a substance supposedly given off by decaying organic matter, sewer gas, caused the epidemic diseases which plagued all large cities. Cities built massive new water and sewer systems to eliminate these problems. Since these utilities worked most efficiently when water flowed through them by gravity, sanitarians demanded properly graded streets to facilitate operation of the subsurface pipes. As a general rule, paved streets drained better than unpaved ones, so sanitarians also called for better pavements on streets and the location of pavements where none existed before for fear that stagnant pools of water or muddy ground might produce the dreaded gas. Most cities introduced water supplies long before sewer systems. The water was supposed to drain on street surfaces, but, in the absence of pavements, often created public nuisances in the form of pools or puddles. For the first time cities began to hire their own engineers, and many cities allowed them to specify the grades of new streets and to veto pavement selected by abutters to make sure that they would conform to drainage requirements.[4] [. . .]

The persistence of traditional urban values concerning streets may be judged from the failure of a number of attempts to use steam vehicles to improve the slow pace of horse-powered intra-urban transportation. Episodically throughout the nineteenth century, inventors developed steam autos, primarily for use in cities. Despite numerous claims of successful prototype operation, none of the vehicles ever reached the production stage. Urban residents, fearing the destruction of neighborhood street environments by the steam cars, bitterly complained to experimenters, obtained municipal by-laws prohibiting steamers and, in one case, mobbed the vehicle. Attempts by horsecar companies to use steam power encountered similar resistance. [. . .] Residents along their routes complained that noise, smoke, fear of explosion, and high speed destroyed the enjoyment of the street space in front of their homes.

Perhaps the power of the concept of abutter control is best illustrated by the famous *Elevated Railroad Cases* of 1883. [. . .] After repeated suits from abutters who claimed that els had destroyed the value of their property, the New York State Court of Appeals finally decided against the els. They ruled that the law entitled abutters to "light, air and access."[5] John Forrest Dillon, the leading nineteenth-century commentator on the law of municipal corporations, interpreted the decision to mean that abutters held "an equitable easement in fee," thus strongly asserting neighborhood control over streets.[6] The prospect of heavy damage payments implied by this decision effectively stopped further el construction in built-up sections of New York City, and elsewhere in the country for many years thereafter.

In sum, by the mid-1880s, it appeared that the older sections of cities had won their battle against new street uses and design. By controlling streets residents could prevent local governments from turning their only open social and recreational spaces into arteries for suburban travel. Some changes in pavement design did take place, largely for health reasons. Suburban boosters [optimistic promoters] and commercial interests had their way on some streets, but between 1880 and 1890 the proportion of urban streets that was paved actually declined (see Table 14.1).

The triumph of the new perception of urban space

Yet, by 1900, a decade later, political reformers, usually based on suburban wards, and their bureaucratic allies, the municipal engineers, had routed the forces of local control, and modern concepts of street paving held almost complete sway. This happened for two major reasons. The first grew out of a series of changes in the structure of cities and the new transportation patterns of the late nineteenth century. Most, importantly, cities allowed horsecar companies to electrify their lines. Higher speed led to more suburban sprawl. As increased numbers of urbanites moved to outlying, low-density neighborhoods, they adopted new attitudes toward streets, which

came to serve them primarily as arteries for travel rather than open social spaces. Also, increased trade led urban traffic to grow three to six times more rapidly than burgeoning populations (see Table 14.2). Trucking interests thus became much more potent lobbying groups. In addition, another new form of traffic, the bicycle, emerged in the 1890s. Bicyclists preferred smooth asphalt pavements.

Second, the new suburban middle class created major new political reform movements. The reformers urged that cities pursue policies to encourage suburbanization, not only to serve the immediate interests of their own constituencies, but also as a general policy to relieve housing pressures and the social pathology which they associated with tenement districts close to downtown. Better paving was one such remedy. The civil service reforms pursued by advocates of the new politics also strengthened the hand of the municipal engineers, who usually believed in smooth pavements and especially urged the adoption of asphalt. The municipal engineers were a

Table 14.2 Traffic increase in American cities, 1870–1930

	% increase in no. of teamsters 1870–1900	% increase in population 1870–1900	% increase in prime movers 1900–1930 (1900 + horses & mules 1930 − motor vehicles)	% increase in population 1900–1930
New York	311.4	92.8	458.7	75.1
Chicago	626.5	237.6	598.1	91.2
Philadelphia	350.7	52.7	305.3	59.6
St. Louis	243.8	64.1	484.8	47.4
Baltimore	157.6	53.1	762.8	64.2
Boston	412.6	54.5	562.9	44.4
Buffalo	422.3	227.2	916.2	63.4
San Francisco	303.2	46.5	828.2	123.3
Cincinnati	125.7	27.1	936.5	37.1
Median	311.4	54.5	598.1	63.5

Sources: Census Office, Interior Department, *Statistics of Population in the United States, Ninth Census (June 1, 1870)* (Washington, DC, 1872, 775–804; Bureau of the Census, Department of Commerce and Labor, *Occupations at the Twelfth Census* (Washington, DC, 1901), 428–479; Bureau of the Census, Department of Commerce, *Thirteenth Census of the United States, Taken in the Year 1910, Agriculture* (Washington, DC, 1913), 441–446; National Automobile Chamber of Commerce, *Automobile Facts and Figures* (1930), 50–51.

crucial group who would come to dominate municipal administration in the early twentieth century because they provided much of the personnel for the city manager and city-planning movements.[7] [. . .]

New programs led to the emergence of a substantial mileage of smooth, durable pavements by 1900. Many cities began for the first time to lay pavements paid for out of general funds, especially after the growing teaming companies of the 1880s and 1890s lobbied for such improvements (see Table 14.2). These transport interests had changed their traditional opposition to smooth pavements when engineers successfully demonstrated that horses did not slip any more often on asphalt than on granite blocks, despite the supposedly superior foothold characteristics of the latter. Bicyclists, a new and numerous group in cities, also agitated for better paving.

Two powerful groups led the political pressure on cities. First, public health reformers, alarmed by the increasing menace of horse manure on streets, began to lobby for easily cleaned asphalt pavements, especially in densely crowded sections of cities where the death rate was high. George Waring, supervisor of the 1880 Urban Census and the leading public advocate of the sewer gas theory of disease, became Street Cleaning Commissioner in New York City during a reform administration in the 1890s. Waring argued that sanitation workers could clean asphalt streets more completely and cheaply than the traditional block pavements. Public health and engineering journals viewed Waring's program with approval, and public health reformers joined teamsters [hauliers] as potent lobbyists for general fund paving programs and increased use of asphalt.

Second, municipal engineers began to use several new powers to limit the power of abutters to select pavements. Over the last quarter of the century, many cities had given their engineering offices the mandate to veto the pavements selected by abutters. Municipal

reformers believed that this precaution would prevent unscrupulous contractors from soliciting petitions from abutters for inferior pavement types, thus avoiding a repetition of the Nicholson block fiasco. Engineers drew increased support from reformers during the 1890s for other reasons as well. One corporation, the Barber Asphalt Company, held a monopoly on the excellent natural deposits in Trinidad. The municipal engineers, led by Victor Rosewater, Omaha's Municipal Engineer, and Allen Dow, Inspector of Pavements for the District of Columbia, launched an attack on this "asphalt trust," which was one of the pet targets of muckraking journalists. Through a variety of scientific tests, these engineers demonstrated that artificial asphalt made with oil from California's newly discovered fields was equal or superior to the "trust's" product. Thus, their scientific research not only led to the ultimate domination of asphalt as a pavement surface, but also provided important political propaganda in support of engineering autonomy.

Some municipal engineers, feeling secure from "political" interference, announced that they would now use the power to veto all pavements, except properly laid smooth pavements. In effect, the engineers used their veto power to dictate policy and eliminate those traffic-slowing pavements which abutters had traditionally favored. Cities also began to exercise more control over new suburban plots, forcing subdividers to conform to municipal street policy before engineers approved new residential developments for settlement. While, ordinarily, engineers used those powers to regulate street layout and grade, they may have included demands for asphalt and brick.

The new paradigm for streets

It is useful to examine the values used by these newly powerful municipal engineers in judging pavement types to more fully understand why they opposed the macadam and cobble pavements which best served the interests of those who resided on the streets in densely packed, older neighborhoods.

In 1901, George Tillson, one of the best known of the municipal engineers, published the first major work of the new century on pavements. Tillson prepared a scale to show the attributes which engineers should value in pavements. He then took this ideal scale and weighed actual paving types against it. Table 14.3 shows the results of his evaluation. Both the ideal scale and the results merit discussion. Tillson assigned 45% of the ideal pavement's value to direct cost standards like cheapness, durability, and ease of maintenance. He next concerned himself with ways in which the pavement facilitated traffic in addition to these basic construction standards. These qualities included little resistance [to] traffic, non-slipperiness (horses frequently fell and became disabled on urban streets) and favorableness to travel. These received a weight of 27%. Finally, variables related to sanitation (ease of cleaning and sanitariness) received a weight of 28%.

On this scale, which seems a good reflection of the values held by most of the new municipal engineers, asphalt was clearly the best pavement. Its greatest advantage over competitors was its cleanliness and, to a lesser extent, its aid to traffic. The worst pavements were macadam (gravel) and cobblestones, primarily because they were hard to clean. Tillson (and, presumably, other engineers) loaded the deck in several ways. He grossly overstated the cost and difficulty of maintenance of macadam, and, more importantly, he overstated the values associated with cleanliness.

The emphasis on sanitary qualities for health reasons served several purposes. Such sanitary precautions were an appropriate concern for engineers, whose water supply and sewer systems urbanites often credited with the massive decline in urban death rates during the late nineteenth century. Hence, urbanites were willing to grant some autonomy to engineers on health questions. Members of the profession often used specious evidence to warn about the threat of street filth, seeking to reinforce their power over pavement decisions. There can be little doubt that animal wastes on streets contributed to the high urban death rate. Any cut that came in contact with the manure in the streets might lead to tetanus. More importantly, dust in the air due to horseshoes pounding the filth may have provided the vector for the bacillus carrying tuberculosis, the leading urban killer.[8] [. . .]

Table 14.3 Evaluation of paving types

Quality	Maximum possible value on a scale of 1 to 100	Granite rock	Asphalt	Brick	Macadam	Cobblestone
Cheapness	14	2	4	3	7	14
Durability	21	21	15	13	7	15
Ease of cleaning	15	11	15	12	5	2
Little resistance to traffic	15	7	15	12	6	4
Non-slipperiness	7	6	3	6	7	5
Ease of maintenance	10	10	6	6	3	2
Favorableness to travel	5	3	5	4	5	0
Sanitariness	13	9	13	11	5	2
Total	100	69	76	67	45	44

Source: George W. Tillson, *Street Pavements and Paving Materials* (New York, 1901), 167.

The arguments made by engineers for smooth pavements to serve public health needs seem to be largely political propaganda. For example, Tillson used far more space in his book discussing the difficulties of providing a favorable operating surface for wagons, carriages, and bicycles than street cleaning. Incidentally, he did not mention autos. Yet, his table, the keystone of the treatise, valued cleanliness as much as mobility. If we view this table from the perspective of abutters, the political reasons for the emphasis on cleanliness become clear: abutters primarily sought the cheapest pavement possible; they were not concerned with maintenance, which the municipality generally paid for out of general tax funds. Ease of travel might even be a negative attribute to a householder seeking to keep traffic off the streets where his children played. By these standards, cobblestones were one of the best pavements, asphalt one of the worst. But if the abutters accepted the value which engineers placed on cleanliness, asphalt appeared superior. Thus, by emphasizing their supposed expertise on health problems, engineers might more easily overcome abutter resistance to smooth pavements.

While better health was important, especially to the general public, those who sought most actively for improved pavements seem more concerned with easing travel. Increasingly in the late nineteenth century, engineers and reformers like Tillson had begun to use physical or mechanical metaphors to describe cities. For example, they used the word, "machine" to describe certain highly successful urban political organizations. Tillson and others described cities as organisms, streets as arteries. In this imagery streets were vital, but only because of what they carried. The arterial metaphor was widespread in the engineering literature. It emphasized mobility and ignored other uses of streets. But to the abutters who paid for most paving, mobility was a negative, not a positive characteristic, of pavements.

Fast-moving vehicles made streets, the only open spaces in most older neighborhoods, too dangerous for social gatherings or children's games.

[. . .]

The engineers did not act alone, of course, but operated with strong political support. Interest groups like teaming [haulage] companies and bicyclists supported them. In many cities the settlement of outlying wards with suburban homes provided new support for their politics. Suburbanization diminished the clout of inner-city wards, where abutter resistance was strongest. Progressive reformers favored both technological advance and increased power for municipal bureaucracies. Such reformers were unlikely to question decisions by engineers, the key agents of both changes.

By 1900 the new cultural values and the innovations in pavements which they implied had taken firm hold. Engineers had created a new paradigm for pavement design that most city dwellers accepted. These changes in streets literally paved the way for the automobile, whose rapidly rolling wheels finished off the old pavements. Thus, in their headlong search for modernity through mobility, American urbanites made a decision to destroy the living environments of nineteenth-century neighborhoods by converting their gathering places into traffic jams, their playgrounds into motorways, and their shopping places into elongated parking lots. These paving decisions effectively made obsolete many of urban America's older neighborhoods.

Urban thinkers of the teens and twenties almost universally forgot the important social functions of street space. The suburban mentality dominated the political, social, and physical reformers who planned cities and dominated the policy issues until the 1960s. [. . .]

Street pavements were a minor aspect of the technological innovations which dramatically transformed late nineteenth- and early twentieth-century cities. Technological innovation is a complex phenomenon and this example indicates some of the complexities. Increases in scientific knowledge (industrial chemistry), the needs of a complementary technological innovation (the auto), and economic considerations (by 1910, asphalt cost less to lay than block pavements) all played a role in the adoption of smooth pavements. But the causes of innovation in this case may be more properly attributed to shifts in the cultural and political climate of nineteenth century cities, shifts which antedated and led to the scientific and economic changes. In the absence of effective social organization, late nineteenth-century Americans sought technological solutions to social problems. The two problems which perhaps most concerned nineteenth-century urbanologists were high death rates and the emerging middle class's desire for its own segregated neighborhoods, isolated from the chaos its members perceived downtown. Water supply and sewer construction provided the most important technological solutions to health problems. Smoother, more easily cleaned pavements were also part of the solutions. Class segregation depended on greater mobility which could open up land suitable for subdivision into tracts in the proper Victorian, middle-class taste. Here the crucial technological solutions were first the horsecar; then, in rapid succession, the trolley, the bicycle, and the auto. Smooth asphalt or brick pavements also facilitated mobility. Thus, transformation in pavements primarily grew out of a demand for technological solutions to social problems.

Notes

1 Artificial asphalt used asphalt-based petroleum and added sufficient sand or gravel to it to form a firm surface capable of supporting vehicular weight. Engineers had experimented with such pavements in the 1850s and 1870s, but they had proved unsuccessful because they used the paraffin-based petroleum found in Western Pennsylvania. Paraffin froze during the winters, causing such pavements to crack and ultimately deteriorate. The natural deposits consisted of asphaltic tars containing mineral impurities which could support weight. However, such deposits also contained a variety of other organic impurities, which producers had to burn off. The amount and kind of impurities varied from source to source and even from batch to batch within the same source. As a result, natural asphalt pavements were notoriously unpredictable. Some might last for decades, some might fail within months. The artificial asphalts proved superior not so much because they were cheaper, but because their quality was more predictable. Portland cement was an improvement on natural cement for much the same reason. Manufacturers could guarantee that each batch that they mixed from pure ingredients would behave exactly like every other. Natural cements, although often cheaper, varied enormously and unpredictably in quality

2 Lewis Mumford, *The Culture of Cities* (New York, 1938), 97-98

3 George Rogers Taylor, "Beginnings of Mass Transportation in Urban America, Part 1," *Smithsonian Journal of History* 1 (Summer 1966), 41-47; Glen Holt, "The Changing Perception of Urban Pathology: an Essay on the Evolution of Mass Transportation in the United States," in Stanley K. Sehultz and Kenneth T. Jackson, eds., *Cities in American History* (New York, 1972), 134-135; Clay McShane, *Technology and Reform, Street Railways and the Growth of Milwaukee 1887-1900* (Madison, 1975), 1-4. For an important contemporary view of horsecars, see John Noble *et al.*, *Facts Respecting Street Railways: The Substance of a Series of Official Reports from the Cities of New York, Brooklyn, Boston, Philadelphia, Baltimore, Providence, Newark, Chicago, Quebec, Montreal and Toronto* (London, 1866)

4 Attempts to use concrete as a street surface awaited the triumph of the automobile since constant pounding from horseshoes caused concrete to disintegrate

5 *Story vs. New York Elevated Railroad Co.*, 90 NY 122 (1883)

6 John F. Dillon, *Treatise on the Law of Municipal Corporations* (Boston, 1890), 834-856

7 Stanley K. Schultz and Clay McShane "To Engineer the Metropolis: Sewers, Sanitation and City Planning in Late-Nineteenth-Century America," *Journal of American History* 65 (September 1978), 389-411

8 Joel A. Tarr, "Urban Pollution: Many Long Years Ago," *American Heritage* 22 (October 1971), 65-69, 106; Lawrence H. Larsen, "Nineteenth Century Street Sanitation: A Study of Filth and Frustration," *Wisconsin Magazine of History* 52 (Spring 1969), 239-247

15

THE DECENTRALIZATION OF LOS ANGELES DURING THE 1920s

by Mark S. Foster

Source: Mark S. Foster, "The Model-T, the Hard Sell, and Los Angeles's Urban Growth: The Decentralization of Los Angeles during the 1920s", *Pacific Historical Review* 44 (1975), pp. 459–84

[. . .] Leading scholars have suggested that nineteenth-century American cities were shaped by forces significantly different from those that influenced the growth patterns of twentieth-century cities. Perhaps the most important changes occurred in the technology of urban transportation. In the last half of the nineteenth century, the rise of heavy industry and rapidly growing populations in America's older, eastern cities created enormous pressure for horizontal expansion. Until the late nineteenth century, however, their expansion capabilities were severely limited by inadequate urban transportation systems. Although primitive street railways were in operation in several American cities prior to the Civil War, the electric trolley was not in widespread use until the late 1890s. While the electric trolley permitted some outward expansion of these older cities, it appeared too late to alter significantly their highly centralized character.

In contrast, rapid technological advances in transportation – particularly the automobile – permitted twentieth-century cities to spread out to a degree which was previously inconceivable. [. . .]

Although the symptoms of Los Angeles's decentralization were evident to all by midcentury, the causes were not. They have sparked controversy among knowledgeable observers for years. Several scholars have suggested that the impetus for the region's horizontal growth developed prior to the 1920s. Reyner Banham, for example, believes that it can be traced to the emergence of the first street railways in the 1870s, while Robert M. Fogelson emphasizes the importance of the interurban railways in promoting decentralization during the first two decades of the twentieth century. Fogelson noted that the critical trend-setting annexations occurred between 1910 and 1920; in that brief period the city's size more than tripled. Expansion of city boundaries, claims Fogelson, encouraged expansion of city services, often in anticipation of, rather than in response to, actual demand. Thereafter, especially during the 1920s, the influx of thousands of retired farmers and small-town dwellers from the Midwest gave further impetus to decentralization in the region. The antiurban attitudes of the newcomers led them to demand homesites containing ample elbow room. Like Fogelson, Richard Bigger and James D. Kitchen emphasize the importance of annexations to the city's horizontal growth, but they also suggest that much of the impetus for regional expansion came from the suburbs. Between 1900 and 1920 such towns as Glendale, Pasadena, and Long Beach launched "empire building" annexation drives of their own, which contributed to the region's decentralization. The dramatic rise of new types of industry from the 1920s onward also encouraged horizontal growth in the region. A sophisticated land-use study by Frank G. Mittelbach notes that such twentieth-century enterprises as the movie and aircraft industries require vast amounts of space; their tendency to choose suburban locations augmented Los Angeles's horizontal development. [. . .] More recently, Sam Bass Warner has explored the automobile's effect upon Los Angeles's decentralization, and he suggests that its major impact was exerted after local planners had committed themselves to a freeway system in 1939.[1]

Despite their valuable insights, earlier studies have placed insufficient emphasis upon three important facets of Los Angeles's decentralization. First, a single decade, the 1920s, was a more dynamic period of horizontal growth than heretofore believed. During the decade, widespread acceptance of the automobile over the trolley as a preferred mode of urban transportation emphatically confirmed a tendency toward decentralization which had been emerging for some time. Second, the real estate boom of the 1920s profoundly affected the overall pattern of regional growth. By that decade, developers were no longer confined to building residential subdivisions within walking distance of streetcar lines. They responded by promoting thousands of homesites located miles from the nearest railway facilities. Third, the 1920s was a period when city leaders consciously and confidently committed themselves to a decentralized pattern of development as a positive goal. [. . .]

[M]ost local planners perceived encouragement of horizontal growth as the only viable response to changing realities. They believed that their policies would be a positive force effecting new patterns of urban development which other cities would eventually imitate. [. . .]

Although Los Angeles was founded in 1781, it remained a small provincial town for over a century. By its centennial, the community boasted only 11,183 residents. Although the land boom of the 1880s brought many newcomers to the region, Los Angeles's population of 102,379 in 1900 hardly ranked it among the nation's major cities. Not until the 1920s did Los Angeles experience the growth which perhaps most profoundly influenced its present day appearance. During the decade its population more than doubled, jumping from 576,673 to 1,238,047.[2]

The population boom of the 1920s coincided with the start of the decline of the electric railways and with the widespread adoption of the mass-produced, low-priced automobile as the primary mode of urban transportation. The fact that the region's two street railways – the Pacific Electric and the Los Angeles Railway – had begun their decline by 1930 escaped virtually every local observer at the time and can be thoroughly understood only in retrospect. In some ways, the two lines appeared healthier in 1930 than they had a decade earlier. Both lines realized net gains in patronage, as their combined yearly passenger totals experienced a fifteen percent increase between 1920 and 1930. Nevertheless, certain unmistakable signs of decay were evident. The 1930 combined patronage figures lose much of their healthy glow when one notes that the region's total population more than doubled during the 1920s. Even more important, during the last half of the decade, when the region's population was still growing rapidly, both lines lost patronage. [. . .] Just as important, the street railway companies were not expanding fixed facilities into the vast new real estate tracts which were opening up during the 1920s. Instead, they initiated bus lines into many of the newer areas – in itself tacit realization that fixed lines were outdated in the eyes of growing numbers of citizens. On the other hand, the 1920s witnessed the public's conversion to the automobile on a massive scale. Though Los Angeles County's population doubled between 1919 and 1929, the number of registered automobiles in the county multiplied five-and-a-half times, from 141,000 to 777,000.[3]

The concomitant rise of the automobile and decline of the trolley in the United States generally and Los Angeles in particular have been traced to a multitude of factors. A general increase in trolley fares may have driven away some public transportation patrons. In addition, automobiles became easier to acquire during the 1920s. [. . .]

The rise of the automobile may well have fostered a feeling of apathy on the part of both the general public and elected officials about the future of the street railways. This certainly appeared true in Los Angeles by the 1920s; people simply took the trolley for granted. At the same time, overconfidence on the part of local street railway officials probably contributed to that casual attitude. [. . .]

In retrospect, it is surprising that Los Angeles's worsening traffic situation during the 1920s did not severely temper this optimism. One of the earliest and most noticeable effects of the conversion to the automobile was traffic chaos, particularly in the downtown area. Several factors contributed to the problem. Automobiles required far more street space to transport a given number of passengers than did trolleys.[4] As more and more automobiles jammed downtown streets, all traffic, trolleys included, was inevitably slowed. Los Angeles's downtown traffic congestion was aggravated by the fact that more automobiles entered its central district each day than entered the central districts of other cities, and its streets were also among the narrowest of any city in the United States. A 1931 traffic study of ten major American cities revealed that over twice as many vehicles entered the downtown area of Los Angeles in a twelve-hour period than entered the downtown area of any other city studied.[5] Another traffic expert discovered that Los Angeles's business district "had less street proportionately than any of the larger American cities." It was also his opinion that the

traffic congestion in Los Angeles was the worst in the United States.[6]

Even as the 1920s opened, downtown traffic congestion had reached crisis proportions. Traffic snarls created by shoppers during the 1919 Christmas season inspired front-page newspaper coverage. Predictably, those merchants having a large economic stake in the preservation of the central area pressured public officials to find a solution to the problem. Traffic officials reasoned that since most downtown streets contained four lanes, elimination of on-street parking during business hours would effectively double the rate of traffic flow. In hasty response to rapidly mounting pressures, the city council on February 7, 1920, enacted the required ordinance; it went into effect on April 10.

The strict no-parking ordinance was a drastic, short-lived failure. Downtown merchants were flabbergasted at the negative impact of the ordinance upon their sales. Deprived of the convenience of parking at the doorsteps of their favorite stores, large numbers of shoppers avoided excursions to the central district. Commenting upon the immediate and sharp drop in downtown retail sales, the *Los Angeles Times* noted that "traffic has been expedited [. . .] but – NOBODY wants to cripple trade or turn Los Angeles into a Sleepy Hollow."[7] One merchant estimated that half of all downtown businesses would be bankrupt in three years if the ordinance remained in effect. Another merchant stated that unless the ordinance was quickly rescinded, it was "very likely that promoters would build numerous stores in suburban residence sections and elsewhere."[8] [. . .] Bowing to overwhelming opposition to the no-parking ordinance, the city council drastically revised the statute on April 26; thereafter, the ordinance applied only during the evening rush-hours. The original no-parking ban had withstood the test of public opinion for only sixteen days.

This was but the first of numerous

attempts during the 1920s to solve the grow-ing problems of downtown traffic conges-tion. Public officials and concerned civic groups engaged in protracted deliberations over the feasibility of providing some sort of comprehensive rapid transit system. While these discussions were unquestionably well intended and in some ways productive, they delayed positive action to solve the problem. This delay abetted the triumph of the automobile. [. . .]

Their expectations were not realized. Early in 1925, Kelker, De Leuw and Company pre-sented its comprehensive transit proposal. The firm recommended that the city spend over $133 million to build a system which would contain 26 miles of subway lines and 85 miles of elevated tracks. However, instead of receiv-ing enthusiastic public support and resolving the rapid transit question, the plan sparked a rancorous debate that continued for the rest of the decade. The proposal drew strong criti-cism from several civic organizations. [. . .] They contended that huge public outlays for subways and elevateds in such older cities as New York, Philadelphia, and Boston had inten-sified downtown congestion. The majority report further noted that widespread use of the telephone and public acceptance of the automobile had emerged after rapid transit systems were constructed in some eastern cities. By permitting almost instantaneous interpersonal communication, the telephone largely eliminated the need for businesses to remain in close proximity to each other. The automobile served to facilitate movement among outlying districts not directly con-nected by public transportation. Thus, the majority report suggested that these advances in technology presented Los Angeles with a golden opportunity to develop in a wholly different pattern. The committee envisioned the emergence of a decentralized metropolis, suggesting that

the great city of the future will be a harmoniously developed community of local centers and gar-den cities, a district in which need for transporta-tion over long distances at a rapid rate will be reduced to a minimum.[9]

The majority report emphatically rejected the notion that rapid transit should be used for promoting and perpetuating a centralized city with a large downtown area.

These criticisms of rapid transit generally coincided with the views of local city planners. Gordon Whitnall, one of the City Planning Commission's more articulate spokesmen, emphasized the inflexibility of traditional rapid transit systems. He admitted that construction of a new system might make sense in cities where population growth had leveled off some-what and where work patterns were relatively stable. Since it was clear that Los Angeles had not yet reached that stage, he favored reliance upon the automobile. According to Whitnall, when large numbers of Americans began to rely upon the automobile,

cities began to grow with it. Instead of the auto-mobile conforming itself to the limitations of the cities, the cities began to conform themselves to the necessities and services of the automobile. That is the BIG thing that has been happening here in the Southland. . . . So prevalent is the use of the motor vehicle that it might almost be said that Southern Californians have added wheels to their anatomy.[10]

[. . .]

Mounting criticisms of the rapid transit con-cept did little to ease the apprehensions of those who wanted to preserve the central dis-trict. Downtown merchants had hoped that a modern rapid transit system would stimulate patronage of their stores. By the mid-1920s some merchants believed that public policy was directed toward virtual abandonment of the downtown area. Even worse from their standpoint was the endorsement by public offi-cials of a massive street and highway building program in the outlying areas. As later events would conclusively demonstrate, they had every reason to fear that an ambitious highway building program would deprive them of patronage.

Their growing concern was caused in part by the deflation of earlier hopes. In the early 1920s, downtown merchants had proposed large-scale street widening as one of the more promising solutions to the problem of traffic congestion. Public agencies and civic groups presented a variety of specific proposals. Local planners soon discovered, however, that widening existing streets in the downtown area would be prohibitively expensive; one estimate placed the cost at a million dollars per mile. Therefore, some traffic planners suggested that the city council would employ its limited street funds more wisely by providing adequate streets and highways in outlying areas. [. . .]

The city council, pressured on all sides by competing interest groups, turned to professional planners for help. The nationally renowned firm of Olmsted, Bartholomew, and Cheney accepted the assignment of making a comprehensive regional traffic survey. The firm's report – labeled the Major Traffic Street Plan – was presented in 1924; it offered little comfort to downtown business interests. The plan suggested large expenditures of funds for construction of highways in outlying areas. More important, the plan implicitly endorsed the concept of decentralization. It envisioned the downtown area's future role in limited terms: as a center for theaters, government offices, corporation headquarters, and some ''specialty'' stores. The bulk of retail trade would be transacted in local neighborhood centers located miles from the central district.

Timing may have been a crucial factor affecting approval of the Major Traffic Street Plan and the rejection of the rapid transit proposal. The former plan was submitted, approved, and funded several months before presentation of the latter proposal. In November 1924, local

voters approved the traffic plan, along with $5,000,000 for funding its initial projects. This recent commitment to a comprehensive highway proposal helps explain why the rapid transit plan encountered so much opposition when introduced in the spring of 1925.

[. . .] The plan envisioned a checkerboard pattern of one-hundred-foot-wide thoroughfares that would eventually blanket the whole metropolitan area; they would be about one mile apart. A supporting network of secondary streets would include eighty-foot-wide streets every half-mile, and sixty-foot-wide streets every quarter mile. The president of the Los Angeles County Planning Commission explained why the plan emphasized highway development in the outlying areas:

> When we faced the matter of subdivisions in the County of Los Angeles[,] . . . subdivisions which were coming like a sea wave rolling over us . . . [,] we reached the conclusion that it would be absolutely necessary to go out into the country and try to beat the subdivider to it by laying out adequate systems of major and secondary highways at least, thus obtaining from the subdivider the necessary area for highways and boulevards.[11]

[. . .]

This clear pattern of new street and highway development, combined with the decline of the trolley and intolerable traffic congestion downtown, convinced many people that central Los Angeles's future prospects were dim. During the 1920s, the area lost much of its vitality as almost all new businesses and professional offices located outside of the central district. The latter part of the 1920s witnessed the first stage of A.W. Ross's Miracle Mile commercial development and the opening of a second Bullock's department store. Both of these new enterprises were located on Wilshire Boulevard, several miles from downtown. In 1920, only 16.1 percent of the city's dentists and 21.4 percent of its physicians maintained their offices outside the central district; by 1930 the figures were fifty-five and sixty-seven percent, respectively. Banks, department stores, movie theaters, and real estate offices all followed the dramatic trend toward location outside of the downtown area during the 1920s. As ever increasing numbers of businesses and professional offices located in outlying areas, a smaller proportion of local residents entered the downtown area. In 1924, sixty-eight percent of all residents within a ten-mile radius of the central district entered it daily; by 1931 the percentage had declined to fifty-two.

This clear trend toward decentralization also represented a reaction to new directions of population flow. Businesses and professional offices followed the people to suburban locations. The influx of thousands of newcomers into the Los Angeles basin during the 1920s created a heavy demand for residential property. Between 1904 and 1913 approximately 500 new subdivisions were opened annually. [. . .] The 1920s witnessed an enormous increase in the number of subdivisions opened. In 1920, only 346 new subdivision maps were recorded; the number rose to 607 the following year. In 1922, the boom really developed momentum as 1,020 new tract maps were filed, and it reached its peak in 1923, when 1,434 plans were presented. In 1924 the number dropped slightly to 1,306 and then declined steadily during the rest of the decade.

The real estate boom of the 1920s and the subdividers' scramble for maximum profits unquestionably exerted a strong influence upon the city's decentralization. [. . .] Between 1917 and 1927, property values in outlying areas rose twice as fast as did property values in the central district.

[. . .] Where the subdividers went, so did the building contractors. Construction activity dramatically revealed the extent of Los Angeles's decentralization during the 1920s. [. . .] In 1924, only 13.4 percent of the ring of land located between 8.6 and 10.3 miles from the downtown area was devoted to urban use; a decade later the percentage had more than

doubled to 27.8 percent. The primary reason for the remarkable increase in urban land use in outlying areas was the construction boom in single-family homes. Fully three-fourths of the area between 8.6 and 10.3 miles from the downtown area that was devoted to any type of urban use contained single-family homes.

This evidence also suggests that although the interurban railway provided the initial impetus toward real estate development in outlying areas, the public's preference for the automobile confirmed and intensified the direction of that growth. While the trolley promoters established a number of subdivisions miles from the downtown area, they had developed only a tiny fraction of the land in the Los Angeles area by 1920. Pre-World War I residents were so dependent upon the trolley for transportation that developers made few attempts to promote single-family homesites more than a half-mile from the lines. When a thirty-man syndicate, headed by Harry Chandler, purchased 47,500 acres of land in the San Fernando Valley in 1909, it immediately arranged for interurban railway service to the area. Although the valley's early development generally took the form of farms and small ranches, the few single-family homes on lots of an acre or less were almost invariably located within a block or two of the interurban railway lines. Proximity to streetcar lines continued to be an important prerequisite for successful development until the 1920s. City maps drawn in 1902 and as late as 1919 show few streets more than five or six blocks from streetcar lines.

[. . .] Prior to 1920, even the least pretentious property notices usually emphasized accessibility to public transportation. Modest three- and four-line advertisements almost always mentioned proximity to trolley lines, though relatively few of these advertisements mentioned garages. Before 1920, realtors apparently assumed that the average home buyer did not own an automobile, and that he relied chiefly upon public transportation.

As southern Californians became increasingly dependent upon the automobile during the 1920s, developers promoted property more and more remote from streetcar lines. [. . .] [T]he automobile's triumph exerted a dramatic effect on the remote areas which were not so well served by the trolleys. The development of the San Fernando Valley during the 1920s was, perhaps, the most spectacular example. The real estate boom of the 1920s witnessed the promotion of thousands of lots, many located miles from the nearest trolley lines. [. . .]

The public's changing taste in transportation during the 1920s was clearly revealed in the changing emphasis of real estate advertising. At the outset of the decade, advertisements for subdivisions usually stressed the proximity and convenience of local trolley service. By 1923, however, the beginning of a marked shift of focus was evident. Although some advertisements mentioned accessibility to trolley lines, most emphasized the automobile as the most convenient form of local transportation. [. . .]

By the end of the decade, realtors throughout the region were convinced of the automobile's indispensability. Advertisements seldom mentioned proximity of developments to trolley lines, but only rarely did advertisements fail to praise the accessibility of a particular area to the highways. [. . .] Improvements of local highways were important considerations in predicting success for any subdivision by the end of the decade. [. . .] By 1930, the automobile had gained such widespread acceptance that it had become nearly indispensable even to those with modest incomes. For example, inexpensive bungalow homes located on 50 by 130 foot lots frequently featured two car garages.

[. . .] Despite the myriad factors influencing Los Angeles's decentralization, that pattern of growth might not have been so spectacular had it not been that so few challenged the viability of population dispersal during the 1920s. To be

sure, local planning agencies were not sufficiently well established to exert a strong countervailing influence against decentralization. The city planning commission was not founded until 1920; organizational difficulties hampered its first few years of operation. Although Los Angeles County created the nation's first regional planning commission in 1923, not until the late 1920s did that organization and the city planning commission begin to work together effectively. Even then, both commissions operated with small professional staffs and limited budgets throughout the crucial period of the 1920s.

Despite such handicaps, local planners might have exerted more influence upon the pattern of Los Angeles's growth had they disapproved of the effects of horizontal development. Such was not the case. They favored decentralization at least in part because they had had the opportunity to study the massive problems created by a high degree of centralization in many older eastern cities. Speaking on behalf of the Los Angeles delegation before a national conference of city planners in 1924, Whitnall commented:

> We know, too, the mistakes that were made in the east and to a degree the things that have contributed to those mistakes. We still have our chance, if we live up to our opportunities, of showing the right way of doing things. It will not be the west looking back to the east to learn how to do it, but the east looking to the west to see how it should be done. This is our regional ambition for Southern California.[12]

Two years later, Clarence Dykstra, an efficiency expert with the Department of Water and Power, expressed the opinion of many local planners when he asked:

> Is it inevitable or basically sound or desirable that larger and larger crowds be brought into the city's center; do we want to stimulate housing congestion along subway lines and develop an intensive rather than an extensive city [and] must all large business, professional and financial operations be conducted in a restricted area; . . . as a matter of fact are all these assumptions, which

were controlling in the past generation, being severely arraigned by thoughtful students?[13]

Since Los Angeles's core-city problems did not begin to compare with those of many older, highly centralized cities, local planners appeared justified in the belief that their most exciting opportunities for future planning lay in more remote areas. [. . .]

By approving decentralization, planners in Los Angeles certainly challenged what they believed to be old-fashioned concepts. At the same time, they probably knew that the automobile was encouraging horizontal growth in many other American cities during the decade, and that Los Angeles's decentralization was not in itself unique. [. . .] Nevertheless, local planners took pride in being leaders in the nation in officially promoting horizontal urban growth. Late in the decade, they noted with obvious satisfaction that regional planners in New York, clearly one of the most centralized cities in the world, had determined to reverse the trend toward centralization in that city.

In retrospect, two factors distinguished Los Angeles's decentralization from that in other cities during the 1920s. No other major city in the United States approached Los Angeles's growth rate between 1920 and 1930. During a decade when the automobile won widespread public acceptance, Los Angeles's population more than doubled. In contrast, such "established" older cities as New York, Cleveland, Boston, and Pittsburgh all experienced population increases of less than fifteen percent. Those cities in the United States which rivaled or exceeded Los Angeles in size by 1930 had been shaped largely by the limits of nineteenth-century technology and had generally experienced their most dynamic growth before the 1920s. Consequently, the impact of the automobile upon those cities was far less dramatic than its impact upon Los Angeles. In contrast, cities such as Atlanta and Kansas City, which expanded in a manner similar to Los Angeles,

had growth rates during the 1920s and populations by 1930 which were a fraction of those in the latter city.[14] Just as important, residents of Los Angeles purchased more automobiles per capita than did residents of any other city in the country. By the end of the 1920s there were two automobiles for every five residents in Los Angeles, compared to one for every four residents in Detroit, the next most "automobile oriented" American city.[15] Thus, the impact of the automobile upon Los Angeles's urbanization process compared to that in other cities is distinguished chiefly by its magnitude. The size and timing of the region's population boom, and the suddenness with which local residents adapted themselves to the automobile were major factors shaping Los Angeles into its highly decentralized form. Both critics and defenders of Los Angeles's decentralization generally concede that by 1930 the city was in many respects the prototype of the mid-twentieth-century metropolis.

The factors affecting the pattern of growth in any region as big and as complex as metropolitan Los Angeles defy simple explanation. Natural factors aside, harbor improvements and the interurban railways unquestionably provided the initial impetus behind the area's horizontal growth. The rise of defense industries in the 1940s and 1950s, and the space industry in the 1960s further encouraged that pattern of development. Yet Los Angeles is basically an automobile city. The automobile assumed the primary transportation role in the region in a single decade. [. . .]

Notes

1 It is impossible to cite every work that has tried to explain the decentralization of Los Angeles, but some of the more important are Reyner Banham, *Los Angeles: The Architecture of Four Ecologies* (New York, 1972), 77-79; Robert M. Fogelson, *The Fragmented Metropolis: Los Angeles, 1850–1930* (Cambridge, Mass., 1967), 85-107, 226-227; Richard Bigger and James D. Kitchen, *How Cities Grew: A Century of Municipal Independence and Expansionism in Metropolitan Los Angeles*

(Los Angeles, 1952), 220–222; Frank G. Mittelbach, "Dynamic Land Use Patterns in Los Angeles: The Period 1924-1954" (Real Estate Research Institute, University of California, Los Angeles), unpublished manuscript in possession of the author; Ashleigh Brilliant, "Some Aspects of Mass Motorization to Southern California, 1919-1929," *Southern California Quarterly*, XLVII (1965), 191-208; Sam Bass Warner, Jr., *The Urban Wilderness: A History of the American City* (New York, 1973), 192-195

2   U.S. Dept. of Commerce, Bureau of the Census, *The Fifteenth Census of the United States;* Volume I, *Population* (Washington, D.C., 1931), 18-19

3   Brilliant, "Mass Motorization," 191

4   A 1924 study argued that in terms of passengers carried versus street space occupied, streetcars were 14.3 times more efficient than automobiles. Frederick L. Olmstead, Harland Bartholomew, and Charles H. Cheney, *A Major Traffic Street Plan for Los Angeles* (Los Angeles, 1924), 16

5   Donald M. Baker, "A Rapid Transit System for Los Angeles, California: A Report to the Central Business District Association, November 15, 1933," pp. 36-39, mimeographed report, available in the Los Angeles Public Library. According to Baker's 1931 figures, a twelve-hour cordon count showed that 277,000 automobiles entered the central business district of Los Angeles, compared to 113,000 in Chicago, 66,000 in Boston, and 49,000 in St. Louis during identical twelve-hour periods. He also stated that, in 1924, only 21.4 percent of Los Angeles's downtown area was devoted to streets, compared to figures ranging from 29 to 44 percent for other large American cities

6   Clarence R. Snethin, "Los Angeles Making Scientific Study to Relieve Traffic Congestion," *American City*, XXXI (1924), 196-197. Other traffic experts shared Snethin's view that Los Angeles's traffic congestion was the worst in the United States. For example, see C.A. Dykstra," Congestion DeLuxe - Do We Want It?" *National Municipal Review*, XV (1926), 394-398

7   *Los Angeles Times*, April 23, 1920

8   *Ibid.*, April 21, 1920

9   *Supplement City Club Bulletin: Report on Rapid Transit, January 30, 1926* (Los Angeles, 1926), 3-4, 9

10 *Bulletin of the Municipal League of Los Angeles*, V (Nov. 30, 1927), 3-6. Whitnall also opposed rapid transit on economic grounds. He believed that the city's low population density would make it impossible for even the most efficient rapid transit system to be self-supporting. Whitnall, "Zoning," mimeographed paper in Whitnall Scrapbook, City Planning Commission Library, Los Angeles City Hall

11 *Proceedings of the 16th National Conference of City Planners, April 7-10, 1924, Los Angeles, California* (Baltimore, 1924), 10

12 Whitnall, untitled mimeographed paper, April 7, 1927, Whitnall Scrapbook

13 C.A. Dykstra, "Congestion DeLuxe - Do We Want It?" *National Municipal Review*, XV (1926), 397

14 Los Angeles's population growth rate of 114.7 percent during the 1920s dwarfed the rates of the other top ten cities in the United States. For the decade, Detroit's growth rate of 57.4 percent was a distant second. U.S. Dept. of Commerce, Bureau of the Census, *Fifteenth Census: Volume I*, 18-19

15 In 1929 there were 228 automobiles per 1,000 residents in Milwaukee and 231 per 1,000 in Kansas City, Missouri. Older, more densely settled eastern and midwestern cities had much lower automobile-to-people ratios. For example, there were only 129 automobiles per 1,000 residents in Chicago; comparable figures for other cities included Boston's 125, Pittsburgh's 78, and New York's 74. Miller McClintock, "Trends in Urban Traffic," in R.D. McKenzie, *The Metropolitan Community* (New York and London, 1933), 275; Brilliant, "Mass Motorization," 191-208.

16

THE MINIMUM HOUSE

by Greg Hise

Source: Greg Hise, *Magnetic Los Angeles: planning the twentieth-century metropolis*, Baltimore, MD, The Johns Hopkins University Press, 1997, pp. 56–85

Scientism and housing

In volume 3 of his encyclopedic survey, *The Evolving House* (1936), Albert Farwell Bemis appealed to readers for a "new conception of modern houses . . . better adapted not only to the social conditions of our day, but also to modern means of production: factories, machinery, technology, and research."[1] Bemis, a successful manufacturer, [. . .] advocated setting aside outdated, "preindustrial" structures for alternatives engineered to current standards in materials, mechanics, and mass production. The means he endorsed, research informed by the scientific method and conducted according to industrial tenets, would, Bemis believed, rationalize the entire domain of housing and secure his penultimate objective, improving individual and collective well-being.

Bemis's wide-ranging project encapsulates the aggressive interwar campaigns to isolate, codify, and manufacture a standard, low-cost, minimum house that the majority of American wage earners could afford. [. . .]

Of course Bemis and his contemporaries were not the first to consider small-house planning, production techniques, and questions of use, value, and markets. Wage earners often built their own houses; and for at least three decades prior to Bemis's tome, catalogue and mail-order companies had been providing customers with manufactured dwellings. Along with national merchandisers such as Sears, Roebuck and Company, regional firms like Pacific Ready-Cut in Los Angeles offered a range of plan types sold as kits-of-parts. The latter's Pacific System offered buyers a predictable product and quality assurance at reduced costs by passing along savings the firm achieved through quantity production and the use of standard parts. At the house site, builders or buyers with basic skills could readily assemble these precut dwellings. Pacific Ready-Cut's 1925 catalogue included a 715-square-foot, four-room-plus-bath dwelling with a roughly square floorplan (Fig. 16.1). Over the next decade government agencies, researchers allied with the public and private sector, and home builders would adopt variants of this basic house type as an ideal minimum house.

Two factors, urgency and ethos, set Bemis and other principals engaged in these investigations apart from suppliers of manufactured housing. While they were interested in personal profit, many were motivated as well by a social agenda and set out to provide housing for those in need and to increase the number of home owners. In terms of method, they were, generally speaking, searching for a universal solution, the one best way to provide a unit plan, structural system, and construction technique appropriate to new living patterns. This is not to suggest that theirs was a highly coordinated project with intrinsic purpose and

Style 263 ~ *Pacific Ready-Cut Home* ~ Specifications

The following specifications briefly cover the materials furnished. See Price List. Cost of constructing this home on your lot ready for occupancy, including all carpenter labor, painting labor, cement work, plastering, plumbing, etc., quoted on request.

Foundation—Floor 1'-10" above ground. Wood steps for rear door. 2" x 6" redwood mudsills; 2" x 4" underpins on outside walls; 4" x 4" girders; 4" x 4" underpins on piers.

Frame—Douglas fir. 2" x 6" floor joists 16" o.c. 2" x 3" studding 16" o.c.; 2" x 4" rafters 24" o.c.; 2" x 4" ceiling joists 16" o.c. Double headers for all openings. Hood in kitchen over stove.

Floor—Sub-floor 1" fir boards covered with ⅜" x 1½" oak flooring in all rooms except kitchen, breakfast nook, screen porch and bath room, which are 1" x 4" tongued and grooved vertical grain fir flooring.

Walls and Partitions—Framed for lath and plaster. Ceiling height 8'-2¼". Outside of building covered with insulating felt and ⅝" x 4" rabbetted and beveled redwood surfaced siding.

Roof—1" x 3" or 1" x 4" surfaced fir sheathing covered with No. 1 "A" cedar shingles laid 4½" to the weather, every fifth course doubled. 18" projection finished with boxed cornice as shown.

Terrace—With segment roof over door as shown. Masonry not included.

Doors—Front door 3'-0" x 6'-8" 1¾" thick special. All other doors No. 210 except No. 303 from kitchen to screen porch.

Windows—Casement and double hung as shown.

Screens—14-mesh galvanized wire. Full hinged screens for all casement sash, half sliding screens for double hung windows. No. 551 screen for rear door.

Interior Finish—Baseboard No. 1; casings No. 1; picture moulding No. 1; continuous head casing in kitchen, breakfast nook and bathroom.

Built-in Features—Mantel shelf No. 913; linen closet No. 501; cooler No. 402; kitchen cupboard No. 202; drainboard prepared for tile or composition; sink cabinets No. 305 and No. 306; breakfast nook No. 702; medicine cabinet No. 602.

Hardware—Solid brass door knobs, escutcheons, drawer pulls, cupboard turns, etc. Nickel finish in kitchen, breakfast nook and bathroom. Dull brass for all other rooms. Cylinder lock for front door. Door butts, hinges, sash locks, etc., plated steel.

Paint—Exterior and screen porch two coats of paint, either white or color. Roof to receive one coat of shingle stain. Interior—Floors, oak floors to receive paste filler, one coat of shellac and wax. Screen porch floor and rear steps to receive two coats of floor paint. Interior—Three coats throughout, two of flat and one of enamel.

Refer to pages 135 to 155 for illustrations of trim, doors and built-in features.

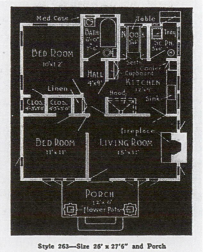

Style 263—Size 26' x 27'6" and Porch

Figure 16.1 A catalogue minimum house, Pacific Ready-Cut, 1925.

clearly identified goals. On the contrary, despite common objectives and a shared belief in the value of a scientific method, optimization, and large-scale operations, it was a disparate set of technicians and productionists who participated in this enterprise. [. . .]

Technicians and productionists shared an abiding faith in a technological fix they believed would address issues as diverse as job creation and retention, cycles in production and consumption, and shelter provision commensurate with an American way of life. [. . .]

Surveys to standards

During the interwar years technicians employed by industry, philanthropic institutions, and multiple branches of the federal government worked individually and jointly to establish spatial standards for a minimum dwelling. [. . .] An assessment of existing housing informed these strategies. When analyzing house plans, technicians noted that room layouts and the space assigned for daily activities was based on tradition, which they regarded as mere fashion. Second, their evaluation of existing codes and regulations revealed that spatial requirements typically reflected unexamined or outmoded standards rather than current knowledge concerning health and safety or the possibilities offered by new materials. From these analyses they argued for jettisoning the traditional and replacing "outdated" house types with scientific designs drawn from a rational assessment of how people actually used space, designs that would incorporate up-to-date equipment and technologies. [. . .]

John B. Pierce, a vice president at American Radiator Company, established a foundation in 1924 to support survey research and publications in the fields of heating, ventilation, and sanitation. Over the next five years the foundation systematically broadened its purview to focus on the dwelling and, more broadly, housing rather than on specific technologies. [. . .]

As a first step the foundation, in cooperation with the American Public Health Association's Sub-Committee on Occupancy Standards, conducted a study of 131 families in the New York area [. . .]. Interviewers and respondents completed a twenty-five-page form, families kept chronological records of selected activities, and researchers documented each dwelling, measuring and recording room sizes and storage space and cataloguing possessions. The guiding assumption was that a "detailed list [of] scientifically verified" data could serve as a baseline specification for unit design.[2] [. . .]

Laboratories for living

For technicians, the interpretation of quantitative data and user preferences could only serve in constructing a benchmark, a metric of existing conditions. This may have been adequate for modifying existing housing, but they were interested in creating something new, a minimum house which, they reasoned, required basic research, an investigation of people in space. Investigators employed by the Pierce Foundation recognized that a static analysis might reveal what people did in a house and when, derived from daily logs and interviews, but could not offer insight into the space required for any given activity. To this end, they made a deliberate effort to begin programming and design with an agreed-upon set of activities, such as eating, sleeping, dressing, and washing, rather than with a set of predetermined spaces, such as dining room, bedroom, and bath. To determine minimum functional requirements, they developed space-and-motion studies. These began with the "human being [. . .] wrapping around them space, equipment, and environment," an approach they contrasted with the "usual one of beginning with four walls and a roof and then subdividing this shell into rooms, which are later 'furnished.' " In a laboratory setting, researchers used still photography to record everyday activities, mounting a series of cameras to capture subjects in motion and the volume of space their actions consumed.[3]

There are significant methodological parallels between the foundation's space-and-motion investigations and Frederick W. Taylor's better-known time-and-motion studies. And even though twenty years separated these experiments, Taylor's "one best way" informed the functional home studies. However, there were considerable differences between these two projects. Where corporate managers capitalized on Taylor's studies of workers performing routine tasks to subdivide labor processes into their component motions, increase task fragmentation, and exert control over the workplace and production, technicians used the data gleaned from space-and-motion studies to construct an alternative to the fixed coupling of tasks and space. They proposed a flexible arrangement of multipurpose zones. In the minimum house, tasks would be performed in loosely defined "activity areas." This diversity is in sharp contrast to scientific management and labor specialization.

Pierce Foundation researchers also constructed experimental housing to conduct performance tests, refine standards, and elicit users' evaluations before scheduling a unit for production. Pierce Heights functioned as a laboratory for construction technology, building maintenance, and the spatial requirements of family living. Here, technicians refined a twenty-four-by-twenty-eight-foot four-room-plus-bath unit, a house type that became a community building standard.

Performance testing was not confined to private sector initiatives. Engineers employed by the U.S. Housing Authority's (USHA) Technical Division constructed a full-scale, four-room study house with exterior and interior walls mounted on movable tracks. After placing essential equipment and furniture in each room, researchers adjusted the walls' position to determine spatial minimums for fixtures and families. When evaluating kitchen design, technicians cranked in the walls incrementally as demonstration homemakers baked, cooked, and cleaned. [. . .] Nathan Straus, the USHA administrator who authorized the study house, had a predilection for minimum standards. In an article in the *Washington Evening Star* he cited over a dozen material economies the administration had achieved through "degadgeting," including the combination of cooking and dining spaces, elimination of basements, and the use of concrete as a finished floor.[4] [. . .]

Better living in fewer rooms

Among the liabilities that technicians and housing reformers identified were single-purpose rooms with limited use, such as the dining room, and unnecessary zones within the dwelling, such as the basement and attic. [. . .]

Reformers had been crafting their critique of the basement in home improvement literature throughout the 1920s. [. . .] Robert T. Jones, technical director of the Architects' Small House Service Bureau, [. . .] called the "millions of dollars invested in cellars under American homes" a buried treasure. He noted that recent practices, such as placing a central heater on the ground floor, offered considerable savings and quoted leading architects who found that basements represented 15 percent of the total cost in small dwellings. The Dwelling Construction Committee of the President's Conference reported that builders and design professionals interested in eliminating needless parts of the house should focus on the basement and the attic, for which the "original needs no longer exist or are rapidly disappearing."[5] [. . .]

For advocates of the minimum house, the dining room was anachronistic as well. The authors of *How to Own Your Home: A Handbook for Prospective Home Owners* informed their readers the dining room was no longer a functional necessity. "Where it is used but three times a day, it is the most expensive room in the house."[6] [. . .]

While a decision to include a dining room or to substitute an alcove or eating nook might be made solely for economic reasons, the move away from a discrete room assigned a single activity was also understood as one aspect of an emergent ethos of informality. [. . .]

The FHA played a key role in disseminating these principles and guiding the adoption of an efficient, four-room-plus-bath minimum house. [. . .] "The small house is not a large house compressed and trimmed down. It may not be created by determining what can be left out of a large house, but only by analyzing the essential functional requirements."[7] [. . .]

The FHA's minimum-house prototype was a 624-square-foot dwelling, which it presented simply as the "minimum house" or "basic plan" (Fig. 16.2). The accompanying text informs readers that "in the design of small, low-priced houses, the principles of efficiency, economic use of materials, and proper equipment, which are important in any class of dwellings, become paramount."[8] [. . .] The kitchens were small, planned for efficiency, and stocked with up-to-date appliances. A utility room with an integrated mechanical system replaced the basement heating plant and coal storage. A wet wall separated the kitchen from the bath, a configuration that permitted reductions in the material, time, and labor required to plumb each unit. [. . .]

Manufacturing the minimum house

[. . .] While technicians conducted surveys and tests and strove to craft an efficient house type, productionists were engaged in a complimentary endeavor, formulating a standard house design amenable to high-volume, quantity output. They targeted their efforts at revolutionizing building practice and the industrial organization of residential construction. This effort to rationalize housing and transform home building into a modern industry extended beyond unit design to encompass quantity production and large-scale operations. [. . .]

A.C. Shire, an engineer and technical editor of *Architectural Forum*, gave these arguments against the home-building status quo their full rhetorical voice in "The Industrial Organization of Housing."

> In an age of large-scale financing, power, and mass production, we have the anachronism that the oldest and one of the largest of our industries, concerned with the production of one of the three essentials of life, is highly resistant to progress, follows practices developed in the days of handwork, operates as a large number of picayune businesses, is overloaded with a whole series of overheads and profits, is bogged down by waste and inefficiency, is unable to benefit by advancing productive techniques in other fields, and is tied down to an obsolete and costly system of land utilization.[9]

In Shire's estimation, correcting these shortcomings would free up the available expertise and technique necessary for the quantity production of quality, low-cost housing.

Productionists championed regional or national firms practicing twelve-month construction and building for less-specialized markets. Most home-building concerns were not structured or capitalized for this level of internal management and external coordination. Increasing output while reducing production time demanded an operation tooled to the principles and strategies of the modern business enterprise. These included close oversight throughout the production process, from estimating and cost accounting to the sequencing of trades on a job site and the marketing and sales of multiple units. Implementing these management practices, productionists believed, would reform home building and allow a steady expansion in the construction of inexpensive, up-to-date dwellings, an important step toward creating a nation of home

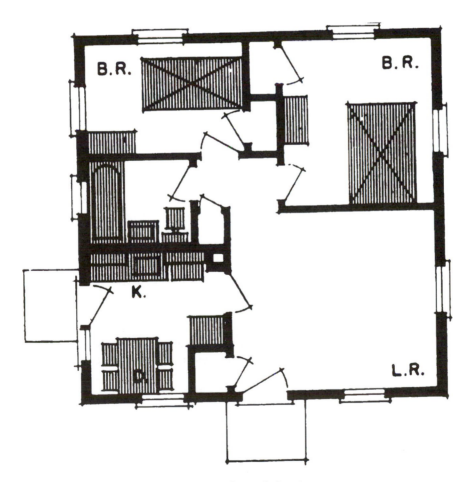

Figure 16.2 The FHA's basic minimum house, plan and elevation.

owners. Government analysts and policy makers agreed. The FHA identified four "fronts" in an "attack" on the high cost of housing production: first, factory fabrication of components and the design and engineering of multipurpose building parts to reduce on-site assembly; second, the development and use of materials better suited to factory fabrication; third, purchasing materials and equipment in larger quantities from fewer sources; fourth, improved labor and hiring practices, specifically a promise of continuous employment in exchange for hourly wage reductions.

The FHA's interest in factory fabrication underscores the degree to which pundits, the press, and the public had linked housing progress with mass production. During the interwar years technicians, design professionals, and builders shared an abiding fascination with factory-assembled dwellings and panelized housing, which they imagined would propel building into the modern era. Manufactured housing appeared in numerous editorials and articles in the home-building, architectural, and popular press. [. . .]

For many observers, prefabrication offered the single solution that would jump-start the building industry. *Fortune* alerted its readership to manufactured housing as a promising new industry, "perhaps the greatest single commercial opportunity of the age." [. . .]

> It is now past argument that the low-cost house of the future will be manufactured in whole, or in its parts, in central factories, and assembled at the site. In other words, it will be produced in something the same way as the automobile[.] . . . Design will dictate the form of thousands of units instead of the form of one. And the designer will necessarily consider not only the appearance, convenience, and efficiency of the completed product, but the feasibility of its production in mechanized plants and its distribution by modern transportation.[10]

However, some questioned whether the automobile analogy was an oversimplification or even inappropriate. Housing reformer Catherine Bauer articulated this concern in an unpublished review of *Fortune's* compendium, *Housing America*. Here she challenged the commonly held assumption that factory-based prefabrication could meet the current crisis in housing inventory and affordability. Bauer uses the editors' bias toward a new "big industry" to make her point: "The very reason the dwelling business has been so backward is the fact that mechanization has so little to do with it. Shelter is inseparable from problems of land, government, and income distribution." In her view, *Fortune's* editors understood housing solely as an inefficiently produced good. Their solution, a factory-fabricated dwelling set up on individual lots by individual entrepreneurs and sold through the same "indefensible Own Your Own Home propaganda," succeeded simply in shifting production from contractors to the "more nauseous cunning of high-pressured national [firms engaged in] cosmetic advertising."[11]

Bauer feared stories detailing housing "built in a day" would lull the public into a false complacency regarding the depth of need and the breadth of the response necessary for solutions. In its place, she preached the gospel of quantity production, reducing costs through large-scale operations and rationalized building practices. Bauer advocated design and planning that anticipated savings through a judicious siting of units and systems, component standardization and preassembly, and the on-site application of jigs and other techniques adopted from belt-line processes.

As this abstract of an extended, vitriolic exchange suggests, *prefabrication* was a catchphrase for an array of processes that might be applied or adapted to home building. Contemporary usage was imprecise; pundits and critics used the term for practices as diverse as factory-based continuous-flow principles modified for the construction site, the off-site production of wall, floor, and roof panels, or the factory production of an entire house with furnishings and appliances, ready for shipment to a finished lot. Robert L. Davison, director of the Pierce Foundation's housing division, offered a concise, inclusive definition: "Prefabrication is the assembly [. . .] of parts or subassemblies into sections [that are then] assembled into a structure, as distinguished from the assembly of parts during the erection of the building."[12]

Here Davison articulates an essential distinction between manufactured units assembled in the field and the on-site manufacture of a dwelling. [. . .]

A final distinction concerned locale, whether assembly takes place away from or at the building site. Using this criteria, prefabrication can be defined typologically as either factory assembled, ready-to-assemble, or assembled on-site. Factory-assembled housing would roll off the line like Ford's Model T, the entire dwelling manufactured in a central facility and then transported to the site either in sections or complete. Once at the lot, all workers had to do was set each dwelling on its foundation and connect the utilities. A number of firms, including Acorn and General Panel, produced factory-assembled units during the 1930s. The second type, ready-to-assemble, also relied on off-site fabrication, but in this case subassemblies such as floor and wall panels, storage and cabinet units, and mechanical systems were delivered to the site for final assembly. On-site prefabrication, or horizontal building, required a reorganization of the project area. Work crews would move through the site sequentially, completing carefully orchestrated, precisely timed, and repetitive tasks. Often a temporary staging area served as a shop, a centralized, occasionally enclosed facility for the multiple sizing and shaping of parts such as rafters and the quantity assembly of components such as wall panels, which could then be distributed across the job site.

Although horizontal building shares many techniques and practices with traditional build-

ing, there are critical distinctions. Theorists and practitioners modeled horizontal or serial production on the belt-line factory. Treating the building site as a continuous process meant that each unit would be assembled uniformly, from sitework to finish, and all units would be produced sequentially. For individual workers, horizontal building accelerated the advent of dedicated operatives whose labor increasingly was augmented by portable power tools. Skilled and semiskilled labor replaced the crafts person who had performed a variety of tasks within an assigned field. Serial production workers performed tasks repetitively as construction proceeded through the site. Although task organization was similar to a belt-line factory, at the building site operatives moved from station to station, whereas at the factory an assembly line brought each unit to the worker. [. . .]

Interchangeability [. . .] was indispensable for a minimum house and modern community planning. Building practice offers a demonstration. For example, an individual prefabricated panel was a module, a unit in a space-forming system. A vertical panel might be used interchangeably for an exterior or interior wall, a horizontal one for a floor or ceiling. Within a wall, each panel could be placed in a variety of positions: modified with a door or window, it could become an entry or a segment in a fenestrated wall. Each panel served as a replicable module in an intricate, interlocking system. [. . .]

Once materials and parts were in place, the builder or general contractor had to organize a team of craftspeople and coordinate independent subcontractors, each of whom wanted to complete a designated segment of the project in the shortest time allowable within a critical path. [Virgil] Tobin captured the dissimilarities between this method and the iconic auto assembly line. "When you buy an automobile, you don't hire a mechanic to shop around for an engine, wheels, transmission, [and] body, and then assemble those thousands of various parts into a finished car. Yet in this age of assembly lines and mass production, that is exactly what takes place when you buy a house." The author lamented the fact that there was no "integrated responsibility for materials, design, quality, or price of the product through the stages of manufacture until it is finally delivered to the consumer."[13]

Tobin's corrective was a belt-line with subassemblies progressing along a track. In Figure 16.3 a series of wall panels are proceeding along a conveyor in Gunnison's New Albany, Indiana, plant. The Gunnison prototype led many to believe that factory prefabrication would set home building on the desired track. However, residential construction was not amenable to this production type. Instead, quantity or large-batch production and the use of jigs and templates revolutionized home building.

Industrial management, simplified practice, and standardization

The Forest Products Laboratory (FPL) and the Pierce Foundation had been experimenting with quantity production since the 1920s. Through research and testing, technicians perfected a range of manufacturing innovations that builders subsequently adapted for residential construction. Forest Products Laboratory, a Department of Agriculture division established to promote timber utilization, began studying woods and glues for plywood in 1921. Following the aircraft industry's World War I advances, particularly in adhesive technology, engineers at the lab eventually shifted their focus from furniture to plywood's structural applications. Over time, FPL engineers fixed on a stressed-skin panel, a composite of waterproof plywood sheets bonded to lightweight framing. Manufacturing this structural box

Figure 16.3 The productionists' ideal. Beltline operations at the Gunnison Corporation's New Albany, Indiana, plant.

required the introduction of phenolic resin glues under carefully monitored conditions. Finished floor and roof panels maximized plywood's sheet form for a two-way membrane with high compressive and tensional strength. Preassembled wall units were slipped into slotted, four-by-four-inch verticals set at eight-foot centers. The FPL premiered its first stressed-skin unit in 1935, the most widely known experimental house from this period.

Initially, the Pierce Foundation directed its research efforts toward monolithic wall materials infilling structural steel frames. The first prototype, produced in 1932, featured Microporite, manufactured from cellular glass and a cementitous fiber. In 1935 the foundation redirected its research program to plywood. In plywood, this system resulted in a single or double wall infill panel set between four-inch-square posts at twelve-foot centers. By 1941 the foundation had released its system to twelve licensees. Commercial prefabricators such as Gunnison, Plywood Structures (Los Angeles), and Bates Prefabricated Structures (Oakland, California), as well as federal agencies such as the Farm Security Administration, adopted the stressed-skin panel. During the defense emergency the Pierce house, modified by the addition of a Celotex asbestos-surfaced insulating board, saw extensive service.

[. . .] Although manufacturers and builders continued to experiment with factory and other off-site house-building systems, these methods found limited applications. However, over time progressive builders' alternative site fabrication techniques and the use of plywood as a thin and lightweight sheet material applied to a scaled-down wooden frame, the basis for either factory fabrication or on-site construction, became standard practice.

Builders adopting horizontal operations had to enhance and restructure their advanced planning. Engineering high-volume production required precise coordination between the supply and delivery of material inputs and the sequence of work on the site. Implementing these changes proved a challenge. During the 1920s the Department of Commerce and the Federated Engineering Societies issued reports that assigned more than one-half of total construction costs to inefficiency, specifically poor management. [. . .]

The federal government, through the Department of Commerce, encouraged the adoption of industrial engineering principles in its effort to improve efficiency and reduce waste. [. . .] Secretary of Commerce Herbert Hoover was instrumental in organizing the Division of Building and Housing and the Division of Simplified Practice, voluntary associations for industry, trade groups, professional societies, and government representatives striving to forward these objectives. [. . .]

Consistent with Hoover's interest in better housing and home ownership, the Division of Building and Housing began addressing key home-building issues such as standardizing materials and housing components, industrial production and efficiency, and financing. [. . .]

Housing analysts considered building cycles wasteful and inefficient. [. . .]

While conceding that weather was an important factor in a production process that occurred to a great extent out-of-doors, critics such as Robert Lasch condemned the industry for letting contracts primarily in the spring and the subsequent shutdown of operations come fall. Lasch ascribed this to builders' reactionary embrace of conventional methods and rejection of modern practice, such as off-site subassembly, which could take place under cover. Analysts pointed optimistically to alternative procedures such as an early installation of mechanical heaters, which would allow for interior work during inclement weather, new materials including concrete additives, and the educational program under way by national trade associations. Productionists believed twelve-month construction would reduce labor costs on a by-project basis, since tradespeople and laborers would agree to reduced hourly wages in return for a guarantee of uninterrupted employment. However, while all these changes might reduce seasonal variances, home buying remained subject to longer-term cycles driven by financing, debt costs, and consumer mobility.

Another avenue productionists pursued when engineering cost reductions centered on material standardization and simplified practice. For productionists, standardization meant fixing the basic mechanical and physical properties of materials, their process of fabrication, or their applicability and usage. After identifying fundamental qualities and use, standardization could focus on establishing ratings governed by law, consent, or general use. Advocates for simplified practice focused on means for rendering industrial practices, in this case home building, less complex through an absolute reduction to the fewest possible variables. Efficiency experts at the Department of Commerce viewed simplified practice as a first step toward standardization. Proponents concentrated on eliminating "excess dimensions, types, and qualities of a specific article of commerce by mutual agreement between producers, distributors, and consumers."[14] [. . .]

In the home-building field, simplified practice was credited with significant reductions in the number and variety of basic materials, equipment, and construction tools. Edwin W. Ely cited common brick, which went from forty-four sizes to one, and the elimination of twenty-nine of forty rebar cross sections as representative.

Simplified practice accelerated a process that was already under way. The FHA acknowledged its benefits for high-volume construction: "The standardization of the parts of a house offer economies in construction without a loss in flexibility in planning. Based upon a common unit of measurement, parts can be made interchangeable. This form of standardization simplifies manufacturing and production and ease of erection. Cutting of materials on the job is reduced and waste eliminated."[15] [. . .]

Robert Lasch identified the degree of coordination necessary for a revolution in building practice.

> This enterprise demands a special type of producer organization bringing together a variety of specialized skills. There must be a real estate expert to handle land acquisition and related problems; a land planner and staff to lay out the general scheme of the subdivision; an architectural staff to design the houses; a construction engineer to organize production; [and] financial experts to manage the loans. The size of the organization, and the degree of specialization which can be called upon, depends upon the scope of operations.[16]

Compare this ideal to the small-scale building of a single house or a handful of units, a system A.C. Shire captures in this passage.

> If we examine an ordinary house in the process of construction we find delivered to the site . . . twenty to forty thousand parts or pieces of material to be handled, assorted, rehandled, fitted, and fastened together. We find over a thousand kinds of materials to be ordered, received, checked, invoiced, and paid for; there may be a hundred men working directly on the design and construction of the house; that these hundred men represent twenty different kinds of skills and specialized work; that as many as one hundred different firms studied the plans and prepared estimates of the cost of certain parts of the work; and that about fifteen different contracts were drawn up with the fifteen successful bidders.[17] [. . .]

[. . .] A typical housing project required the services of 13 subcontractors, 5 utility organizations, 15 material supply dealers, 8 material and equipment jobbers, 1 retail dealer, and 108 manufacturers. If operative builders would only change their operations, the argument continued, they could begin developing comprehensively planned communities. However, regardless of the extent to which productionists reengineered home building for volume production, their efforts would eventually reach a ceiling in terms of cost reduction. Fabrication and construction of an individual housing unit's shell accounted for just over one-third of the total cost. After a point, further improvements in production would garner diminishing returns.

Although this discussion of management, standardization, and innovation has focused on housing production, there was an equally vigorous debate concerning marketing and consumption. Quantity output required volume sales. The means for achieving this objective and opening up the market for low-cost dwellings was not self-evident.

In 1932, General Houses, a Chicago corporation, advanced one strategy that had direct implications for modern community planning. Howard T. Fisher organized the firm to market a panelized steel house of his design. In its choice of material and product the company was not exceptional. What was innovative about Fisher's organization was his marketing strategy. By establishing a network of licensed dealers and service units associated with local builders, Fisher intended to replicate the auto manufacturers' sales system and "Fordize" the building industry. In other words, Fisher and General Houses would act primarily as merchandisers. Paralleling the auto industry's strategy, Fisher planned to operate General Houses as an assembler of parts, not a prime producer. As suppliers he enlisted a Who's Who of American manufacturing, including Thomas A. Edison, Inc., General Electric, Pittsburgh Plate Glass, American Radiator, and the Pullman Car and Manufacturing Corporation. [. . .]

Technicians and productionists, along with decentrists and pragmatists, cooperated in the formulation of modern housing and, by extension, modern community planning. Technicians and productionists engaged in identifying the minimum ideal, restructuring building practice, and instituting management innovations participated in the creation of a new kind of city. They contributed to the efficient, high-volume production of standard, low-cost housing, an ideal that intersected with the spatial, formal, and social concerns of the neighborhood and region outlined by the decentrists and pragmatists. [. . .]

Notes

1 Bemis 1936, viii
2 APHA 1939
3 The quote is from Callender 1944, 12–13
4 "USHA Cuts Costs by Eliminating all Frills and Gadgets," *Washington Evening Star*, May 25, 1940
5 Jones 1925 and 1926
6 Department of Commerce 1923, 15
7 FHA 1936, 2
8 *Ibid.*, 2
9 Shire 1937, 37–38
10 MacLeish 1932, III
11 "When is a House Not a House?" C.B. Wurster Papers, carton 3
12 Davison 1943, 6
13 Tobin 1947, 336
14 Priest 1926
15 FHA 1936, 31
16 Lasch 1946, 156
17 Shire 1937, 37

References

APHA (American Public Health Association) (1939) *Basic Principles of Healthful Housing*, Committee on the Hygiene of Housing, New York, APHA

BEMIS, A.F. (1936) *The Evolving House*, Vol. 3, *Rational Design*, Cambridge, MA, MIT Press, Technology Press

CALLENDER, J.H. (1944) "The Scientific Approach to Design", in DAVISON, R.L. CALLENDER, J.H. and MACKEY, C.O. (eds), *The Engineered Dwelling: The Pierce Foundation*, New York, John B. Pierce Foundation

DAVISON, R.L. (1943) "The Engineered Dwelling", *Prefabricated Homes* 5 (9), pp. 6–11

DEPARTMENT OF COMMERCE (1923) *How To Own Your Home: A Handbook for Prospective Home Owners*. Prepared by John M. Gries and James S. Taylor, Division of Building and Housing, National Bureau of Standards, Washington, DC, US Government Printing Office

FHA (Federal Housing Administration) (1936) *Principles of*

Planning Small Houses, FHA Technical Bulletin 4, Washington, DC, FHA

JONES, R.T. (1925) "Omitting the Cellar to Cut Building Costs", *Small Home* 4 (9), p. 22

—— (1926) "Fifty Ways to Lower Building Costs", *Small Home* 4 (12), pp. 3-4, 28-30

LASCH, R. (1946) *Breaking the Building Blockade*, Chicago, University of Chicago Press

MACLEISH, A. (ed.) (1932) *Housing America: By the Editors of Fortune*, New York, Harcourt Brace

PRIEST, E.L. (1926) *Elimination of Waste: A Primer of Simplified Practice*, Washington, DC, US Government Printing Office

SHIRE, A.C. (1937) "The Industrial Organization of Housing: Its Methods and Costs", *Annals of the American Academy of Political and Social Science* 190 (Mar.), pp. 37-49

TOBIN, V. (1947) "Some Hope in the Housing Industry", *Consumer Reports* 12 (Sept.), pp. 336-38

17

LIGHT, HEIGHT, AND SITE: THE SKYSCRAPER IN CHICAGO

by Carol Willis

Source: Carol Willis, ''Light, Height, and Site: The Skyscraper in Chicago'', in John Zukowsky (ed.), *Chicago Architecture and Design, 1923–1993: reconfiguration of a metropolis*, Munich, Prestel; Chicago, The Art Institute of Chicago, 1993, pp. 119–39

In Chicago in the summer of 1893, the critic Barr Ferree addressed the annual convention of the American Institute of Architects on the practical problems of building in the modern commercial city. In this competitive arena, he observed, the profession had little control over the basic decisions of design because factors such as height, massing, and budget were determined not just by the client, but by a mathematical equation of costs and profits.[1] [. . .]

Indeed, no one on the congested streets of downtown Chicago in 1893 could ignore the awesome effects of the profit motive. A boom in speculative construction that preceded the World's Columbian Exposition that year had raised the standard level of new office buildings to sixteen stories, and some even higher. At twenty-one stories, the majestic Masonic Temple was the tallest office building in the city and, briefly, the world. Its gabled roof supported an observation deck that allowed the public to view the urban panorama from an altitude of 302 feet, and a glass skylight covered a large interior court that penetrated the full height of the building. The lower nine floors of this atrium were planned for retail shops, while the upper floors were divided into about 540 offices. [. . .]

[T]he history of the Chicago skyscraper has not been a steady ascent from small to tall. In fact, during the first half of this century, municipal restrictions on height shaped the city's skyline far more than bold feats of construction. In 1893, in reaction to the towering height of the Masonic Temple, the city council set a limit of 130 feet, the equivalent of ten or eleven stories, and for the next thirty years the maximum height moved up and down between 130 and 260 feet. In 1923 the city enacted a zoning law that allowed towers, but from the 1940s until 1956 the city again reduced heights, effectively to twelve stories. These height and zoning restrictions, along with other factors such as land-use patterns and rental markets, directly affected the formal development of the Chicago high-rise and of the city's skyline (Figs 17.1 & 17.2). [. . .]

Skyscrapers are designed from the inside out and from the smallest cell to the complex whole: a standard office unit is multiplied and arranged in an efficient floor plan. Even when the space is not partitioned, but pooled into larger areas, the ''phantom'' office governs the design from the dimensions of the structural bay to the fenestration pattern on the facade. Because of their dependence on natural light, interiors changed little from the first metal-skeleton buildings in the 1880s until the 1940s. The rule of thumb for first-class build-

Figure 17.1 View of Chicago, c. 1935.

ings dictated that no space be more than 25 to 30 feet from the windows to the innermost wall, which was the distance that daylight could penetrate an interior. From this constraint evolved both the dimensions of offices and the typical perimetal floor plan.

Such fundamental factors of function and economics make tall buildings in all places take similar forms, while local conditions such as land-use patterns, municipal codes, and rental markets differentiate them. This play of forces can be seen by comparing the skyscrapers and skylines of Chicago and New York. Most previous histories have emphasized the cities' differences. Commercial architecture in both, however, is formed by the same fundamental factors, although modified by a specific urban context. They represent, therefore, two variations on a vernacular of capitalism.

The "typical Chicago office building," 1893–1923

"If you look at our streets, you will find that the typical building of Chicago is the hollow square type," asserted architect George C. Nimmons in a paper prepared for the Zoning Committee of the Chicago Real Estate Board in 1922.[2] He was referring to the many structures erected from the 1880s through the teens that had a large light court at their centers. [. . .] These atrium buildings were not, in fact, the predominant type; U-shaped plans and other variations were more common, and there were, of course, many buildings on small lots. They did, however, represent many of the city's most impressive landmarks.

The hollow-square plan was the logical solution, given the city's height restrictions and the

Figure 17.2 View of Chicago, c. 1927–28.

characteristically large building sites in the Loop, because in order for adequate illumination to reach all interior spaces, light courts had to be cut into the mass, usually at the center or rear of the building. In New York, by contrast, interior courts were rare because lots were generally small, and there were no restrictions on height. As a result, the typical New York high-rise was a tall, generally solid block, often towerlike in height and slenderness.

In both cities, though, the same standard office unit and the same planning strategies governed commercial development. In "Some Practical Conditions in the Design of the Modern Office Building," published in the *Architectural Record* in 1893, George Hill outlined seven basic elements of a successful building: good location; good light; good service; pleasing environment and design; maximum rentable area consistent with true economy; ease of arrangement to suit tenants; and minimum cost, consistent with true economy.[3] Summarizing these criteria in another article, Hill noted that "an office building's prime and only object is to earn the greatest possible return for its owners, which means that it must present the maximum of rentable space possible on the lot, with every portion of it fully lit."[4] To achieve the utmost rentability, Hill advised that the plan should be flexible enough to allow division for many single tenants or for large companies that might rent an entire floor. He explained that the "economical depth" for an office was determined by the fact that "after a certain point is reached, no more money can be obtained [. . .] no matter what its depth," and he reported that "the generally accepted requirement of good lighting is that every portion of the office be within 20 to 25 feet of a window."[5]

A developer received the most profit by leasing a large number of small offices. [. . .]

Although the early 1890s was still a period of

experimentation with floor plans, later in the decade and especially in the early years of the new century, an efficient office layout became quite standard and changed little through the 1920s. The depth of the space from exterior window to interior corridor was subdivided for use by multiple employees. The smallest offices had only a single room, and furniture was used to define different work areas. For larger suites, one entered from the public corridor into a large room that could serve as reception and staff areas. Off this space were the private offices, which were generally about 12 to 14 feet deep and 8 to 10 feet wide so that two fit within a standard structural bay (when used most efficiently, steel columns measured about 18 to 20 feet on center).[6] This arrangement allowed for an anteroom, or reception area of 8 or more feet in depth, which was large enough for a stenographer, files, and a waiting area, and for two private rooms, each with a window. Examples of this layout can be seen in the plans of the Marquette (Fig. 17.3), Old Colony (Fig. 17.4), and Peoples Gas Company (Fig. 17.5) buildings. [. . .]

The hierarchy of the office was reflected in the spatial arrangement. Executives occupied corner offices or at least a windowed room, while secretaries and other staff were relegated to the deeper, darker spaces. The arrangement was described by Earle Shultz and Walter Simmons in *Offices in the Sky*:

> Of course the boss had to have his private office next to the window with the light coming in over his shoulder. In some cases his secretary worked in the office, too, but usually she and other clerical help used the reception room space between the private office and the corridor wall. To get maximum light into the reception room, the partition dividing it from the private office was glass. Sometimes this glass was opaqued to prevent people waiting in the reception room from seeing into the private office.[7]

This plan could be expanded laterally with ease, allowing the company to add as many units as it wanted on one floor. Small companies generally averaged only one or two staff members per private office; larger firms requiring open areas to accommodate many clerks or agents often preferred unpartitioned spaces. Over 80 percent of tenants throughout this period, however, leased suites of less than 1,000 square feet, which usually contained four or five individual offices.

Tenants were willing to pay higher rents for shallow, well-lighted spaces. A 1923 survey of values in Boston, for example, showed that offices that were 15 feet deep leased for $3.00 per square foot, while space that was 25 feet deep cost $2.60 per square foot and a 50-foot space averaged $1.65 per square foot.[8] Since deep spaces cost the same to build and operate, but netted lower earnings, the logic of producing only better-quality space was clear. By 1900 the norm for first-class space was 20 to 25 feet deep, and it stretched a bit more in the later 1920s to about 28 or 30 feet.

The main reason for this standard was the dependence on daylight to illuminate work areas. Although electric wiring was universal from the 1880s, incandescent light was weak and inefficient. Desk-top lamps produced only about three to four footcandles (one footcandle is a measure of the amount of light on a surface held one foot away from a burning candle), while the average for a room with good daylight exposure was around ten, which was sufficient for reading and working.[9] In some interiors, however, daylight levels could reach from fifty to one hundred footcandles. Measures of adequate lighting varied significantly over the decades: one sanitary survey in New York in 1916 recommended eight to nine footcandles, and in the 1920s, ten to twelve footcandles was advised.[10] Indirect lighting from large ceiling lamps was considered the most desirable type of illumination since it produced the fewest and faintest shadows. In the 1930s, spurred by aggressive sales tactics of large

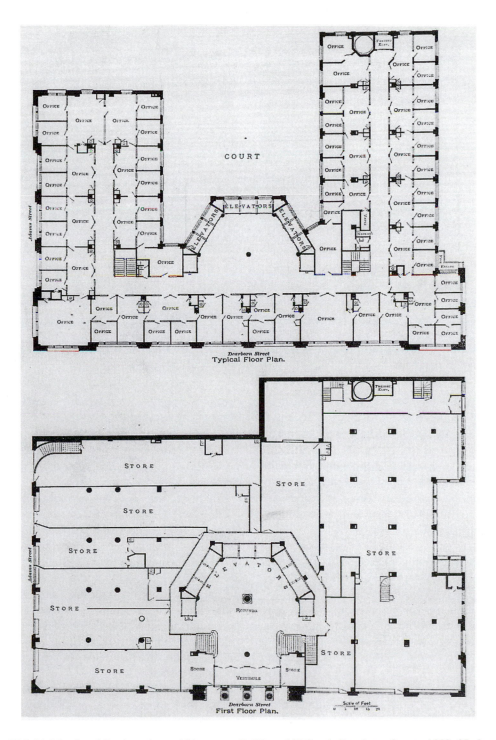

Figure 17.3 Holabird and Roche, plans of Marquette Building, 140 South Dearborn Street, 1893–95; from *Prominent Buildings Erected by the George A. Fuller Company* (Chicago, *c.* 1895), p. 42.

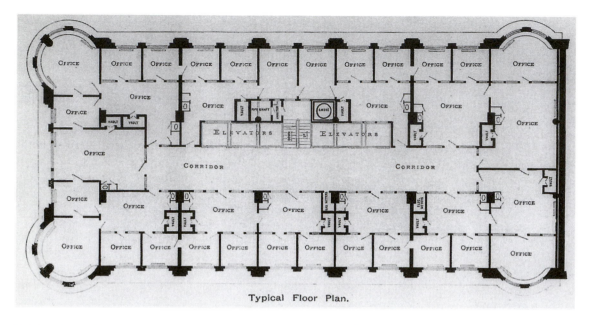

Typical Floor Plan.

Figure 17.4 Holabird and Roche, plan of Old Colony Building, 407 South Dearborn Street, 1893–94; from *Prominent Buildings Erected by the George A. Fuller Company* (Chicago, c. 1895), p. 30.

power companies, some experts urged a new standard of twenty-five footcandles. The drawback to using more bulbs, however, was that they added heat to the room (a 500-watt bulb gave off the equivalent of one pound of steam per hour).[11] Fluorescent lighting, introduced in 1939, eliminated the problem of excessive heat, and in modernist buildings, the ceiling often became a plane of light.

Before this new technology, however, in order to allow sunlight to penetrate as deeply as possible into the workspace, ceilings had to be high (at least 10 to 12 feet) and windows large – though not too big and heavy to open (Fig. 17.6). The "Chicago window" with its large central pane and side sashes solved that problem simply. Single windows usually measured about 4 to 5 feet wide and 6 to 8 feet high. They were often paired or grouped in multiples, which gave rhythm to the facade, but the main reason for the spacing was to allow for the partitioning of the interior. Win-

dows were also used for ventilation, for although most buildings had mechanical systems, outside air and breezes were vital to comfort, especially in summer months (air conditioning was not available until the 1930s).

Together, the optimal economic depth for office space of about 25 feet and the dimensions of typical Chicago blocks and lots produced two characteristic floor plans favored by developers until the mid-1920s: a solid, rectangular mass with a double-loaded corridor; and a light-court building – either a hollow square, a U-shaped plan, or a truncated variation such as an L shape for corner lots. Chicago's original grid had been platted with blocks of about 320 feet square and streets 66 or 80 feet wide. Most of these blocks were bisected by an alley, running either north–south or east–west, which was a public right-of-way and therefore preserved some light and air on the interior of blocks and

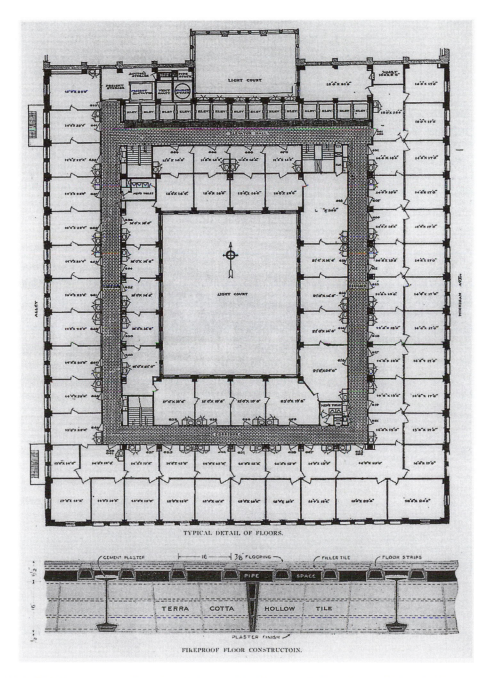

Figure 17.5 D.H. Burnham and Co., plan of Peoples Gas Company Building (now 122 South Michigan Avenue), 1910–11; from *Construction News* 30 (July–Dec. 1910), p. 100.

Figure 17.6 Holabird and Roche, interior of an office in the Cable Building (now 57 East Jackson Street), 1899; from Cable Building brochure.

the rear of many buildings. In the 1890s there were many parcels of a full quarter-block; smaller lots were sold by front footage and were generally 160 feet deep. [. . .] [N]ear the railroad stations, the area [. . .] was divided into long and narrow north–south blocks, which afforded through-block lots of about 70 feet and of various lengths.

A building with a double-loaded corridor – two rows of offices about 25 feet deep, plus a generous central circulation space – fits perfectly onto these narrow blocks. All offices had excellent light and views onto streets, rather than courts or alleys. Many of the famous early skeletal structures of [. . .] Chicago [. . .] occupied these southern sites, including [. . .] the Old Colony Building (68 by 148 feet; see Fig. 17.4). [. . .]

Light-court plans offered the best solutions for lots of at least 80 by 100 feet – that is, where buildings could not cover the full site because the interiors would be too deep to rent as first-class space. The hollow-square was the most logical plan for the largest sites. [. . .]

In the hollow-square plan, offices could be arranged either on a double-loaded corridor

with an outer and an inner ring of suites, or simply with a circulation space on the court's inner perimeter. The first solution could increase the space on each floor by 50 percent or more; the court-side offices were of secondary value, however, especially on the lower stories. The building's open center was often developed as a commercial atrium, sometimes several stories high. [. . .] A more common use for the central court space was as a second-floor banking hall of grandiose proportions.

For smaller sites, the U-shaped plan with the light court at the rear of the lot, often abutting the public alley, produced the most efficient layout. With one open end, the U allowed more light to enter the court throughout the day and saved the high costs of another exterior wall. For tenants, this plan offered better proximity to elevators. It also offered great flexibility, since a single company could take a half or full floor and incorporate the corridors into the firm's space leaving public only the area near the elevators. [. . .]

New York skyscrapers provide a useful counterpoint through which to understand the distinctiveness of Chicago during this period. Because building sites were generally smaller and land values were higher in New York than in Chicago, and because the city imposed no restrictions on height or bulk until 1916, towers proliferated in New York (Fig. 17.7). In the Central Business District in Lower Manhattan, the area of the original colonial settlement, real estate had originally been sold as lots, rather than as frontage, with the result that ownership was broken up into many parcels, which for well over a hundred years had been occupied by profitable low-rise buildings. Large sites were difficult and expensive to assemble (no great fire conveniently cleared the way for redevelopment just as a period of rapid corporate expansion was beginning, as had happened in Chicago in 1871). Farther uptown in the regular grid of avenues and cross streets established by the 1811 Commis-

Figure 17.7 Postcard view of the skyline of New York City, 1911.

sioners' Plan, the standard blocks measured 200 feet deep (north–south) and about 600 to 800 feet wide (east–west). Lots were generally 100 feet deep (half of the block) and conveniently divided for homes or shops into 25-foot frontages or multiples thereof. Despite the regularity of the blocks, therefore, parcels tended to be small.

Very tall buildings rose on constricted sites, stretching high above their neighbors to capture sunlight and views. [. . .] Many [. . .] skyscrapers were erected on small lots in the 1890s: the Manhattan Life Insurance Company (67 by 119 feet); the Commercial Cable Building (48 by 142 feet); and American Surety tower (85 by 85 feet). All exceeded 300 feet in height, surpassing Chicago's Masonic Temple. [. . .] Floor plans for buildings on such small sites were highly efficient because all offices could be placed on the outer perimeter, compactly arranged around a central core of circulation and services. Larger lots could also profit from a compact plan, since the additional elevators required by taller towers could occupy the otherwise unrentable center of the building.

In Chicago, however, municipal restrictions vitiated the logic of the compact core. From

1893 until the passage of a zoning law in 1923, the height limit ranged between 130 and 260 feet, or between ten and twenty-two stories.[12] This regulation, in combination with the typically large lot sizes in the Loop and the nearly inviolable custom of designing buildings that rose straight above the sidewalks (i.e., with no light courts on major facades), meant that high-rises had to be cut away at the rear or in the center in order to let adequate light into all interiors. Thus, the characteristic plan in Chicago was a central court while in New York it was more often a central *core*. Although the layout of an individual office was essentially the same in both cities, the full-floor plans and their projection into three-dimensional massing were entirely different due to historical patterns of land use and local codes.

The skylines of the two metropolises were likewise shaped by these conditions. Many Chicagoans had believed that height restrictions would eventually produce a harmonious streetscape as blocks were developed to the maximum. In fact, the fluctuating limit created a particularly motley array of flat-topped boxes, and, despite the visions of Daniel Burnham and a generation of civic efforts, the skyline never developed a strong visual order. In contrast, the distinctive silhouette of Manhattan bristled with towers. In 1913 the island boasted 997 buildings of between 11 and 20 stories, and 51 buildings ranging from 21 to 60 stories in height; the Woolworth Building soared to 792 feet. Even ten years later, Chicago had only 49 buildings of between 11 and 16 stories and 43 more of between 17 and 22 floors.[13] New York's laissez-faire environment seemed to spawn towers, for, while the city had about double the population of Chicago, it contained more than ten times the number of tall buildings. [. . .]

Between the Great Fire in 1871 and 1923, about 14 to 16 million square feet of office space was created in the Loop. During the same period New York constructed about 74 million square feet. Certainly one important factor in Manhattan's success as a business center in the early twentieth century was its vast supply of commercial space, most of it built by speculators, which ensured that rents were highly competitive. In Chicago, where height restrictions at times discouraged speculative development, the rate of construction – compared to the population and economic growth – was much lower than in New York. [. . .]

Twenties towers

The Chicago skyline changed dramatically during the 1920s as slender towers punctured the old 260-foot limit. Tallest of these was the 612-foot Board of Trade, but there were also eight buildings of more than 500 feet and eleven exceeding 400 feet. This proliferation of towers was the product of two factors: the real estate boom that began in 1923 and continued in force until 1929, and the passage of the city's first zoning ordinance, also in 1923.

Historically, the Chicago real estate market was characterized by cycles of boom and bust that were "like tidal waves in their magnitude."[14] In his classic study of 1933, *One Hundred Years of Land Values in Chicago*, Homer

Hoyt identified five major cycles of growth and decline since the city's founding in the 1830s. The cycle that saw the rise of the first skyscrapers began in 1879 and reached its peak between 1889 and 1892. The oversupply of office space produced during those years, along with the national depression of 1893, resulted in a decline of construction and land values that reached a nadir in 1898. Recovery began around the turn of the century, and until 1918, real estate values in Chicago showed an uncharacteristic pattern of gradual and sustained growth. When World War I diverted materials and labor to the military effort, development was stanched at the same time that businesses were expanding, putting great pressures on existing rents, which increased by 80 to 100 percent between 1919 and 1924.[15]

This situation fueled a new cycle of development. As building costs began to fall after the war, many developers rushed to fill the demand. [. . .] What began as an answer to a real demand, ended in frenzied speculation and a collapse in land and building values. By 1928 the financing of many projects was already in trouble, and after the stock market crash in October 1929, the situation grew steadily worse.

From the perspective of the mid-1920s, though, the phenomenal growth seemed a continuous spiral of prosperity. Land values in the Loop doubled between 1920 and 1928, which in turn put pressure on owners to capitalize land costs with bigger buildings. During the peak years of construction from 1923 to 1929, the supply of office space in the city nearly doubled. [. . .]

The 1923 zoning ordinance responded to the pressures of expansion by increasing the cubic volume permitted for high-rise buildings. In this sense, it had the opposite aim of the New York City law of 1916, which, after a completely laissez-faire condition, greatly reduced the height and bulk of a building that could be

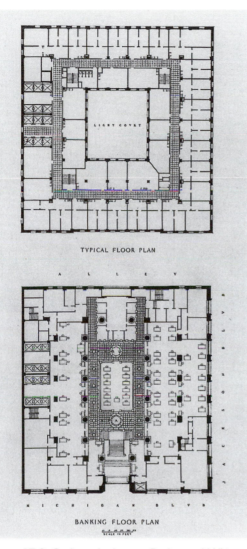

TYPICAL FLOOR PLAN

BANKING FLOOR PLAN
SCALE IN FEET

Figure 17.8 Graham Anderson, Probst and White, typical floor plan and banking floor plan of Straus Building, 1923–24; from *The Architectural Work of Graham, Anderson, Probst and White* (London, 1933).

erected on a given site. Although Chicago's guidelines were modeled on New York's, the specific formulas differed substantially. [. . .]

The major effect of zoning in the Central Business District was that towers were allowed. [. . .] Under the new zoning laws, the vertical limit above the sidewalk was 264

feet for the Loop and North Michigan Avenue. Above that height, a tower could be erected on up to 25 percent of the lot, although this upper section could not exceed one-sixth of the maximum cubic area of the main building. The volume restriction severely limited the height of commercially viable towers. About seventeen to twenty stories of additional tower floors was the maximum number possible for a quarter-block site. [. . .] . While the 264-foot limit for the base section of the building was the highest in any American city with regulations, the limit on the volume for the tower kept it quite small, especially compared to the buildings in New York, where a tower could rise to unlimited height on 25 percent of the lot. The seventy-story Chrysler Building, for example, stands on a lot of 200 by 205 feet.

The new regulations tended to produce an awkward hybrid – a large blocklike base with a puny tower. [. . .]

The combination of the central court plan with a tower was unusual, but not unique. The Straus Building of 1923–24 (now the Britannica Center) was an imposing twenty-one-story block with a nine-story tower surmounted by a solid stepped-pyramid and heraldic sculpture that extended its total height to 475 feet (Fig. 17.8). The building combined corporate headquarters for the investment banking company S.W. Straus and an income-producing property. [. . .]

Their analysis, which was documented in detail, supplied the economic logic behind nearly every decision of the design. Their problem was to achieve the highest ratio of rentable area to the cubic contents of the building and enclosing walls. [. . .] The plot solution that fulfilled this goal was a hollow square (Fig. 17.9). [. . .]

Another key decision was the number of elevators, since the quality of service directly affected rents. The standard for a first-class building was a maximum waiting time during peak periods of twenty-five to thirty seconds.

Figure 17.9 Graham, Anderson, Probst and White, Straus Building (now the Britannica Center), 310 South Michigan Avenue, 1923–24.

By projecting the total population of the building (some 4,000 to 5,000 tenants and their visitors) and the daily traffic pattern, experts determined that twelve cars were needed for public use. These were placed on the south side of the property, principally because of the possibility of future expansion, and arranged in two banks with three cars on each side of the corridor. The separate banks served different floors, which was the first time such a system had been adopted in Chicago. The nine-story tower was truly a separate section of the building, accessed by its own elevators that began on the twenty-first floor. [. . .]

The Palmolive Building [Fig. 17.10] was unique among Chicago high-rises for its "New York-style" massing. Under the formulas of

Figure 17.10 Holabird and Root, Palmolive Building (now 910 North Michigan Avenue), 1927–29.

New York's 1916 zoning law, the first setback began at a lower level than in Chicago, generally about 150 or 200 feet. Above that point, the building was required to step back within a prescribed diagonal plane projected from the

center of the street. This base could be surmounted by a tower which, as in Chicago,
could cover no more than 25 percent of the
lot; however, there was no limit on the height
of the tower. The typical New York skyscraper
thus had a base section, a transitional zone
consisting of a series of shallow setbacks, and
a tower that was slender and often very tall. A
comparison of the Palmolive Building and 500
Fifth Avenue illustrates the resulting differences in proportion under the two cities' zoning laws (see Fig. 17.11). Although the site of
500 Fifth Avenue (101 by 208 feet) was in fact
smaller than that of the Chicago building, its
tower rose twenty-two stories higher. In addition, the New York building sacrificed symme-
try to profit; the setbacks were different on its
east and west shoulders because the owner
wanted to exploit the maximum bulk allowed
by the formulas that governed each side of the
corner lot.

The typical massing solutions that evolved
from Chicago's zoning ordinance were, as

described earlier, the composite of a big base
plus a small tower and the integrated central
tower with flanking wings. There was also the
persistent popularity of the traditional interior-
court plans, in part because they were familiar,
but also because they could be highly profitable. Increasingly, though, in the later twenties, the compact-core arrangement became
the preferred plan, especially for its efficiency
of elevator access and service. Of the two formal approaches, the central tower was the one
adopted for buildings that were stylistically
most modern. [. . .]

Postwar postscript

The more than twenty-year halt in construction, caused by the oversupply of the 1920s
and made worse by the Depression economy,
was sustained by shortages of materials and
labor during World War II. These conditions
affected all cities, but the hiatus lasted longer
in Chicago than in most places. In New York,

Figure 17.11 Postcard view of mid-town Manhattan, with 500 Fifth Avenue building in left foreground.

recovery began almost immediately after the war, and by 1959, 54 million square feet of new office space had either been completed or was in development. During the same period, only 2.6 million square feet of new space was constructed in Chicago.

What caused the slow growth? It was not low demand. Although the vacancy rate had been 19.2 percent in 1931, and the market bottomed out in 1937, by the early 1940s, not only had the space been absorbed, but 2.7 million square feet in semiloft buildings had been converted to office use. Stringent zoning restrictions seem to have been a major factor inhibiting new construction. In 1942 the city again reduced the maximum size of a building; under the new formula, high-rise buildings were limited to a gross volume of 144 times the area of the lot, or if the building covered the full site, up to twelve stories.[16] To pile up several dozen floors in a high-rise, therefore, the site had to be very large, and most of it would be left empty. [. . .]

In an analysis of the market conditions in 1959, Earle Shultz maintained that the new limit on height and volume was lower than what most developers or companies believed to be the economic height. Without the hope of an attractive return, their money found more lucrative areas of investment. Shultz argued further that corporations that would have preferred to build Chicago headquarters in income-producing structures like the Straus or Palmolive buildings were discouraged from doing so by the restrictions and instead moved to other cities, in particular to New York. Although his theory cannot be tested, the data are consistent with such an explanation. In any case, the most significant fact about the Chicago skyscraper in the postwar decade was its absence.

New York high-rises constructed in the late 1940s and the 1950s were [. . .] architecturally undistinguished. They followed the setback massing of the twenties (which remained standard until the revision of the zoning law in 1961), but the pyramidal forms were clad, not in stone, but in an International Style aesthetic of banded windows or curtain walls of metal and glass. The more important change in these buildings, though, was in their plans, which featured larger floor areas – some 50 feet or more deep.

Businesses had prospered during the war years and needed additional space for their growing workforce and new types of office machinery. From 1940 to 1960 employment of office workers doubled and the size of the average office increased twofold, as did rents. Companies often preferred to consolidate operations on large, full floors, and improved technology made the deep floors viable. Fluorescent lighting provided high levels of illumination without excessive heat to all interior spaces, and air conditioning ensured ventilation. In 1950 Lee Thompson Smith, president of the Real Estate Board of New York, summarized the advantages of postwar structures as follows:

> These buildings are modern. Primarily because they are air conditioned. But one salient characteristic of the new buildings that cannot be adapted to old buildings at any price is their basic planning. They provide large blocks of space on one floor, with great glass areas, better lighting, fewer courts, less waste space and new automatic elevator arrangements, with fewer cars and faster service. Deeper floor areas, among the other developments in design, result in as much as 80 percent of the space on each floor being rentable space, as compared with 65 percent in the buildings that were conventional 20 years ago.[17]

Such buildings were modern in terms of program and function, but less so in terms of style. Yet the advantages they offered were attractive to large companies, and, as a result, pre-Depression buildings became less desirable to many tenants.

In the postwar era, technology liberated the interior from its dependence on external light and air. [. . .]

Although this essay has argued that, above

all, economics determined the forms of skyscrapers, the influence of aesthetics and ideology, especially in the postwar period, cannot be denied. But the contributions of great architects and engineers or the importance of schools or styles should be understood and appreciated in the context of the parameters of commercial architecture. In 1894 the straight-talking Barr Ferree noted:

> The design of a high building is a definite problem which may be expressed in a very few words. It is the arrangement of the largest number of windows of the greatest possible size in a limited space, which is fixed by external circumstances, such as the width of the lot and the number of stories high the owner is intending to build.[18]

The art of skyscraper design, one might say, is no more and no less.

Notes

Many thanks to Robert Bruegmann for his comments on this essay

1 Barr Ferree, "Economic Conditions of Architecture in America," *Proceedings of the Twenty-Seventh Annual Convention of the American Institute of Architects* (Chicago, 1893), pp. 228–41

2 Charles M. Nichols, ed., *Studies on Building Height Limitations in Large Cities with Special Reference to Conditions in Chicago* (Chicago, 1923), p. 70. Similarly, in their 1959 history of the office building industry, Earle Shultz and Walter Simmons noted that the majority of the city's buildings erected up to 1920 were of the O-shaped plan; see *Offices in the Sky* (Indianapolis, 1959), p. 132

3 In 1893 George Hill published a number of articles on economical design, which were among the first to appear in professional journals; they included "Some Practical Conditions in the Design of the Modern Office Building," *Architectural Record* 2 (1893), pp. 222–68; and "Wasted Opportunities, No. III," *Architectural Record* 3 (1893), pp. 436–38

4 Hill, "Wasted Opportunities" (note 3), p. 436

5 Hill, "Some Practical Conditions" (note 3), p. 471; and *idem*, "Wasted Opportunities" (note 3), p. 437

6 Shultz and Simmons (note 2), p. 130

7 *Ibid.*

8 *Ibid.*, p. 131

9 Statistics are quoted from an article by Kirk M. Reid, an illuminating engineer at Nela Park, Cleveland, a research center for General Electric; see "Artificial Light in Office and Stores," *Buildings and Building Management* 25 (June 8, 1925), pp. 43–46

10 *Ibid.*, pp. 43–44

11 Shultz and Simmons (note 2), pp. 202–203

12 The city council moved the limit up or down chiefly in response to real estate pressures – that is, the limit was raised when there was high demand for office space and lowered when there was an oversupply. The best source on height regulation in Chicago is Nichols (note 2); see also Shultz and Simmons (note 2)

13 New York statistics come from the *Report of the Heights of Buildings Commission* (New York, 1913), p. 15. For figures on Chicago, see Nichols (note 2) and Shultz and Simmons (note 2), p. 284

14 Homer Hoyt, *One Hundred Years of Land Values in Chicago* (Chicago, 1933), p. 372

15 *Ibid.*, p. 238

16 Shultz and Simmons (note 2), pp. 286–287

17 *Ibid.*, p. 247

18 Barr Ferree, "The High Building in Art," *Scribner's Magazine* (March 1894), p. 300

DECISIONS ABOUT WASTEWATER TECHNOLOGY, 1850–1932

by Joel Tarr

Source: Joel Tarr, *The Search for the Ultimate Sink: urban pollution in historical perspective*, Akron, OH, University of Akron Press, 1996, pp. 111–30

[. . .]

This paper is focused on a specific collection of environmental problems that were first addressed by the United States over a century ago. [. . .]

The duration of this study is conveniently divided into three relatively well-defined periods: 1850–80, 1880–1900, and 1900–32. Three technological system controversies will be examined, one for each time period. The resolution of each of these selected controversies led to the emergence of new problems. [. . .]

During each of these periods, the different technological, institutional, economic, and medical factors had varying weights and interacted to produce a particular orientation and policy regarding waste disposal and water and wastewater problems. Each of the periods was marked by controversy among competing systems and approaches, as adherents of one system or technology strove to dominate the field. By the end of the period, however, a particular approach was normally dominant. In each case, however, it was the application of this system or approach that led to the major policy controversy of the next period, as the dominant system developed problems and anomalies, some of which were unanticipated.

Selection of sewerage based on use of water for transporting domestic wastes

The conventional explanation for the adoption of water as the transport system for waste removal in American cities during the nineteenth century is related to public health. Historians and others have argued that cities adopted this new technology in order to escape from the hazards of the cholera and yellow fever epidemics that they experienced at periodic intervals throughout the century. This explanation, however, is based almost entirely on the writings of public health authorities, as well as special cases like Memphis, Tennessee. While this explanation has some validity, it is so general that it does little to help us fully understand the causes of the breakdown of the previous technological system, or the costs of this breakdown to society or individuals. [. . .]

Presewer cesspool and privy vault system

The period of 1800–80, what we might call the presewer period, was marked by a relatively primitive system in regard to the technology of human-waste removal and disposal. Human wastes were primarily deposited in cesspools and privy vaults. In many cities, when the cesspool or privy vaults became full, they were cleaned by municipal or by private scavengers, usually at the owner's cost. The wastes were then utilized as fertilizer on nearby farms or dumped in adjacent water courses and on vacant land. For most of the period, there were no public sewers for human-waste removal; the only sewers that existed were for the removal of storm or surface water.

Government involvement in waste disposal was minimal. Although there were many municipal regulations concerning the building and cleaning of cesspools and privy vaults, they were seldom enforced. City health departments, when they existed, had little money and limited powers. Usually only the threat of an epidemic could stimulate them to action. Medical theory in regard to the cause of epidemics was divided, with some physicians arguing that disease was the result of miasmas produced by decaying organic matter (anticontagionists), while others maintained that specific contagia, probably animate and usually imported, were responsible (contagionists). Reflecting this divided opinion, the response of city officials to epidemics was usually twofold: (1) clean the city and (2) institute a quarantine.

Introduction of running water at household tap

A major technological change was introduced during the 1800–1865 period – a continuous supply of water, freely running at the household tap. Until the beginning of the nineteenth century, all American cities obtained their water from local sources such as wells, cisterns, and springs. Population growth, the pollution of local sources, and a rising demand for more water for drinking, cleaning, and fire fighting resulted, however, in the construction of water systems that obtained their supplies from outside the city. Philadelphia was the first city to install such a system, followed by other large cities such as New York, Boston, and Baltimore. By 1860, the nation's sixteen largest cities, as well as many smaller ones, had water-

works of some kind. In total, there were 136 American waterworks in 1860, 58 percent of which were owned privately and the remainder public, although the privately owned waterworks were more characteristic of the small rather than the large communities.

The introduction of running water into cities resulted in a huge increase in water consumption: per-capita water consumption increased from about 2–3 gallons per day (gal/day) to between 50–100 gal/day. Boston, for instance, introduced water from the Cochituate Aqueduct into the city in 1848, and by 1855 the system had nearly 18,000 household connections, over 1,200 hydrants, and over 110 miles (180 km) of distribution pipes. [. . .] [A]lthough many cities introduced running water during this period, no city simultaneously made provision for means to remove the water. It was expected that the previous means of water disposal – street gutters or cesspools – would deal with the problem.

Installation of water closets

Lack of provision to handle the increase in the daily load of wastewater from kitchen and household use could be expected to add to cesspool or privy-vault overflows. Difficulties such as these were amplified with the introduction of another new technology – the water closet. This device was an old idea, but it had to await the availability of running-water supplies to enable its widespread adoption. [. . .] No water closet patents were issued in the United States until 1833. Once running water was available, however, many affluent urbanites installed the "modern" convenience of a bathroom with a water closet. In Boston, for instance, in 1864 (population approximately 180,000), there were over 14,000 water closets; while in Buffalo in 1874 (population approximately 125,000), there were over 3,000 water closets in use. By 1880, roughly one-third of urban households had water closets.

[. . .] The wastes from the water closets were now run into the cesspools and the privy vaults that retained their use as collectors. Even though some cities had stormwater sewers, the law usually forbade the placing of human excrement in them, although householders occasionally made surreptitious connections. The result of directing the flow from water closets into cesspools and privy vaults was to overwhelm the capacity for infiltration of wastewater into the soil around the cesspools or simply to overflow the privy vaults. This happened in city after city in the nineteenth century. As late as 1894, Benjamin Lee, the secretary of the State Board of Health of Pennsylvania, complained that [. . .] "Copious water supplies [. . .] constitute a means of distributing fecal pollution over immense areas [. . .] and no water closet should ever be allowed to be constructed until provision has been made for the disposition of its effluent in such a manner that it shall not constitute a nuisance prejudicial to the public health."[1]

Incentives for sewers based on water transport

[. . .] The previous analysis illustrates how the introduction of a new technology, running water, resulted in the adoption of another new technology – the water closet. The unanticipated result of the adoption of these technologies was to upset the existing system of waste disposal and to create problems of both sanitary nuisance and higher out-of-pocket costs for cesspool cleaning. There were attempts to preserve the existing system by using "odorless evacuators" (vacuum pumps) to empty cesspools, but this technology was both costly and undependable. Increasingly, in the period after the Civil War, as more cities installed running-water supplies and householders adopted water closets, urban policymakers became convinced that the only method to deal with the wastewater problem

was to adopt another new technology – the installation of sewers that transported wastes by water carriage.

Variations in sewer technologies: separate or combined sewer debate, 1880–1900

[. . .] [T]here were important debates about sewer technology, sewage disposal, the etiology of disease, and the role of government. Utilization of water-carriage removal, for instance, caused a shift in concern about the locus of pollution and media that were being polluted. With the cesspool–privy vault system, pollution was localized – the earth around cesspools was saturated, and wells and groundwater supplies were endangered with contamination. With the water-carriage system, the locus of pollution was shifted to streams and rivers often distant from the vicinity of the household. In addition, the disposal of one city's wastes often created problems for other downstream cities. Whereas in the period before the Civil War boards of health had only been local, it now became evident that extralocal authorities – either metropolitan or state boards of health – were needed to deal with pollution questions.

Origins of separate versus combined sewer debate

[. . .] The debate over separate and combined sewers was an old dispute that went back to the 1840s in Great Britain and essentially involved the question of whether human wastes and surface runoff from storms should be carried in the same pipe channels. The original urban sewers were for stormwater only. In Great Britain and Europe, storm sewers became combined sewers when households connected up to the existing sewers after the installation of running water and water closets.

The improper working of these systems,

however, due to poor design, led to the idea of a separate pipe system for household and human wastes. In addition, some sanitarians argued that the separate system would make possible the retrieval of human wastes for fertilizing purposes – an opportunity for resource recovery in addition to simple waste removal. A few cities in Great Britain and Europe did construct sanitary sewers and allowed their combined sewers to revert back to their original stormwater purpose. But the great majority of cities, faced with the problem of providing for both stormwater and household waste removal at the cheapest cost, built combined sewerage systems.

To a great extent, American sanitary engineers looked to Great Britain and Europe for models on which to base their designs. From 1857, the date at which the first sewerage system designed to handle sanitary wastes as well as stormwater was constructed, in Brooklyn, to 1880, combined systems were typically constructed in American cities. [. . .]

Waring's separate system and its impact

The first major application of a separate system came in 1880 in Memphis, Tennessee. Conditions there were unusual in that the system was built as a direct result of a severe yellow fever epidemic that occurred in 1878–79. Another unusual circumstance was the involvement of a federal agency, the short-lived National Board of Health, which entered the situation as the result of an invitation from local officials. After surveying the city and considering a number of sewerage plans, a commission appointed by the National Board of Health recommended the plan for a separate sewerage system conceived by one of its own members, Colonel George E. Waring, Jr. [. . .]

Waring's small-pipe system, in contrast to other sewerage systems, was based on a theory of disease etiology – a version of the filth theory – rather than engineering design principles.

Waring maintained that unless human fecal wastes were quickly transported out of the city in "fresh condition" they would "undergo putrefaction and give off objectionable gases." These sewer gases would be lethal to the extent that fecal matter had decomposed. [. . .] Waring criticized combined sewers as too large and as having insufficient velocity to prevent sewer gas-producing deposits from forming. Stormwater removal, he insisted, was only a secondary function of sewers, and stormwater could safely be permitted to run off along surface channels. A further feature of his system, said Waring, was that, in contrast to the combined system, it permitted land treatment of sewage where neighboring streams provided insufficient dilution for disposal. This was an opportunity to select an alternative method of disposal at the end of the pipe.

[. . .] By 1892, approximately twenty-two towns and cities had constructed Waring systems. [. . .] [T]he relatively inexpensive cost of the separate ($7,000/mile [$4,300/km]) sanitary sewer only, compared to the combined system ($28,000/mile [$17,000/km]), recommended it to many city officials. Most cities that installed a Waring system made no provision for the underground removal of stormwater. Some, however, did provide for treatment of their sewage by broad irrigation or sewage farming.

Typical late-nineteenth-century sewerage systems

Despite the alleged sanitary virtues of the Waring system, no large city installed a separate system of any design in the years up to 1900. Most engineering opinion followed the course laid down by Rudolph Hering in his famous report of 1881 to the National Board of Health. Hering took the position that the combined and separate systems had equal sanitary value and that the construction of one system or another depended on local conditions. Generally, he maintained, the combined system was best suited and cheapest in large, densely built-up cities where surface drainage of rainwater was unacceptable. The separate system, on the other hand, was suitable for cities and towns that did not require underground removal of stormwater and where sewage required treatment before discharge to a receiving waterway.

By the last decade of the nineteenth century, therefore, an accepted model had emerged in regard to the building of sewer systems. [. . .] The most important factors in the choice of one system over another were local conditions of population and traffic density, the costs of the alternative systems, and availability of water courses for disposal.

Impacts of combined sewer technology

[. . .] As sanitary engineer Moses N. Baker commented in the 1905 *Social Statistics of Cities*, "The general rule observed by American cities of all sizes is to discharge their sewage into the nearest available water until the nuisance becomes intolerable to themselves, and then to divert it from their own shores, resting content with inflicting their wastes on neighbors below, until public protest or lawsuits make necessary adoption of remedial measures."[2] [. . .]

The fact that sewage disposal in water courses could, if populations rose, cause pollution problems cannot be called an unanticipated result. There were early cases of stream pollution in Great Britain in the 1850s and 1860s, and Massachusetts actually took action against such pollution in the 1870s. A number of well-known sanitary engineers such as Rudolph Hering, Moses N. Baker, and Colonel Waring warned of the threat of dangerously polluting inland rivers through sewage pollution. Waring's solution to the problem was to build separate sanitary sewers and treat the collected sewage through land disposal. During the 1890s, a number of sanitary engineers took

the same position. But while Waring's motivation in urging the separate system was his concern over sewer gas, as well as the need for sewage treatment, engineers such as Baker and George Rafter who took the same position in the 1890s were primarily concerned with bacterial contamination.

[. . .] Until approximately 1890, chemical analysis was the only means to test water quality. In the late 1880s, however, bacteriological research played a more prominent role in defining water quality and gradually resulted in the substitution of the germ theory of disease for the filth theory. In the early 1890s, the work of William T. Sedgwick at the Massachusetts Board of Health laboratories clarified the etiology of typhoid fever. The clear identification of the waterborne nature of communicable disease produced an increased concern with methods of sewage "purification." And, given the nature of existing technology and the high volume of wastewater produced by combined sewers, purification required the installation of separate sanitary sewers.

Needs to modify the nineteenth-century system

[. . .] The 1880s and 1890s witnessed rising typhoid death rates in downstream cities that drew their water supplies from rivers in which upstream cities discharged their sewage. Bacterial analysis confirmed the link between sewage pollution of rivers and typhoid fever. The system of waste disposal that had replaced the cesspool–privy vault system – a system that primarily used combined sewers to remove both household wastes and stormwater and that relied on dilution as a disposal method – had broken down and recreated the health hazards of the old system. Increasingly, the health hazards caused by the sewage pollution of rivers initiated the development of new institutions – state boards of health with enforcement powers – that stood ready to force change in the system.

Debate over wastewater treatment and water filtration: 1900–1932

Evolution of state regulations

The sewage pollution of streams and the resulting typhoid death rates focused the attention of public health officials and sanitary engineers on the power of the state to prevent disposal in streams. Massachusetts established the first state board of health in 1869, and this example was followed by a number of other states in the late nineteenth century. More significantly, during the 1890s and the beginning of the twentieth century, state legislatures in states like New Jersey, Ohio, and Pennsylvania equipped their state boards of health with power to forbid sewage pollution of streams by municipalities.

In Pennsylvania, for instance, the State Board of Health was established in 1885 as the result of a severe typhoid fever epidemic that swept through the mining community of Plymouth. The epidemic had clearly demonstrated that state authority in public health matters was required because the state had sole power to enforce regulations beyond municipal boundaries. The board's role, however, was limited to the investigation and abatement of nuisances and the recommendation of new legislation, and it received no specific powers in regard to stream pollution.

In 1903, however, a severe typhoid epidemic caused nearly 1,400 cases of illness and 111 deaths in the town of Butler. The Pennsylvania legislature responded to the Butler epidemic by passing a stream pollution law. This law forbade the discharge of sewage into the state's waterways without treatment, although cities that had sewerage systems prior to the passage of the law were exempted from its effects. The law, however, did apply to extensions of existing sewerage systems. In addition to the stream pollution legislation, the legislature also replaced the State Board of Health with the Department of Health with the authority to enforce the legislation.

By 1905 there were several other states, such as Minnesota, New Jersey, and Ohio, that had laws comparable in strength to the Pennsylvania law. High typhoid death rates, although not necessarily epidemics, were the motivating cause. [. . .]

Progress in treatment technologies: sewage and water

The passage of laws forbidding sewage pollution of streams assumed the existence of a technology available at a reasonable cost capable of preventing such pollution. At the end of the 1890s and the beginning of the twentieth century, both sanitary engineers and public health officials were quite optimistic in this regard. In 1898, for instance, the Pennsylvania Board of Health stated that the "purification of the sewage of large towns, is a matter of not the slightest difficulty" and recommended that all cities possessing public water-supply systems install sewage-treatment facilities.[3] [. . .] This optimism, however, was clearly unwarranted.

[. . .] [B]y 1902 ninety-five cities treated some portion of their sewage, but only eleven of these had populations of more than 30,000, and all but a small fraction had separate sewerage systems. The major sources of river pollution, however, were the large cities and, as of 1905, New Orleans was the only city with a population above 30,000 to have a separate system. The remainder of the large cities, all with combined systems, disposed of their sewage by dilution. They followed this practice because it was the cheapest form of disposal and because sewage pollution of water supplies was considered primarily a problem for downstream cities.[4]

Even if these cities had wanted to treat their sewage and reduce the danger of typhoid epidemics, their combined sewerage systems and the uncertain and changing state of sewage-treatment technology would have made this immensely costly. The first methods utilized in this country were broad irrigation and sewage farming. In the early 1890s, however, experiments at the Massachusetts Board of Health's Lawrence Experiment Station had proven the feasibility of intermittent filtration, and this method became the dominant technology utilized in the 1890s. Intermittent filtration, however, required both the availability of soil with a high infiltration rate and large tracts of land. Both factors restricted its feasibility for many cities. In the beginning of the twentieth century, other sewage-treatment technologies such as household septic tanks and central treatment plant trickling filters were introduced.

Simultaneous with the changes in sewage-treatment technology, important developments were taking place in the area of water filtration. Here, as with sewage treatment, the Lawrence Experiment Station of Massachusetts led in the development of technology. In 1893, a slow sand filter designed by Hiram F. Mills of the Lawrence Station was installed in the city of Lawrence, and in five years the typhoid death rate was observed to decrease by 79 percent. The development of other types of filters, such as the mechanical filter and the rapid sand filter, followed and were installed for water-supply treatment in a number of cities. By 1900, 28 percent of the urban population was drinking filtered water, and typhoid death rates in cities were dropping quickly.[5]

The striking success of water-filtration processes in lowering typhoid fever rates, as compared with the slow progress in sewage-treatment development, caused a crucial shift in engineering opinion. As *The Engineering Record* editorialized in 1903, "'Many engineers believe today that in some cases it is often more equitable to all concerned for an upper riparian city to discharge its sewage into a stream and a lower riparian city to filter the water of the same stream for a domestic supply, than for the former city to be forced to put in sewage-treatment works.'"[6] Not only was water filtration more certain in terms of lowering typhoid death rates, but it also could be provided at a far cheaper cost than sewage treatment.

Conflict over enforcement: the Pittsburgh case

While engineering opinion was shifting toward the choice of water filtration over sewage treatment in order to protect drinking-water supplies, state legislatures and the courts were moving in a different direction. By 1909 a number of states, such as California, Indiana, Minnesota, New Jersey, and Ohio, had passed laws forbidding the discharge of raw sewage into streams from new municipal systems. These laws were designed, in most cases, to be administered by state boards of health staffed, to a large extent, by physicians. The medical position, in contrast to that of most sanitary engineers, was that the public health would best be protected if no municipality was permitted to discharge raw sewage into streams. And they stood ready to use the authority of the state to enforce their directives.

The critical clash between the state boards of health and the sanitary engineers came in the city of Pittsburgh. In 1910, Dr. Samuel G. Dixon, the Pennsylvania commissioner of health, required Pittsburgh to submit within one year a comprehensive plan providing for sewage treatment. [. . .] The city hired two well-known sanitary engineers, Allen Hazen and George Whipple, to make recommendations. [. . .]

The Hazen and Whipple report concentrated on the economic feasibility of constructing separate sanitary sewers and a sewage-treatment plant in Pittsburgh. They estimated that replacing Pittsburgh's combined sewers with a separate system and building a treatment plant would cost Pittsburgh taxpayers a minimum of $46 million with no provision for the costs of disruption caused by construction of the system. They also noted that Pittsburgh's water-filtration plant had cost

$7 million and calculated that the twenty-six towns downstream from Pittsburgh on the Ohio River could provide filtered water for their residents for far less than the $46 million it would cost Pittsburgh to build its treatment plant. No precedent existed, they argued, "for a city's replacing the combined system by a separate system for the purpose of protecting water supplies of other cities."[7] [. . .]

Engineering opinion overwhelmingly supported the Hazen and Whipple report and viewed the controversy as an issue of "how far engineers are at liberty to exercise their own judgment as to what is best for their clients and how far they must give way to their medical colleagues."[8] [. . .]

Heritage of decision to treat water but not sewage

In the period from approximately 1900 into the 1930s, engineers were again confronted with major policy choices in regard to wastewater treatment and the protection of public water supplies. [. . .]

The pollution problems produced by combined sewers, however, spawned not only sewage- and water-treatment technologies but also water pollution laws enforced by activist state boards of health. These boards of health, often staffed by public health doctors representing the so-called "New Public Health," concerned themselves with controlling the causes of disease at the source. For them, this meant keeping all human wastes from contact with public water supplies. In a number of states these boards of health compelled small cities to stop discharging raw sewage into streams and to build treatment plants. The crucial question, however, involved the major cities that had combined sewer systems. Could the state force them to convert their combined sewer systems to separate systems and build treatment plants, often at a cost that exceeded their bonded debt limits? Engineering opinion took the position that such demands were "radical if not [. . .]

quixotic." Dilution, they insisted, was "sound in principle and safe in practice if carried on with proper restrictions," especially since it was "physically impossible to maintain waterways in their original and natural condition of purity."[9] In addition, economics militated against so-called "purification" of sewage that could only take place at enormous expense.

The focus on bacteriological pollution associated with sewage wastewaters and the cost-effective alternative of treating water at the point of use was expedient and proper for a limited set of water-quality concerns. It neglected, however, chemical pollutants associated with industrial sources and initiated an overdependence on dilution in streams. In addition, it did not anticipate the need for waste-treatment technologies for waste constituents that proved refractory to water-filtration plants. [. . .]

Conclusion

The adoption of a technology and its development in each historical period strongly influenced choices in subsequent periods. Decisions had impacts, many unanticipated, in a variety of technological, environmental, public health, and governmental areas. [. . .]

The examination of the record shows the complexity of past decision making in the area of wastewater collection and disposal. Choices made were often based on a combination of local needs and current scientific ideas. Although the adoption of a new collection technology caused problems to become regional rather than local, financing remained available solely on a municipal basis. Only the intervention of either regional authorities or the federal government, with its higher funding basis, could begin to cope with the pollution problems that were the heritage of the original adoption of a water-carriage system for domestic waste removal.

Notes

1 B. Lee, "The Cart Before the Horse," *Papers and Reports of the American Public Health Association* 20 (American Public Health Association, 1895): 34-36

2 U.S. Department of Commerce and Labor, *Statistics of Cities Having a Population of over 30,000: 1905* (Washington, DC: U.S. Government Printing Office, 1907)

3 *Annual Report of the Pennsylvania State Board of Health, 1898* (1898), 1: 5

4 U.S. Department of Commerce and Labor, *Statistics of Cities Having a Population of over 30,000: 1909* (Washington, DC: U.S. Government Printing Office, 1913)

5 A. Hazen, *Clean Water and How to Get It* (New York: John Wiley and Sons, Inc., 1907)

6 "Sewage Pollution of Water Supplies," *Engineering Record* 28 (August 1, 1903): 117-118

7 "Pittsburgh Sewage Disposal Reports," *Engineering News* 67 (February 29, 1912): 398-402

8 "Compulsory Sewage Purification," *Engineering Record* 65 (February 24, 1912): 198

9 "A Plea for Common Sense in State Control of Sewage Disposal," *Engineering News* 67 (February 29, 1912): 412-413; G.W. Fuller, "Is It Practicable to Discontinue the Emptying of Sewage into Streams?" *The American City* 7 (March 1912); and G.C. Whipple, "Sewage Treatment vs. Sewage Purification," *Engineering News* 68 (October 3, 1912): 388-389

19

REFUSE POLLUTION AND MUNICIPAL REFORM: THE WASTE PROBLEM IN AMERICA, 1880–1917

by Martin Melosi

Source: Martin Melosi, ''Refuse Pollution and Municipal Reform: The Waste Problem in America, 1880-1917'', in Martin Melosi (ed.), *Pollution and Reform in American Cities, 1870-1930*, Austin, TX, University of Texas Press, 1980, pp. 105-33

If the entire year's refuse of New York city could be gathered together, the resulting mass would equal in volume a cube about one eighth of a mile on an edge. This surprising volume is over three times that of the great pyramid of Ghizeh, and would accommodate one hundred and forty Washington monuments with ease. Looked at from another standpoint, the weight of this refuse would equal that of ninety such ships as the ''Titanic.''[1]

The graphic imagery of Franz Schneider's remarks dramatically illustrates the severity of the refuse problem in American industrial cities. [. . .] Of course, garbage and other forms of refuse date back to the cavemen, but it was not until the late 1880s that public awareness of the problem brought forth outcries for a permanent solution in the United States. [. . .]

Contemporary statistics on refuse, although incomplete, indicate the magnitude of the problem for most industrial cities. For instance, between 1910 and 1916, each citizen of New York City produced approximately 1,625 pounds of garbage, rubbish, and ashes per year. In other cities, the annual per capita production ranged from 185 to 1,840 pounds during the early twentieth century.[2] Table 19.1 shows the amount of garbage collected in eighteen major cities during 1916. [. . .]

The crude state of collection and disposal practices added to the gravity of the refuse problem. During the late nineteenth century, cities were still grappling with the issue of who was ultimately responsible for providing scavenging service. [In the 1880s, c]ities with a population of less than 30,000 tended to favor private collection more than did larger cities. These findings reinforce the notion that urban growth, in and of itself, had an impact on the extension of city services – in this case, refuse collection.

Figures for street cleaning show a much more significant trend toward municipal responsibility than with refuse collection. [. . .] 70 percent of the cities surveyed in the 1880 census made public provisions for street cleaning. [. . .] The question of ultimate responsibility for street cleaning was determined more easily than that of refuse because, unlike households, streets had no clear territorial limits and transcended the question of individual responsibility.

Compounding the dilemma of ultimate responsibility were problems of waste disposal once it had been collected. The cities had to contend with a vast array of refuse, such as garbage (organic waste), rubbish (paper, cans,

Table 19.1 Tons of garbage collected by city, 1916

City (Population)	Tons of Garbage Collected
Baltimore (593,000)	37,915
Boston (781,628)	52,650
Bridgeport, Conn. (172,113)	19,897
Cincinnati (416,300)	40,692
Cleveland (674,073)	59,708
Columbus, O. (220,000)	20,393
Dayton, O. (155,000)	16,621
Detroit (750,000)	72,785
Grand Rapids, Mich. (140,000)	8,678
Indianapolis (271,758)	23,267
Los Angeles (600,000)	51,062
New Bedford, Mass. (118,158)	10,162
New York (5,377,456)	487,451
Philadelphia (1,709,518)	101,678
Pittsburgh (579,090)	73,758
Rochester, N.Y. (275,000)	30,782
Toledo, O. (220,000)	23,971
Washington, D.C. (400,000)	46,293

Source: Rudolph Hering and Samuel A. Greeley, *Collection and Disposal of Municipal Refuse* (New York, 1921), p. 40.

old shoes), human and animal excrement, dead animals, street sweepings, and ashes. The variety of solid wastes made collection and disposal complex. Collection and disposal methods for one substance, such as garbage, might not be suitable for another, such as dead animals. Not only was the type of waste a factor in determining proper collection and disposal methods, but also the city's geographic location, its climate, economic and technical factors, and local traditions figured in the decision. More often than not, expediency won out over health or aesthetic reasons. [. . .] [M]ethods in 1880 varied in form but were equally primitive. The one method that dominated, if any, was dumping refuse on land or into water, an immediate solution that merely shifted the problem from one location to another. [. . .]

By the mid-nineteenth century, experiments in England and the United States indicated some direct relationship between communicable disease and unattended waste. The work of British sanitarian Sir Edwin Chadwick led the way to new sanitary laws in England and inspired American health officials to regard refuse collection and disposal as a health issue. This attitude was an important step toward improved sanitation in the United States, even though it was not until the late nineteenth or even the early twentieth century that the germ theory of disease supplanted the arcane filth (or miasmic) theory and revolutionized American public health.

After mid-century, environmental sanitation, especially refuse disposal, increasingly became the responsibility of the cities' sanitary authorities. The 1880 census reveals that at least 94 percent of the cities surveyed had a board of health, health commission, or health officer. Of these sanitary authorities, 46 percent had some direct control over the collection and disposal of refuse, and almost all had the power to deal with nuisances created by refuse. As might be expected, the larger American cities granted their health boards and commissions more direct control over refuse collection and disposal than did smaller cities.

It would be naive, however, to assume that the health boards of the late nineteenth century were capable of thoroughgoing refuse reform. For instance, in 1880 few sanitary authorities operated without overt political interference. A large proportion of the boards of health were dominated by city officials, rather than by physicians or sanitarians. In fact, some boards included no physicians as members. [. . .] [T]he politically dominated boards were less able to provide the leadership necessary to improve health conditions on a daily or periodic basis: they could merely react to cataclysmic events such as epidemics – and not very well in those cases.

Despite the limitations, health officials tended to dominate thinking about refuse collection and disposal practices in the 1880s and 1890s. Now considered a health problem, refuse no longer could be regarded simply as a nuisance. In 1887, noting the unsatisfactory condition of refuse collection and disposal throughout the nation, the American Public Health Association (APHA) appointed the Committee on Garbage Disposal, chaired by the eminent sanitarian Dr. S.S. Kilvington. The committee was assigned the task of inquiring into the extent of the refuse problem in the United States. For ten years the group gathered statistics, examined European collection and disposal methods, and analyzed the various methods employed in the United States, in an effort to obtain some practical answers.

Civic organizations that called for improvements in sanitary conditions in the city relied heavily upon the health argument to make their case. [. . .]

The Ladies' Health Protective Association (LHPA) of New York City became a leading civic force in the fight to bring about sanitation reform in American cities. Organized in 1884, the LHPA undertook a variety of projects, including slaughterhouse and school sanitation,

street cleaning improvement, and refuse reform. Although influenced by national trends in public health, the LHPA was a community organization without extensive medical or technical expertise. Thus its protests were often couched in broader, if not more simplistic, terms than that of danger to the city's health. Aesthetic considerations underscored much of the association's interests. [. . .]

The association's successful efforts in lobbying for improved sanitary conditions in New York led to the formation of similar groups in other eastern cities. [. . .]

Public awareness was important, but city governments had to commit funds to improving sanitary services. Furthermore, departments of public health and public works needed to implement efficient and effective methods of collection and disposal. In the 1890s, civic groups, journalists, and sanitarians publicized the need for improved collection and disposal methods, but with few practical results. [. . .]

The emergence of the sanitary (or municipal) engineering profession played a more important role in refuse reform than did municipal initiatives. Sanitary engineers became intrigued with European, especially English, efforts to deal more efficiently and "scientifically" with refuse. [. . .] By 1876, Alfred Fryer built the first municipal refuse "destructor" in Manchester, marking the beginning of large-scale use of incinerators throughout England. The apparent success of British efforts led to the construction of similar "destructors" in the United States. (Americans called them "garbage furnaces," "cremators," and "incinerators".) In 1885, Lt. H.J. Reilly (United States Army) built the first American garbage furnace at Governors Island, New York. In 1886–87, the first municipal cremators were built in Wheeling, West Virginia, Allegheny, Pennsylvania, and Des Moines, Iowa.

To many engineers and sanitarians, the introduction of the incinerator harkened a new day for refuse disposal. Regulated disposal by fire seemed to be the technological panacea to a monumental problem. [. . .]

Soon other technological breakthroughs were being tested and marketed. In 1886 a company in Buffalo, New York, introduced the so-called Vienna or Merz process for extracting oils from city garbage. The reduction process, as it became known, was intended to provide cities with salable by-products such as grease, fertilizer, and perfume base, which would offset part of the high cost of disposal. [. . .]

[T]he appointment of Colonel George E. Waring, Jr., as street cleaning commissioner of New York City (1895) was the major turning point in the development of modern refuse management. Waring was born in Pound Ridge, New York, in 1833. As a young man he studied engineering and agricultural chemistry, and by the mid-1850s he divided his time between farming in the spring and lecturing before farmers' clubs in the winter. A job on Frederick Law Olmsted's Staten Island farm proved to be the springboard into a career as a renowned engineer. Through his association with the pioneer landscape architect, Waring obtained an appointment as drainage engineer for the Central Park project. The Civil War temporarily interrupted his burgeoning career, but his experiences as a combat officer, rather than stunting his professional goals, sharpened them. "Colonel" Waring, as he would be called for the rest of his days, returned to civilian life strongly impressed by the sense of "duty and authority" that he acquired in the service. In fact, his passion for military discipline provided the foundation upon which his administrative style would be based in later years. In the late 1860s, he devoted almost total attention to drainage and sewerage engineering. By the 1880s, he became a well-known figure in sanitation. [. . .]

Waring's appointment as street cleaning commissioner of New York City in 1895 was

the culmination of his career. In his effort to improve city services, reform mayor William L. Strong had appointed Waring and Theodore Roosevelt (police commissioner), hoping that these men of honesty and expertise would reform the city's corruption-ridden departments. [. . .] After years of Tammany Hall domination, the Street Cleaning Department had become heavily laden with spoilsmen [political placemen] who misused the department's funds and provided only marginal service to the city. After expelling the political cronies, Waring selected young men with engineering backgrounds or military training, thus staffing the department with a technical elite. [. . .]

His ideas were not revolutionary or unique; they represented instead an accumulation of the best methods that had been attempted piecemeal throughout the country. It was the commissioner's ability to perceive of the refuse problem broadly that allowed him to establish the most comprehensive environmental sanitation program of that time.

The collection of refuse posed many difficulties. Waring's first step was to initiate a system of "primary separation" at the household level. Garbage, rubbish, and ashes were to be kept in separate receptacles awaiting collection. This method allowed the Street Cleaning Department to dispose of the disparate material using the appropriate methods for each. He also initiated the building of the first municipal rubbish sorting plant in the United States, where salvageable materials were picked out of the discarded rubbish and then resold. Profits from the plant were returned to the city to offset collection costs.

Street cleaning was the commissioner's most serious collection problem. The streets of the city were crudely constructed; littering was acceptable public behavior; and horse manure was everywhere. Relying on his staff of superintendents and foremen to keep watch on the daily cleaning, he enlarged the corps of sweepers as well as improving their competence and attempting to raise their morale. Street cleaning was far from an honored job in the 1890s, but the Colonel gave his men an esprit de corps by increasing their pay, improving work conditions, and impressing upon them the importance of their work. His "White Wings" (as the sweepers were called) were issued white uniforms, which, although clearly impractical for the job, associated the men with cleanliness, on a par with doctors and nurses. Over 2,000 strong, the "White Wings" performed admirably under Waring's guidance and brought unprecedented attention to the department. Citizens of New York and local newspapers quickly took note of the unaccustomed cleanliness of their streets.

Waring's separation program and his improvements in street cleaning were intended to raise public consciousness about the need for broad-scale sanitary improvements. [. . .] Unlike some of his engineering colleagues, he relied more upon human initiative and involvement than upon technological advancements to solve sanitary problems. The most dramatic example of this emphasis was the formation of the Juvenile Street Cleaning League. Initially, more than five hundred youngsters participated in this program to disseminate information about proper sanitation and to inspire community involvement in keeping the streets clean. [. . .] Waring's ultimate goal was the extension of public knowledge about proper sanitation and the development of a sense of personal commitment to cleanliness of the city. [. . .]

Waring faced his greatest test on the question of final disposition of refuse. Characteristically, he employed a combination of old and new techniques. Although he considered dumping wastes at sea to be the simplest and most traditional method of disposal, he realized many of its pitfalls and sought alternatives whenever possible. The only improvement the colonel initiated in this area was to commission new types of dumping scows, which

ostensibly would empty the wastes more efficiently and farther from shore. Although the new dumpers helped retard the pollution of beaches along the New York–New Jersey coastlines, they were only stopgap measures at best.

The commissioner's treatment of garbage was based upon more innovative methods. He sought not only to eliminate the wastes but also to retrieve resalable by-products. For this purpose, he installed a reduction plant on Barren Island that extracted ammonia, glue, grease, and dry residuum for fertilizer from the garbage. These salvageable materials were to be sold in the city's behalf, just as the items from the rubbish plant. Waring also encouraged experimentation to find newer and more economical ways to reduce garbage and, simultaneously, began an extensive land reclamation program on Rikers Island by using waste as landfill.

By the end of his short tenure as street cleaning commissioner in 1898, Waring had generated considerable local and national attention, which led to similar programs in many other cities. [. . .]

His program for New York City indicated an important shift from viewing the refuse problem simply as a question of health to confronting it as a multifaceted urban problem. [. . .] Waring's efforts emphatically demonstrated a nascent urban environmentalism that sought to upgrade the quality of urban life. He also demonstrated a faith in municipal government to lead the way in environmental reform, aided by a technical elite freed from the influences of patronage. His actions mirrored the emerging Progressive reform spirit of his day, as he sought to place the public welfare above private gain. [. . .] He recoiled in moral outrage at the physical and social degradations of city life, envisioning himself as a noble crusader against the forces of evil. Thus, in a real sense, Waring was a product of his times as well as a pioneer in his own field. His example provided a bridge between the arcane refuse practices of the nineteenth century and the increased sophistication of the twentieth century. [. . .]

Debate within the public health community over environmental sanitation contributed greatly to the broadening in perspective about the refuse problem. The acceptance of the germ theory of disease had not only enlivened interest in preventive medicine but had brought into question the importance of environmental sanitation in curbing disease. [. . .]

More than the sanitary implications of street cleaning and refuse disposal was under scrutiny in these early years of the twentieth century. Emphasis began to shift from simply an awareness of the refuse problem to a major effort to determine its extent and thus better gauge what methods could be applied to a solution. Once the extent of the problem was determined, refuse reform began to move along two important lines: (1) a careful analysis of the available technological solutions conducted by sanitary engineers and (2) a program of community involvement generated by civic organizations. [. . .]

The engineering community actually began to take more of a leadership role in finding a technical solution to the refuse problem in the late 1890s. Although engineers claimed that the sanitary considerations of the refuse problem were central, thus clinging to the value of environmental sanitation to improve health, they more often stressed that collection and disposal of waste was primarily a technical problem that should be left to those with specialized training in such areas. Neither the public health official nor the municipal bureaucrat, the argument went, had the breadth and depth of training of the sanitary engineer. Health officials could uncover the source of the refuse problem, but it was the new professional, the sanitary engineer, who would be called upon to find a solution. [. . .]

The sanitary engineer quickly became the logical successor to the health officer in dealing with refuse problems. The ambivalence within the health community toward environmental sanitation was in sharp contrast with the sanitary engineers' dogged adherence to it. The technical training and expertise of the sanitary engineer seemed to promise an efficient and tangible solution to the waste problem. [. . .]

Sanitary engineering became a powerful force in municipal affairs not only because of the apparent local demand but also through the efforts and actions of important national engineering associations and committees. With respect to the refuse issue, one of the earliest and best-known groups was the APHA's Committee on the Disposal of Garbage and Refuse. Its major functions included gathering statistics, inspecting local sanitation operations, and analyzing local and national sanitation trends. The American Society for Municipal Improvements (ASMI, later the American Society of Municipal Engineers), founded in 1894, was another important group that devoted substantial attention to the problem of refuse. It was the first national organization that attempted to unite all municipal engineers. Other engineering groups, with broader interests than ASMI, such as the American Society of Civil Engineers (ASCE), still devoted some attention to the refuse problem. Almost every convention of ASCE included a panel related to sanitation matters. From time to time, very specialized groups formed, such as the Society for Street Cleaning and Refuse Disposal of the United States and Canada. [. . .]

All these groups and organizations helped establish a vital communications network and a means to disseminate ideas about the problem of refuse and how to deal with it. They also played a major role in collecting statistics about the refuse problem from American and Canadian cities as well as from abroad. The data gathered, although often incomplete, provided a reservoir of information not available to refuse reformers in the nineteenth century. Despite expected differences of opinion and local variations, sanitary engineers began to establish a consensus about many important aspects of the refuse problem gleaned from these statistics. A national focus, which had been lacking in nineteenth-century refuse reform, was beginning to emerge in the early twentieth century. [. . .]

Realizing the problems inherent in interpreting disparate data from a large number of cities, the APHA's refuse committee devised the "Standard Form for Statistics of Municipal Refuse" in 1913 to bring some order out of the chaos. Attempts to standardize statistics also led to a universal recommendation from engineering groups for cities to keep better records, especially in such areas as quantities of wastes produced, seasonal variations, and cost factors in collection and disposal. [. . .]

Sanitary engineers also did much to lobby for municipal control of street cleaning and garbage removal and disposal. The question of ultimate responsibility, which was often avoided in the 1880s and early 1890s, received considerable attention in the early twentieth century. In their important book, *Collection and Disposal of Municipal Refuse*, Rudolph Hering and Samuel Greeley state flatly, "The collection of public refuse is a public utility."[3] If there remained any support for the contract system of collection and disposal, it was modified substantially to include more rigorous municipal supervision. Surveys taken from the 1890s through World War I demonstrate the impressive shift to increased municipal responsibility. In the 1880 census, 70 percent of the cities surveyed cleaned their own streets; in a 1914 study the proportion had increased to 90 percent.[4] Great strides were made in municipal responsibility even in garbage collection and disposal. By World War I at least 50 percent of American cities had some form of municipal collection system, compared with only 24 percent in 1880.[5]

Once sanitary engineers began to have a firmer grasp of the refuse problem, they turned their energies to analyzing collection and disposal methods. Almost immediately they began to condemn the practices of the nineteenth century. Dumping on land or in water, open burning, and the use of untreated waste as land or street fill came under the most serious scrutiny. More sophisticated criteria for selection than simple convenience began to emerge. As pragmatists, the sanitary engineers placed increasing emphasis upon the need to examine local conditions before determining the proper collection and disposal methods for a community. In planning for new systems, the engineers felt it necessary to take into consideration not only the types and quantities of wastes but also the quality of local transportation facilities, the character of the body in charge of the work, and physical characteristics of the city that might determine location and type of disposal system, plus the receptivity of the local government and the citizenry to changes in practice. The reliance of the engineer upon a technical solution to the problem of refuse, therefore, was increasingly being tempered by an awareness of the complexity of the problems faced.

[. . .] Although no single type of refuse technology came to dominate, incineration and reduction were most often discussed during the period. Each method had its advocates and detractors, but both methods came under more careful analysis than they had when first introduced. The success of the English destructors had led to the APHA refuse committee's provisional endorsement of incineration and quick implementation of the method in several American cities. Further study of the application of British systems to American cities indicated that some municipalities had made a hasty decision in building untested and questionable incinerators. [. . .] With further study, it became clear to many engineers that British destructors were not designed to handle the kinds of wastes encountered in the United States. For example, American garbage had more water in it, making it more difficult to burn. Consequently, they recommended design changes to take into account the differences in the composition of waste. They also recommended that incinerators should be considered only after a careful analysis of costs and municipal needs.

The reduction process, which appeared in the United States about the same time as incineration, underwent a similar development: impulsive implementation, severe criticism, reevaluation. The 1897 refuse committee report indicated a great deal of interest in reduction because of its promise to return revenue to the city. However, after a period of operation, several undesirable side effects led to an increased criticism of the method. The foul odors emanating from the plants raised the greatest protests. [. . .] In 1916, ASMI's Committee on Refuse Disposal and Street Cleaning recommended a compromise: reduction was fine for large cities where the revenue derived from it might warrant its use, but for small cities incineration appeared to be more sanitary and less costly.

The debate over the merits and weaknesses of incineration and reduction had a positive effect upon their use. Improvements in technology and more careful applications of the methods led to more rational use. As Table 19.2 indicates, both methods gained in popularity throughout the early twentieth century. However, neither method dominated the field of disposal. In fact, primitive methods still represented from 26 to 66 percent of those employed in American cities between 1900 and 1920. [. . .]

[B]y the early twentieth century, an additional approach – energy retrieval from waste – was commanding increasing interest. Once again England and other European countries led the way. The English had adapted many of their destructors to produce residual sources of heat, such as steam, and had even developed a process of generating electricity for running trolley cars. [. . .] Wide application of energy retrieval methods in the United States was retarded because of the high costs of operation of such facilities and the cheaper and more plentiful sources of energy available at the time. [. . .]

Inspired by the efforts of the engineering

Table 19.2 Methods of garbage disposal (percentage and number of cities)

	1880	1890	1902	1909	1915	1918
Dumping on land, used as fill or buried	39 (78)	46.5 (33)	24 (39)	18.5 (25)	11 (21)	27 (29)
Farm use (fertilizer, animal feed)	22 (44)	29.5 (21)	21 (33)	28 (25)	7.5 (37)	20 (21)
Dumping into water	9 (17)	13 (9)	6 (10)	7.5 (10)	1 (2)	1 (1)
Burning	1 (1)	7 (5)	3 (4)	—	—	—
Reduction	—	—	9 (15)	19.5 (26)	10 (19)	20 (21)
Incineration	—	—	19 (31)	26.5 (35)	30.5[a] (58)	13 (14)
Combination	10 (20)	—	12.5 (20)	—	7 (13)	18 (19)
No systematic disposal	4 (8)	—	—	—	—	—
No data	15 (30)	4 (3)	5.5 (9)	—	33 (63)	1 (1)
Total number of cities per survey	199	71	161	133	190	106

Note: [a] includes open burning.
Sources: Gleaned from George E. Waring, Jr., comp., *Report on the Social Statistics of Cities, Tenth Census of the United States, 1880*, 2 vols. (Washington, D.C.); "Street Cleaning and the disposal of Refuse," *Engineering Record* 22 (29 November 1890): 410; "Some Statistics of Garbage Disposal for the Larger American Cities in 1902," in APHA, *Public Health: Papers and Reports* 29 (October 1903): 152–53; "Disposal of Municipal Refuse," *Municipal Journal and Engineer* 35 (6 November 1913): 627–28; "Refuse Collection and Disposal," pp. 728–30; William P. Capes and Jeanne D. Carpenter, *Municipal Housecleaning* (New York, 1918), pp. 194–99.

community and imbued with a compulsion to improve the urban quality of life, a variety of reform groups sought solutions to the problem of refuse as part of their more comprehensive attempts to rid the cities of vice, corruption, and disease. Many of the most important national municipal organizations that formed during this period, such as the National Municipal League and the League of American Municipalities, incorporated refuse reform into their array of goals.

In the tradition of the Ladies' Health Protective Association, several middle-class women's groups assumed the leadership in civic efforts to end the refuse problem. Mildred Chadsey, commissioner of housing and sanitation in Cleveland, used the term "municipal housekeeping" to define those sanitation functions of the city that were previously performed by individuals.[6] [. . .]

[. . .] Women became aggressive promoters of refuse reform. In Boston, Duluth, Chicago, and other cities, women's municipal organizations began investigations into disposal methods. The Women's Health Protective Association of Brooklyn sought and obtained new ordinances for collection and disposal. The Louisville Women's Civic Association published and distributed four thousand pamphlets about the garbage problem and even produced a movie entitled *The Invisible Peril*. Shown to thousands of citizens, the movie depicted how an old hat discarded in an open dump spread disease. The women of Reading, Pennsylvania – some in their sixties – banded together in a street cleaning brigade to protest poor methods. The local alderman, embarrassed by the action, soon discovered that his position was in jeopardy for not addressing the problem himself.

The city cleanup campaigns were the most dramatic and highly publicized form of citizen involvement in refuse reform. They were extensions of Waring's civic awareness efforts on a grander scale. At least once a year, most cities sponsored a "cleanup week" to generate interest in sanitation and other related problems, such as fire prevention, fly and mosquito extermination, and city beautification. Interest in the campaigns often brought together several civic organizations for a coordinated effort. In Philadelphia, the municipal government assumed the leadership of the city's cleanup campaign and transformed it into a major event. The city distributed 3,400 personal letters, 750,000 gummed labels, 260,000 bulletins, 20,000 colored display posters, 750 streamers, 1,000,000 cardboard folders, 300,000 badges, 300,000 blotters, 350,000 circulars, and various other paraphernalia. Some communities, such as Sherman, Texas, used the opportunity to campaign for better collection and disposal ordinances. Too often, however, the cleanup campaigns were merely cosmetic exercises. Any truly substantive and far-reaching sanitary benefits of the efforts are difficult to determine.

The efforts of the sanitary engineer and the zealousness of civic reformers, although great, did not resolve the refuse problem by 1920. The coming of World War I partially explains the delay, because it diverted attention from urban reform by readjusting societal priorities. The war, however, also played a role in the curious resurgence of one of the "primitive" disposal methods – feeding garbage to swine. The wartime food conservation campaign, led by Food Administrator Herbert Hoover, temporarily tightened enforcement of garbage collection ordinances and ultimately resulted in a de facto reduction in the quantity of garbage. What garbage remained, it was argued, could more profitably be fed to hogs instead of being incinerated or reduced at a time when food production had a high priority. The United States Food Administration waged a vigorous campaign to disarm critics who believed that garbage-fed hogs were unsanitary, and prepared massive reports stating the contrary. The efforts of the Food Administration, however, did not have a lasting impact upon the refuse question, nor did the temporary reduction in the amounts of garbage continue to be a trend after the war. The problems of the early twentieth century were still unresolved.

Sanitary engineers had gone a long way toward gaining a firmer grasp on the refuse problem in the first two decades of the twentieth century and, to their credit, helped publicize the extent of the waste inundating the cities. Yet the faith of Americans in the ability of technology and science to resolve the refuse problem was excessive. On the one hand, scientific knowledge had not advanced sufficiently to determine the precise impact of specific disposal methods in broad ecological terms. For instance, with the advent of the automobile as a primary source of transportation, many people believed that the streets would be free of pollution. Unfortunately, no one realized the cruel irony – horse manure would be supplanted by a greater danger in the form of noxious gasoline fumes. On the other hand, few if any Americans gave much thought to the cause-and-effect relationship between the consumption of goods and the production of waste. Some contemporaries bemoaned the wasting of natural resources, but few made the connection with the production of refuse. In the largest sense, then, the uncritical faith in science and technology was not sufficiently tempered with a sense of its limitations.

The growth of civic awareness with respect to sanitation was a step, but only one step, forward in bringing the refuse problem to the attention of the citizenry. The cleanup campaigns were ephemeral, since they were most often an extension of the contemporary refor-

mist mood and hardly a vehicle for long-range solutions. Even efforts to obtain new ordinances or more effective methods of collection and disposal were inherently difficult. It was one thing to write new laws requiring householders to separate garbage from rubbish or to make littering illegal; it was quite another to enforce such laws. Forcing citizens to change their habits, let alone understanding the more complex problems associated with sanitation, could not be accomplished overnight. Furthermore, city officials had to be convinced that improving waste collection and disposal methods was a high priority among voters, and not excessively costly. Cost of improvements quickly became a prime determinant of what kind of system the city should acquire, and no new system would come cheaply. [. . .]

Notes

The author would like to thank the Rockefeller Foundation and Texas A&M University for the financial support that made this study possible

1 Franz Schneider, Jr., "The Disposal of a City's Waste," *Scientific American* 107 (13 July 1912): 24

2 Statistics gleaned from Rudolph Hering and Samuel A. Greeley, *Collection and Disposal of Municipal Refuse* (New York, 1921), pp. 13, 28

3 Hering and Greeley, *Collection and Disposal of Municipal Refuse*, p. 4

4 "Statistics of Cities," in United States Department of Labor, *Bulletin* 36 (September 1901): 880-85; *Municipal Journal and Engineer* 37 (10 December 1914): 836-44

5 "Refuse Collection and Disposal," *Municipal Journal and Engineer* 39 (11 November 1915): 527-730, B.F. Miller, "Garbage Collection and Disposal," in ASMI, *Proceedings, 1915*, pp. 10-11

6 Mildred Chadsey, "Municipal Housekeeping," *Journal of Home Economics* 7 (February 1915): 53-59

20

"THE BEST LIGHTED CITY IN THE WORLD": THE CONSTRUCTION OF A NOCTURNAL LANDSCAPE IN CHICAGO

by Mark J. Bouman

Source: Mark J. Bouman, "'The Best Lighted City in the World': The Construction of a Nocturnal Landscape in Chicago", in John Zukowsky (ed.), *Chicago Architecture and Design, 1923-1993: reconfiguration of a metropolis*, Munich, Prestel; Chicago, The Art Institute of Chicago, 1993, pp. 32-51

Down State Street in October 1926, marching bands paraded and floats rolled through two huge decorative arches, accompanied by a throng of 250,000 people reveling in the "Splendors that Out Babylon Ancient Babylon." [. . .] [I]n the White House, President Coolidge tapped a golden telegraph key and 140 new street lamps in Chicago flickered and grew bright – brighter, it is claimed, than on any other street in the world. [. . .]

What Chicago shares with most Western industrial cities is a lighting style that emerges from three types of demand: security, urbanity, and utility.[1] [. . .] Crime and public order are frequently given as reasons to light the streets, and the order represented in the lighted landscape is one imposed by a central authority on an often eagerly accepting public. Social hierarchy is also implied in much "urbane" lighting; its highly aestheticized forms result from a desire to provide attractive public spaces for social intercourse, or to convey messages about the highest aspirations of the metropolis. The tallest or most central buildings are often singled out for special lighting attention (Fig. 20.1). Utility follows logically from the increased pace of nocturnal activity in an industrialized society. The modern model of lighting provision is, of course, for large central power plants to generate the gas and electricity; for major industrial firms to manufacture the fixtures; and for city governments to contract for or to manage directly the operation of the system, to provide the locations for the fixtures, to settle on their design, and to raise the funds necessary for their emplacement.

As a center of electrical manufacturing, as a source of innovation in power supply, as a fount of inspiration in urban design, as a city with a loud and long tradition of civic boosterism [optimistic promotion] and as a distinctly politicized social environment, Chicago has witnessed a gradually shifting set of priorities relative to the three sources of demand. Where at one time being the "best lighted city" meant drawing heavily on the city's distinguished, urbane local planning tradition, by the end of World War II it had shifted to imply maximum "utility" for automobile traffic with strong element of "security" to inhibit crime. [. . .]

The production of light

Regardless of the use or display of light at night, it has been characteristic of modern lighting supply since the first gas lighting

Figure 20.1 Night view of Palmolive Building, Water Tower, and North Michigan Avenue, 1931.

systems in early nineteenth-century London to separate the consumption of light from the point of production. So to look at the lighted landscape itself is to see only one part of a whole system, and Chicago was one of the first places where this basic insight was made. Both the landscape of gas production and consumption harked back to the nineteenth century, to what Lewis Mumford called the paleotechnic period of industry. Chicago's gas industry, in existence since 1850, relied heavily on coal; retorts, coke ovens, and huge gasometers were prominent features not only of the landscape, but of the city's smellscape. Gas lamps flickered from cast-iron poles not far above the heads of pedestrians.

By the 1920s, gaslight had already acquired the patina of quaintness and was generally considered to be obsolete. Even the gas industry's own survey of its achievements in 1925 did not touch on lighting. Chicago committed itself to a program of replacing gas street lights with electric ones, but the Depression and World War II held up the process for many years. In 1954, the last eighteen gas lamps in the city were removed. But gas lighting lingered in suburban pockets. When, in the late 1970s, Congress prepared to ban outdoor gas lights in the wake of rising energy costs, communities like Riverside, with gas lights of "traditional [. . .] cultural or architectural style," were exempted. In any case, with their low crime rates such communities did not have high demand for the crime-fighting potential of electric lights and did not care to be associated with anything like the bright lights image of Chicago.

The comment by Riverside's village manager that the gas lamps gave the town "that rustic, bucolic, country-like atmosphere" not only reasserts the old rural ideal operative in suburbia, but it is the perfect argument by negation for the urbanity of electric lighting.[2]

But Chicago has had many positive arguments as well. As a center of innovation in electrical production and distribution techniques and in the manufacturing of electrical equipment, Chicago has claim [. . .] to the title of "Electric City." A large local community of talent included such electrical industry luminaries as Elisha Gray, Elmer Sperry, Augustus Curtis, and, not least of all, Samuel Insull, who constructed the first and most famous American regional utility.

In order to compete with gas lighting, Insull found that the cheapest way to produce electricity was to take advantage of economies of scale. Massive central generating stations were constructed in Chicago to house the larger steam turbines that replaced reciprocating steam engines; the great prototype was the D.H. Burnham-designed Fisk Street Station in 1903. To make the most efficient use of such a large capital investment, the power plants had to run as close to capacity as possible. But daily demand for electricity varied, with nocturnal valleys and strong industrially based midday peaks. In order to raise the load factor of his generating facilities (i.e., the average proportion of capacity actually used), Insull needed smoother daily demand curves. A more diverse demand could be realized by extending service to the broadest possible area through long-distance high-voltage transmission lines and a growing stable of affiliated power companies. Nocturnal demand could also be raised through increased consumption of domestic electrical appliances, the lighting of buildings, and contracts for street lighting (although Chicago's street lighting demands were serviced by the Metropolitan Sanitary District's hydroelectricity into the 1950s). In short, demand had to be created for the supplier's benefit and the use of advertising to create this demand has always played an exceptionally important role for local utilities.[3]

Chicago's nocturnal landscape thus comes to rest on an integrated regional network of large-scale power plants, transmission lines (often running on wide rights-of-way), and substations. The high stacks and huge proportions of Commonwealth Edison's coal-fired plants such as Fisk Street, Quarry Street, Northwest, Crawford Avenue, and State Line clearly show the building's function, though facades of red brick, terra-cotta inlays, arched windows, and some small details such as keystones and quoins are occasionally added for "architectural dignity" (see Fig. 20.2). Easily the most impressive of the power stations was Graham, Anderson, Probst and White's State Line Generating Company (1921–29; Fig. 20.3), conceived as a "citadel of electricity." [. . .]

The urbanely lighted growth machine

Chicago's urbane nocturnal landscape was based not only on the pioneering work of Samuel Insull, but also on that of other civic leaders who insisted on the city's leadership in commerce and culture. Light played a major role at both of Chicago's World's Fairs; it figured as a key image in the local physical plan-

Figure 20.2 Graham, Anderson, Probst and White, Commonwealth Edison Co., Crawford Avenue Generating Station, 1924–25; from *The Architectural Work of Graham, Anderson, Probst and White* (London, 1933).

Figure 20.3 Graham, Anderson, Probst and White, Chicago District Electric Generating Co., State Line Generating Station, Hammond, Indiana, 1921–29; from *The Architectural Work of Graham, Anderson, Probst and White* (London, 1933).

ning tradition; it was used as a tool to establish and maintain the dominance of the retail core; and it figured heavily as an aesthetic and advertising tool in the city's burgeoning skyline. The result was a highly localized urbane landscape: a Central Business District aglow with White Ways, lighted skyscrapers, monumental bridges and plazas; light beribboned boulevards linking the city center to elegant public parks; and scattered outposts of urbanity in the outlying "bright lights" districts. World's Fairs had been closely associated with electricity at least since the display of the Corliss Engine at the 1876 Centennial Exhibition in Philadelphia. A rising tide of illumination peaked at the 1893 World's Columbian Exposition [in Chicago], where the formal spaces and grandiose buildings of the "White City" used more lamps than any city in the country then had.

The "official face" of the night, as eventually codified in the 1909 *Plan of Chicago*, was derived from this classical, formalist, intensely illuminated tradition. The ideal Beaux-Arts nocturne is depicted in several of Jules Guérin's renderings in the Plan. Plate 127,

Figure 20.4 Jules Guérin, "Bird's-Eye View of Grant Park at Night"; from Daniel H. Burnham and Edward H. Bennett, *Plan of Chicago* (Chicago, 1909), pl. 127.

for example, entitled "Bird's-Eye View at Night of Grant Park, the Facade of the City, the Proposed Harbor, and the Lagoons of the Proposed Park on the South Side," depicts a skyline twinkling grandly, and a harbor full of boats enjoying a nocturnal sail (Fig. 20.4). The harbor design includes lighted obelisks, beacons, or lighthouses at the corners [. . .] Another Guérin drawing, "View Looking West, of the Proposed Civic Center Plaza and Buildings" shows how City Beautiful lamp standards envelop and define the formal space at night, with ribbons of light extending into the darkening countryside.

The *Plan of Chicago* therefore formalized one of the great conventions of American urban form: that the American downtown is as much a "central illuminating district" as a Central Business District. Hierarchy is written in lumens as well as in land values, and at the very apex of both is the Great White Way. New

York's Broadway, the American prototype, had its Chicago counterpart in the theater district centered on Randolph Street, but Chicago's White Way was really State Street, its pulsating retail heart. State Street had been the site of some of the earliest uses of arc lights and the Edison incandescent system in Chicago in the 1880s. By the turn of the century, according to Harold Platt, "the bright: lights shining from the palaces of consumption along State Street would have been difficult to miss."[4]

In 1926, with lighting provided by the State Street Lighting Association to provide "facilities for evening promenade," State Street was to become, as we have seen, the "World's Greatest White Way." The light poles themselves, designed by Graham, Anderson, Probst and White, were formally approved by the Chicago Plan Commission. Each block had four poles on a side directly opposite each other. Atop the poles were two 2,100-watt

Figure 20.5 View of State Street, looking north from Madison Street, c. 1926.

9,000-candle-power [c.p.] lamps (Fig. 20.5). At the time, the brightest street lights anywhere else in the city had 1,000 c.p. As if to confirm the notion voiced by a lighting advocate fourteen years earlier that "the number of persons passing a point in the evening is in direct proportion to the illumination at that point," within a week of the announcement of the new lighting scheme came the results of a survey finding State and Madison to be Chicago's busiest corner.[5]

Thirty-two years later, however, State Street was in trouble. Declining sales volume at the major department stores and at other retail outlets reflected the growing competition from suburban shopping malls and the suburbanization of the labor force. In 1958, a year in which office construction began to boom, the State Street merchants again made the decision to try to lure more customers with the "world's brightest lights."

In this case, the designer, Robert Burton, set out in true modernist fashion "to design a standard of pure and artistic form, free from ornateness, yet a creation of simplicity and lasting beauty." Each pole had three luminaires that splayed obliquely out of the top of the pole thirty-four to thirty-six feet up; at twenty-four

feet, another luminaire branched out over and parallel to the sidewalk (see Fig. 20.6). Each luminaire contained six fluorescent tubes, with special color properties. According to the special newspaper published for the occasion, "when the General Electric wizards were busy dreaming up these newfangled lights for State Street one of their priority targets was to eliminate the annoying quirk of most fluorescent street lights that tended to change women's rosy complexions to a ghastly green or pasty yellow [. . .] these new lights actually tend to flatter complexions!"[6] Once again bands played and speeches were made, one hundred actors performed the pageant "Light Through the Ages," and President Eisenhower flipped the switch (see Fig. 20.7).

[. . .] [B]oth the number of shoppers and retail volume continued to decline. In the late 1960s, those generators of night traffic – restaurants and entertainment places – continued to disappear. So State Street was remade again in the 1970s. The twenty-year-old lighting fixtures gave way to a new urban design scheme: this time, the lighting plan featured the now-familiar high-pressure sodium lamps on forty-foot-high standards, two to a block front. Proposed low-level accent lighting on twelve-foot-high poles was never built. [. . .]

Outlying commercial districts followed the State Street example, with local business associations purchasing smaller versions of the White Way or ornamental lighting systems. [. . .]

The *Plan of Chicago* yielded other corridors of urbanity: wide boulevards edged by trees and buildings of uniform cornice line. By night, these corridors were to be accentuated by numerous globe-topped white lights; the poles, such as those designed by Daniel Burnham for the South Park Commissioners, were to be exercises in street furniture classicism (see Fig. 20.8). The boulevards were clearly intended for evening promenade, as the Guérin drawings show, and, as such, they represent

Figure 20.6 View of State Street, looking north from Van Buren Street, 1959, showing street lights installed in 1958.

extensions of the central illuminating district into the region beyond. But the boulevards were also intended for vehicular traffic, and it is the genius of the Michigan Avenue Bridge and Wacker Drive that they somehow combine both functions. Wacker Drive itself is a brilliant

Figure 20.7 View of State Street at night, looking south from Lake Street, 1958, showing new street lights.

example of Beaux-Arts city planning, and its ornamental lighting contributes in no small measure. Given how they function, however, most of the bridges over the Chicago River were designed without lighting because of the jarring effects of traffic. Ornamental lamp posts were removed from the Jackson (1916) and Michigan Avenue (1918–20) bridges because vibration destroyed the incandescent fixtures.

Of course, the Chicago Plan envisioned other places for pedestrian promenade, especially Grant Park, with its grand entrance and lighting fixtures by Edward Bennett. The grand centerpiece is Buckingham Fountain (Bennett, Parsons, and Frost; 1927). Now operated by computer, the fountain's changing array of colored lights was originally controlled by Curtis Lighting's "Lumitone" system. It used a paper roll much like a player piano to produce an ever-changing pattern of forty-five-million-c.p. lights.

Figure 20.8 View of Michigan Avenue, looking north from Randolph Street, c. 1933.

What the Plan and its executors did for grandiose horizontal spaces, the skyscraper tower did for the vertical. Few more visually exciting changes have swept across the landscape of the American city than the building of the skyline, especially as punctuated by high-rise towers. From the start, tower construction went hand in hand with illumination, from the start, the romantic landscape thus created became richly symbolic, evocative of America's progressiveness, modernity, and technical precocity, as well as simply being interesting new additions to the scene. [. . .] [I]lluminating engineers began to concentrate more on the use of light for artistic effect than on maximum output. Lighting designer Luther Stieringer pioneered a method he called the "luminous sketch," in which only the outline of the structure was lighted. Some early examples include

the Eiffel Tower (1900), the Electricity Tower at the Pan-American Exposition in Buffalo (1901), and the Singer Building in New York City (1908). [. . .]

Floodlighting was pioneered by Walter D'Arcy Ryan at the 1915 Panama Pacific Exhibition. His technique was to shine lights on the buildings from a distance, to use reds in corners and, perhaps, green on St. Patrick's Day. Ryan's work would not have been possible without a device that could direct the light's rays on to the subject instead of diffusing it uselessly into the night sky. Chicago happened to be the home of the man known as the "archenemy of glare" and the "pioneer of indirect illumination," as Curtis Lighting (formerly the National X-Ray Reflector Company) referred to its founder, Augustus Curtis. The company was "devoted to the design, manufacture, and installation of engineered lighting equipment which provides the correct amount of illumination for proper architectural and lighting effect." Two basic methods were employed: one set the projectors far from the building and attempted to project as much white light as possible upon the building slab; and the other placed the projectors directly on the building, accentuating through careful placement, and, perhaps, color, the building's silhouette, its ornamentation, setbacks, and accents.[7]

Publicists for Curtis Lighting described floodlighting as "one of the most emphatic advertising mediums. It establishes the location of a structure much more convincingly than the street designation." Building lighting did indeed play an increasingly important advertising role in the great wave of skyscraper construction that followed on the end of World War I. In 1922, the *Chicago Central Business and Office Building Directory* made only one reference to exterior lighting. [. . .] But in 1932, three buildings [. . .] called attention to their lighting schemes, and three others [. . .] were depicted in nocturnal photographs or artistic renderings; by 1942, the number had risen to four. [. . .]

The earliest major nocturnal landmark in Chicago was the Wrigley Building (Graham, Anderson, Probst and White; 1919–24) whose location, brilliant floodlighting, and white-enameled terra-cotta cladding made the building an instant favorite. The combination of light and materials had a precedent in New York's Woolworth Building (Cass Gilbert; 1911–13), the first commercial building to be evenly floodlighted, with lights made by National X-Ray Reflector. No doubt the other most recognizable landmark in the interwar years was the two-billion-c.p., 150-foot-high beacon atop the Palmolive Building (Holabird and Root; 1927–29, Fig. 20.1). [. . .]

The beacon was an important navigational tool for many years. Some airline passengers claimed to be able to read newspapers by it twenty-seven miles away. Beneath it an 11-million-c.p. lamp pointed the way to the Chicago Municipal Airport. [. . .] But that light, rotating twice a minute, became an annoyance when high-rise apartments [. . .] were built to its level in the 1960s. The beacon was partially shielded from its neighbors in 1968 and permanently retired in 1981; a glowing light replaced it in 1990.

Another well-known aviation beacon in the 1930s sat atop Holabird and Root's LaSalle-Wacker Building (221 North LaSalle Street; 1929–30). As in the case of the Palmolive Building, LaSalle-Wacker's location, at the intersection of two important axes from the Chicago Plan, drew attention to an important node in the grid of the downtown area. The building was also floodlighted so that the recessed bays were illuminated, creating a vertical rhythm of light and dark at the top of the structure.

Beacons were not the only method of calling attention to the top of a building. Four searchlights pointed in the cardinal directions from the beehive atop the Straus Building (today

replaced by a blue light). The Trustee System Service Building [. . .] (Thielbar and Fugard; 1929–30) was topped by a ziggurat and lantern. The Jewelers Building by Giaver and Dinkelberg with Thielbar and Fugard (1926) wore a distinctive crown in the Stieringer manner, while Holabird and Root's 333 North Michigan Building of 1928 wore a lighted crown. [. . .]

The Union Carbide company sought a "distinctive and perpetual advertisement" for the occupants of its new regional headquarters building at 230 North Michigan Avenue. Advertisements for the Carbide and Carbon Building (Burnham Brothers; 1929) called attention to its illuminated gold-leaf tower and campanile. [. . .]

The even floodlighting method was wonderfully suited to skyscrapers built after zoning changes encouraged the use of telescoping forms. As architects began to abandon elaborate ornamentation in the late 1920s, the building mass itself and floodlights mounted on the setbacks provided the architectural interest. For example, while the Palmolive Building's elongated fenestration gave a daytime impression of elegant vertically, by night the floodlighting of each of the setbacks emphasized the horizontality of the breaks between them (see Fig. 20.1)

Few buildings could claim as much surface area to be lighted as the Merchandise Mart (Graham, Anderson, Probst and White; Fig. 20.9). Massive facades were relieved by a composition of vertically banded windows set between cornice lines near the base and two slight setbacks near the top. Here the lighting took advantage of cornice lines near the top of the eighteen-story structure, the turrets at the corners, and the massive central tower that rises seven floors higher than the rest of the building. The vertical window bands were lighted from lamps placed above the fourth-floor cornice. The setbacks were brilliantly illuminated. The tops of the corner turrets and the top three floors of the tower and its pyramidal tower

Figure 20.9 Graham, Anderson, Probst and White, Merchandise Mart, 1931.

were also lighted. The effect was surprisingly graceful for so large a structure, with the rhythm of the top of the building emphasized at night, rather than the massive limestone walls. Guidebooks made certain to point out the building to visitors by night. A guide for visitors to the Century of Progress Exposition of 1933–34 found the Mart's nocturnal aspect to be "thrillingly beautiful." The authors of the *WPA Guide* were no less enthusiastic: "concealed lighting of the massive symmetrical structure formed a beautiful composition of light and shadow in the Chicago night scene."[8]

By the time of the Century of Progress Exposition, then, illuminating engineering was in full maturity and prepared to use light to "amaze and delight even the most blasé city dwellers," as Hubert Burnham put it.[9] Ironically, Walter D'Arcy Ryan, the fair's director of illumination, succeeded in that goal because he "leaned more toward the utilitarian than the aesthetic" and sought to maximize foot candles per dollar because of the

fair's tight finances. [. . .] Ryan would employ virtually all the techniques already in use in the downtown streets and skyline (see Figs 20.10–20.12). He would bring the judicious use of neon to the public's attention, and he would play off the highly sculptured architectural forms that were painted in Joseph Urban's vivid palette. Windowless walls were floodlighted in white and outlined in neon; fountains were bathed in changing beams of light; searchlights beamed brightly into the night. Street lights were a mixture of lanterns, mercury-neon pylons, opal glass pylons, tubular incandescent standards, and the unique indirect "mushroom lights." In 1934, cables of the Sky Ride across the lagoon were festooned with small incandescent lamps in the Stieringer manner. From the rocket cars of the Sky Ride one could look down over the whole fairground, and back across Monroe Harbor and Grant Park to the Loop, which from a distance truly became an illuminated landscape.

Figure 20.10 Night view of the Electrical Building, Century of Progress Exposition, 1933–34.

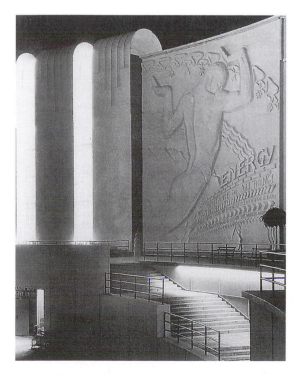

Figure 20.11 Night view of the Electrical Building, Century of Progress Exposition, 1933–34

Figure 20.12 Night view of the Hall of Science, Century of Progress Exposition, 1933–34 (2).

Compositions of Chicago's night scene were so common by mid-century that they threatened to become cliché. But spectacular illuminations soon threatened to become extinct. Older fixtures were often not repaired after breakage, and maintenance was frequently deferred. Bands of lighted office windows shining through glass curtain walls sufficed for nocturnal display in most of the modernist buildings. [. . .] And rising energy prices in the 1970s seemed about to end all such skyline display forever.

Daley night: security and utility in the best lighted city

The State Street celebrations showed a city willing to use light to boost its main shopping street and to promote a bold image of itself to the world at large. But according to city engineers in 1925, the continual demand for better street lighting arose "on account of the automobile problem and public safety protection and the rapid continuous growth of the city." Exactly what was needed was spelled out: "At least twenty more square miles of residential street lighting; the adequate lighting of all street car line streets; the adequate lighting of the downtown loop; the rehabilitation of the underground facilities in many parts of the city; the rehabilitation of much of the steel pole equipment of the city; the adequate lighting of the through streets and reinforcement of lighting in many sections of the city."[10]

Since 1910, the city had bought its electrical power for street lighting from the Sanitary District of Chicago, which produced hydroelectricity at its dam in Lockport. By 1925, 89 percent of the city's 70,873 lamps were electric; the rest were gas or gasoline. The two most commonly used electric lamps were tungsten incandescents: 600 c.p. for commercial districts and 100 c.p. for residential areas. In 1924, the city began a program of adding 1,000-c.p. lamps to widened thoroughfares and upgrading the residential lights to 250 c.p. The 250-c.p. lamps installed between 1924 and 1931 were mounted on posts fourteen feet high, three feet from the curb on alternate sides of the street. The pillar form cast light on street and on sidewalk – and also into the sky.

By the end of World War II, however, the system was in major need of repair. [. . .] [T]he 1920s vintage electric lighting was inadequate to modern needs, and newer sections of town were still unlighted, while older sections showed the effects of a decline in maintenance. All this was a major loss of civic face. [. . .]

Thus began a campaign to upgrade the system that seemed to have no end. From the outset of the rebuilding program, the city's priorities were clearly articulated: to enhance lighting for vehicular traffic and to use lighting to reduce crime. City officials were fond of quoting traffic safety statistics, emphasizing function over aesthetics, and they asserted that "modern street lighting is designed to provide convenience to the public, to safeguard motorists and pedestrians from accidents and molestation at night."[11]

Between 1944 and 1952, $13 million was spent to replace the older lamps. The new lamps were now twenty feet high and were attached to an arm that projected ten feet beyond the curb line in an arc that reminded everyone of a "cobra head." Reflectors ensured that the light was directed on to the street itself, so posts needed to be placed on only one side of the street. And all this lifted lighting far beyond the sphere of the pedestrian sidewalk (see Fig. 20.13). [. . .]

By the time the fifteen-year upgrade program was over in 1959, Richard J. Daley had been mayor for four years. All lamps had been replaced with mercury vapor, and some city officials were saying that the new system would last the city satisfactorily for at least fifty years. But new rounds of street lighting upgrades were to commence as soon as the old were finished. Again, each new program

Figure 20.13 View of the intersection of 73rd Street and Coles Avenue, showing popular harp-shaped lighting fixture of the 1920s and later cobra-head style. *c.* 1950.

was sold to the public on the basis of traffic safety, crime reduction, and the appeal to being the "best lighted city in the United States." Daley and street lights became entwined in the popular imagination. [. . .] Under Daley, the city also assumed responsibility in 1959 for lighting the 200 miles of Chicago Park District boulevards, 40 percent of which had already received mercury-vapor lamps by that time. The Park District itself got into the act with a "security lighting program" for Grant Park that placed modernistic lamps on walkways previously reliant on street lighting for illumination. Having covered its parks, the city turned its attention to the alleys in the 1960s. Designed to "end crime-breeding darkness and promote greater use of alleys," 51,000 mercury-vapor lamps were installed at the ends and in the center of the alleys.

Mayor Daley announced another major upgrade in 1973: $7.5 million would be spent to replace the fourteen-year-old bluish white mercury-vapor lamps with yellowish high-pressure sodium-vapor ones. Some of the first efforts centered on high-crime areas, [. . .] while sodium-vapor advocates cited their higher energy efficiency and greater illumination. In spite of objections, the program moved forward and was completed in 1981 with the installation of high-pressure sodium in all the alleys.

For several years, the debate over the lights sizzled. But it was clear that being the "best lighted city" now meant being the "most lighted city." Even areas that were traditionally treated with great attention to design felt the change. Lake Shore Drive, for example, had been lined for many years with globe-encased Mazda lamps on decorative concrete poles. Yet, Lake Shore Drive was clearly becoming a high-speed throughway, not a park drive. Today, high-pressure sodium-vapor lamps and cobra heads proliferate. Their presence represents the final abandonment of the pretense that Lake Shore Drive is not an expressway. [. . .]

Did all this brightening of the night sky succeed in driving away crime? The Chicago Police Department did report a decline in the incidences of some night crime, but such statistics must be handled with extreme caution. There is little conclusive evidence that street lighting lowers the level of crime; the best that can be said is that street lighting can lessen the *fear* of crime.

The sodium-vapor controversy also reached into the suburbs, especially those with architecturally or historically significant street lighting systems. An explicit rejection of the Chicago aesthetic was heard from Wilmette environmentalists who said that "new lights would make Wilmette look like State and Madison."[12] But the controversy raged most fiercely in Oak Park and Evanston. Both cities boast lighting designs by architect Thomas Eddy Tallmadge. In Evanston, the fluted columns are ornamented with oak leaves and

Figure 20.14 View of Evanston, showing lamps by Thomas E. Tallmadge, 1958.

waves, and a distinctive octagonal luminaire that has become one of the city's architectural symbols (see Fig. 20.14). The city council's efforts to scrap all 6,800 lights in 1976 were met with immediate and widespread protest. Several years later, the city decided to keep the Tallmadge luminaires or replicas on residential streets, to reject crime as a reason for bright, high-pressure sodium-vapor lights, and to allow neighbors to decide on the wattage of bulb to be used. Nonetheless, it proposed cobra-head light standards for major arterial streets for the very reason Chicago did in 1948: to direct the light directly onto the thoroughfare.

Urbanity rekindled

As if to signal a general downtown revitalization and the growing dissatisfaction with the overwhelming emphasis placed on security and utility, old buildings began to be relighted or newly lighted in the 1980s, and new buildings shook off the dark modernist aesthetic for brightly illuminated tops, as a way of drawing attention to the virtues of a particular address. Lighting programs also received a boost from the recommendations of the "Light Up Chicago" committee, formed in 1983 by the Woman's Board of The Art Institute of Chicago. [. . .]

Some notable structures, of course, had never really lost their lights, such as the Kemper Insurance, Wrigley, Chicago Tribune, and Chicago Temple buildings. For others, the "Light Up Chicago" committee recommended that "an overall priority would be to focus on buildings of the twenties and thirties whose dramatic lighting was an integral part of the urban environment."[13] [. . .]

Notes

1 Mark J. Bouman, "The 'Good Lamp is the Best Police' Metaphor and Ideologies of the Nineteenth Century Urban Landscape," *American Studies* 32, no. 1 (Fall 1991), pp. 63-78; *idem*, "Luxury and Control: The Urbanity of Street Lighting in Three Nineteenth Century Cities," *Journal of Urban History* 14, no. 1 (Nov. 1987), pp. 7-37; Harold L. Platt, *The Electric City: Energy and the Growth of the Chicago Area, 1880-1930* (Chicago, 1991); Wolfgang Schivelbusch, *Disenchanted Night: The Industrialization of Light in the Nineteenth Century* (Berkeley, Los Angeles, and London, 1988); David Nye, *Electrifying America: Social Meanings of a New Technology* (Cambridge, Mass., 1990); James M. Tien, Vincent F. O'Donnell, Arnold Barrett, and Pitu B. Mirchandani, *Street Lighting Projects*, National Evaluation Program Phase I Report (U.S. Department of Justice, Law Enforcement Assistance Administration, National Institute of Law Enforcement and Criminal Justice, 1979), p. 3

2 'Riverside Lamps Stay as Most Gaslights Go," *Chicago Daily News* (Jan. 5, 1978)

3 Platt (note 1), *passim*; *idem*, "Samuel Insull and the Electric City," *Chicago History* 15 (Spring 1986), pp. 20-35; Thomas P. Hughes, *Networks of Power: Electrification in Western Society, 1880-1930* (Baltimore, 1983), pp. 201-26; *idem, American Genesis: A Century of Invention and Technological Enthusiasm 1870-1970* (New York, 1989), pp. 226-43; Forrest McDonald, *Insull* (Chicago, 1962)

4 Platt (note 1), pp. 28-39; 91

5 "Congestion Peak Found on Chicago Corner," *Chicago Commerce* 22 (Oct. 2, 1926), p. 11

6 *State Street Special* (Nov. 13, 1958), a special four-page newspaper printed for the occasion, now in the collection of the Chicago Historical Society

7 J.L. Stair, *The Lighting Book* (Chicago: Curtis Lighting, 1930), pp. 25, 235; Glenn A. Bishop and Paul T. Gilbert, *Chicago's Progress: A Review of the Fair City* (Chicago, 1933), p. 83

8 Alfred Granger, *Chicago Welcomes You* (Chicago, 1933), p. 144; Federal Writer's Project, *The WPA Guide to Illinois*, with a new Introduction by Neil Harris and Michael Conzen (New York, 1983), p. 249

9 Hubert Burnham, "Architectural Expression for Chicago's 1933 World's Fair," in Arthur Woltersdorf, ed., *Living Architecture* (Chicago, 1930), p. 130

10 "Survey of Public Improvements," in Chicago Association of Commerce, *Survey of Chicago* (Chicago, 1925), p. 19; "Proper Street Lighting Saves Lives and Reduces Traffic Accidents," in Chicago Plan Commission, *Chicago Looks Ahead: Design for Public Improvements* (Chicago, 1945)

11 Commissioner of Streets and Electricity Lloyd M. Johnson asserted that 65 percent of all traffic accidents and 75 percent of fatalities occur at night, and that "good street lighting and adequate traffic control signals are imperative in reducing night accidents"; "City Sets 250-mile Street Lights Goal," *Chicago Times* (May 14, 1947)

12 "Environment Units Dim Suburban Lighting Plans," *Chicago Tribune* (Aug. 20, 1972)

13 The "Light Up Chicago" Committee, *Light Up Chicago*, A Report to Mayor Jane Byrne, 1982

21

REGIONAL PLANNING FOR THE GREAT AMERICAN METROPOLIS: NEW YORK BETWEEN THE WORLD WARS

by David A. Johnson

Source: David A. Johnson, ''Regional Planning for the Great American Metropolis: New York between the World Wars'', in Daniel Schaffer (ed.), *Two Centuries of American Planning*, London, Mansell Publishing, 1988, pp. 167–96

American metropolitan planning took its modern shape in the two decades between the end of World War I and the beginning of World War II. During the 1920s nascent ideas about planning generated in the 1880s and 1890s came to be accepted and institutionalized. And in the Depression of the 1930s, many of the plans and projects based on those ideas were actually carried out.

The most striking example of this process occurred in the New York Metropolitan Region. [. . .] [T]he 1929 Regional Plan of New York and Its Environs, certainly one of the most important and well-funded efforts in the history of American planning, [. . .] brought together many of the planning and reform ideas, personalities, and forces at work in American society in the early twentieth century. The contradictions and conflicts it generated reflected the internal division within that society and the tension between progressive reformers – those who wished to remake the basic structure of society – and meliorist reformers – those who sought only to remedy its negative consequences. [. . .]

[. . .]

The Port of New York Authority Plan, 1921

One of the most significant events in the development of planning activities in the New York Region was the creation of the Port of New York Authority in 1921. The Port Authority was established after years of promotional effort by prominent business groups, mostly on the New York side of the harbour. Port development had become an obsession in cities on both the east and west coasts following the opening of the Panama Canal in 1914. The Canal symbolized the emerging industrial supremacy of the United States and demonstrated in a dramatic way the practical results produced through the combination of public capital, engineering skills and military discipline, brought together under a government attuned to the needs of business.

The port's importance was recognized by successive New York City administrations, which between 1870 and 1914 had spent over $100 million to improve pier and dock facilities. [. . .] [T]he New York, New Jersey Port and Harbor Development Commission was established by the two state legislatures in 1917. General George W. Goethals, the builder of the Panama Canal, was retained as consulting engineer to the Commission.

The Commission reported back a year later with a recommendation that an autonomous port authority be established for New York Harbor similar to the Port of London Authority. To substantiate the need for such an authority, the Commission devoted the next three years to an exhaustive study of the geography, history and condition of the port. The 500-page report issued in 1920 concluded that rail freight movement was the basic problem. A comprehensive plan was presented to rationalize freight handling around the harbour. Circumferential belt lines were proposed, one along the waterfront, and one at the exurban fringe of development in the New Jersey countryside. A tunnel linking New Jersey and Brooklyn also was proposed to minimize the need for water-borne freight transport across the harbour. And an underground-automated electric goods delivery system was suggested as a way to solve truck congestion on Manhattan streets. While the plan was concerned solely with freight systems, its potential for shaping regional land development patterns was enormous.

[. . .] [T]he Port Authority bill was successfully passed in both states in 1920. The legislation specifically required the Port Authority to adopt a physical plan for port development. A board appointed to propose a plan, chaired by General Goethals, recommended in 1921 one almost identical with the 1916 Comprehensive Plan of the New York, New Jersey Harbor Commission.

Successful implementation of the plan was dependent on [. . .] the cooperation of the region's eleven private railroads [. . .] . As early as 1922 it was evident that the railroads would not support any Port Authority proposals for belt line connections. Each railroad was reluctant to give up the local monopoly it enjoyed along its right-of-way for the benefit of participating in a more efficient overall system. An impasse resulted which prevented the Port

Authority from carrying out its original mission [. . .] . What seemed to be required was a 'Plan of New York Authority' which could look at the region as a whole, an agency whose scope could go beyond the narrow legislative mandate of the Port Authority and yet work closely with it. [. . .]

Two centuries of American planning

The making of the Regional Plan of New York and Its Environs

The most important regional plan for the New York Metropolitan Area, and certainly the best funded, was the monumental Regional Plan of New York and Its Environs, begun in 1921 and published in 1929.[1] Two remarkable men with roots in Chicago were the driving forces behind the New York plan: Charles Dyer Norton, a banker with the House of Morgan and former secretary and counsellor to President William Howard Taft, and Norton's close friend, Frederic A. Delano, an engineer and uncle of an aspiring New York politician, Franklin Delano Roosevelt. Norton and Delano, who had worked closely with Daniel Burnham on the 1909 Plan of Chicago, hoped to generate a similar plan for New York, but on a much broader scale.

Funded for a decade by generous grants from the Russell Sage Foundation totalling $1.2 million, the plan broke new ground but also synthesized many earlier ideas and proposals put forward by the Olmsteds and by the Port Authority. Following the dictum of Patrick Geddes, an elaborate regional survey was undertaken to provide a basis for the regional plan. Economic forces, social trends, and physical conditions were mapped and analysed for their future regional importance. The planning staff, directed by Thomas Adams, a Scotsman who had helped plan Letchworth and Welwyn Garden Cities near London, was assisted by the nation's leading planners: John Nolon, Edward Bennett, Harlan

Bartholomew, George B. Ford, and Frederick Law Olmsted, Jr. In preliminary studies for the regional plan each consultant was assigned a geographic sector and asked to prepare sketch plans and policies.

From these studies Adams was able to fashion a preliminary set of bold and imaginative policy statements to guide the design of the regional plan:

1 Regional zoning should be established on the basis of a regional plan to serve as a guide to local zoning plans.

2 Special wedge-shaped agricultural zones should be established to provide more open space than could be provided through park or forest preserves alone.

3 To ease congestion, new transportation facilities should emphasize circumferential rather than radial movements.

4 Activities that did not require central locations should be decentralized; conversely, functions that could not be decentralized without loss to the region as a whole should be kept centralized.

5 High buildings were generally not desirable as they did not pay the public costs they generated in congestion and transportation requirements.

6 More public open space should be acquired, particularly at river and harbour edges.

7 Large sites should be acquired in advance for future airport needs.

8 Subdivision design and location practice should be improved, particularly to avoid gridiron plots on hilly terrain and premature development.

9 A system should be established to reduce the inequitable distribution of property taxes among industry-rich and industry-poor towns and cities.

10 Development corporations should be established to help industries relocate and to build satellite towns.[2]

Written in 1923, this was an early and comprehensive statement of the concerns which would preoccupy planners and officials in the region for decades thereafter. [. . .]

The final version of the Regional Plan assumed that population in the twenty-two county region would grow from 8.9 million people in 1920 to 21 million in 1965, an increase of more than twelve million. (This forecast turned out to be about 4 million above the actual 1965 population of 17.3 million.) [. . .]

The plan called for a 'recentralization' of industry and business. Carefully selected new areas for industry were proposed while other areas, locally zoned for industry, were recommended for non-industrial uses. Economic studies had shown that the region was greatly over-zoned for industry. Regional survey studies also concluded that while some industries needed to remain in central locations such as Manhattan, others would do better to relocate in industrial tracts outside the core. Recentralization was intended to solve these problems.

The plan and its critics

Although generally well received, the plan was severely criticized by Lewis Mumford, Benton MacKaye, and others largely for its unexamined assumption that population growth was a given and that the problem was simply one of finding the best ways to accommodate it. [. . .] Mumford, a disciple of both Ebenezer Howard and Patrick Geddes, had expressed scepticism about the premises of the Regional Plan as early as April of 1925 when the International Town, City and Regional Planning, and Garden Cities Congress convened in New York. [. . .] Garden City proponents, including Mumford, were disappointed at the absence of proposals for new communities. Raymond Unwin, C.B. Purdom, and the venerable Ebenezer Howard each expressed their concern over the regressive, over-dense growth of the metropolis and reiterated the desirability of decanting excess population into self-sufficient garden cities beyond the fringe of urban development. The Garden City was, they emphasized, not simply a planned suburb. It was an entirely new form of settlement and implied new patterns of life and work. [. . .]

By the time the first volume of the plan was issued in 1929, Mumford's scepticism had grown. He was convinced that the Regional Plan promised a continuation of the worst trends toward over-centralization and concentration. In an article for the *New Republic*, he awarded a 'prize in "Applied Logic" ', one of a number of 'Booby Prizes for 1929', to the Committee on Regional Plan –

> for its admirable demonstration that by providing for a population of 20,000,000 in the New York area, the problems of transportation which are now insoluble would become less so, and park areas and playgrounds, which are now non-existent or impossible to reach, would then be more numerous and easier to reach.[3] [. . .]

[. . .]

Mumford [. . .] expressed admiration for the technical work of the survey but demurred about the basic assumptions underlying the Plan:

> The chief difficulty is not with the Plan's conclusions but with its premises: namely, that continued growth at the present rate in the metropolitan area is inevitable, and that the first duty of the Plan is to facilitate such growth . . . On the basis of past experience, I see no reason whatever for hoping that this growth and vast expenditure will be compatible with a sufficient and timely provision of parks, playgrounds, and housing facilities: so long as growth and the maintenance of land values are the ends in view, it is rather safe to say that these vital facilities will remain in 'embellishments' – scamped and squeezed in order to accommodate the budget [. . .].[4]

[. . .]

Adams argued that to plan for unavoidable growth was not to be construed as supporting or encouraging such growth:

> I can imagine that if you do not believe in the continued growth of the metropolitan area at

some rate approximate to that which has been occurring in recent years, you will not be able to look upon the Regional Plan with favor. We have been forced as a result of our studies to face this growth as inevitable. I deprecate, however, your suggestion that we have regarded as our first duty to facilitate such growth. What we have regarded as our first duty is to facilitate such dispersal of the growth as will make it healthier than it will be if it is left to go on as at present [. . .] I remember when I was the first executive of the Garden City Company in England, I thought it possible – as you appear to do – that the building of garden cities would tend towards stopping the growth of London. What it has done is to make the growth of London better and to spread rather than contract the big city. We all wish it had been otherwise, but that has been the experience.

I agree with you that there is no reason for hoping that the increased growth and vast expenditure will be compatible with a sufficient provision of open spaces and housing facilities, *if* you judge solely from past experience. But we are of the opinion that past experience has little bearing on what will happen under the conditions of the next fifty years. I am surprised that you should say that we have the growth and maintenance of land values as an end in view because one of our chief ends has been to destroy the fallacy that land values matter at all when human values are at stake. [. . .]

I cannot possibly see any benefit from submitting a regional plan that will set up an impossible ideal or fails to accept facts as they are.[5] [. . .]

[. . .]

The most memorable and detailed critique of the Regional Plan came two years later when in June of 1932 Mumford's long-awaited analysis was published in the *New Republic*. It was a devastating treatment revealing the very different premises upon which the two groups were operating. [. . .]

The plan, according to Mumford, contained many internal contradictions. Proponents of garden cities as well as of concentration could find support for their positions at different points in the text. What the plan lacked was a fertile theoretical base upon which concrete proposals might be made. The plan, Mumford thought, was characterized by its drift into the future and its unquestioning assumption of continued growth. There were other shortcomings: the geographic and historical studies which should have been basic to the plan were perfunctory and arranged in no logical order; the choice of the planning region as an area of metropolitan influence based on an hour's commutation time to the centre was, he suggested, arbitrary and too confining; the population projections, modelled on the process of fruit fly multiplication in a closed container, was inappropriately applied to a metropolitan area.

Mumford noted that New York's peak of centralization had occurred about 1910 and was declining thereafter as the effects of automobile and electricity transmission decreased the advantages of metropolitan concentration. The plan and survey, he asserted, ignored this trend and had been erroneously based on growth statistics from the period 1900–1912.

Mumford further criticized the plan for offering solutions capable only of immediate implementation when what was needed was a plan flexible enough to be operative over thirty or forty years. [. . .] He also saw class bias and interest underlying the values on which the plan had been devised:

> It may be more effective, as well as more clear-sighted and honest, to say that no comprehensive planning for the improvement of living conditions can be done as long as property values and private enterprise are looked upon as sacred, than it is to draw pictures of parks that may never be built [. . .] and garden cities that will never be financed [. . .]
>
> The Russell Sage planners did not take advantage of their theoretical freedom; they were so eager to fasten to a viable solution, a solution acceptable to their committee full of illustrious names in financial and civic affairs, to the business community generally, to the public officials of the region, that they deliberately restricted the area of their questions.[6]

[. . .]

Mumford's sharpest criticism was directed at Adams's failure to look at alternative strategies which might require far less capital investment. Mumford proposed an alternative solution: lessen the pressure of congestion in lower Manhattan by recentralizing the metropolitan business districts; lay down new cities and direct the exodus of industries to these new cities outside the Region's congested areas; rebuild the blighted areas and take care of part of the increase in population by a process of 'intensive internal colonization', by which Mumford meant establishing denser neighbourhoods with more open space. By diverting new growth to redevelopment inside the city boundaries, Mumford hoped that the suburban drain of urban resources could be deterred. This alternative conception, an extrapolation of the ideas of Ebenezer Howard, Patrick Geddes, and Henry Wright, was not a denial of the metropolis but of what Mumford deemed to be its unnecessary expansion at the expense of amenity, community, and the vitality of civic institutions.

[. . .]Adams invoked his past association with Patrick Geddes and Ebenezer Howard, asserting that both would have approved the plan's basic ideas. Mumford's ideals, Adams acknowledged, were high, but unworkable. What was essential to progress in reform, Adams asserted, was 'movement'. And for movement, 'one must keep to the road and as nearly the middle of it as possible, if any improvement is to be made'. It was this 'movement' toward progress which Geddes would have applauded, he suggested, and not Mumford's unrealizable Garden City idea. As for Howard, Adams contended, he would have approved the notion of garden suburbs close to the city, for had he not sited his second Garden City, Welwyn, seventeen miles closer to London than the first, Letchworth?

Adams denied categorically that skyscrapers were incompatible with garden cities, that the proposed highways and rail systems were substitutes for housing, and that the boundaries chosen for the region were too close in to allow for new town proposals in the countryside. And, disputing Mumford, he vehemently contended that the plan would reduce congestion, not add to it. Because the criticism was based on a 'wrong diagnosis', Adams asked that Mumford's conclusions be dismissed.[7]

Adams and Mumford, both committed reformers, had sailed past each other like ships in the night. Their differing perceptions and values precluded any communication on fundamentals. Mumford's words still stand tall after fifty years, his vision of human community untarnished by the passage of time. By contrast, Adams's detailed plans and his arguments in defence of them seem more dated, embedded in an era and a place. For Adams the object of the plan was to formulate a usable public agenda acceptable within the bounds of public opinion. To go beyond this was to risk failure and impotence. But for Mumford, the object of a plan was to stretch the limits of opinion itself, and to provide new images of a humane community. Adams sought solid but incremental improvements in the city and region as it existed in reality. Mumford sought to change that political and social reality. Mumford's words seem more durable because the realities are still in need of change. Adams's plans seem dated because they were fitted to the perceived need of the time. [. . .]

Mumford and Adams epitomized two consistent strains of American reform. One radically grounded and visionary, the other conservative and pragmatic, they have often clashed bitterly. [. . .]

The impact of the Regional Plan of New York and Its Environs

Whatever its flaws, the 1929 Regional Plan of New York and Its Environs left a considerable imprint on metropolitan New York. The infrastructure in place today was created in large measure through the successful completion of many of the proposals contained in the plan. Important new agencies and institutions, such as the New York City Planning Commission and the National Resources Planning Board, were established or assisted by the existence of the plan and the work of the Regional Plan staff. Franklin Roosevelt, as President, knew of the value of planning as a result, in part, of his acquaintance with the regional plan. When Frederic Delano accepted his nephew Franklin's invitation to lead the new National Resources Planning Board, the experience of the regional plan of New York was there to guide him.

Planning education was also greatly influenced by the regional plan project. The first graduate degree programme in city and regional planning was established – at Harvard University – directly in response to the need for skilled professional planners which had become apparent in the course of developing the plan.

The plan's physical planning concepts provided the framework for much of the subsequent development of the region. The great bridges and tunnels that knitted together the pieces of the region were located according to the recentralizing principles laid out by Thomas Adams and were accelerated in their construction by the existence of the plan. The circumferential highways in the plan were adopted by engineers and state highway departments. New grade-separated freeways were originated or accelerated by the plan. The acquisition of parklands was also spurred by the proposals contained in the regional plan. City waterfronts were rebuilt along the principles of the plan, though their use for highways took more priority over the amenity and recreational values recognized in the plan. In the growing centres of the cities, especially Manhattan, new structures were spread further apart to allow for better circulation and more light and air, in accordance with the images and ideas of the plan. If many of these projects and concepts did not originate with the regional plan, it nevertheless brought them together in a unified presentation, making them far more widely known and accepted.

The plan resulted in many positive successes, but it also failed in serious ways, both in its fundamental conception and in its implementation. The plan failed to address the long-term implications of vastly increased automobile ownership, despite accurate predictions of the increase. It failed to address the problem of providing housing for low- and lower-middle income families. And it was ambiguous in its support of new communities, affirming the concept, on one hand, and reluctant to indicate where or how new communities might be built, on the other.

The plan also failed to realize its bold rail proposals, thanks largely to the fragmented, private ownership of the railroads, and it failed to realize its too modest proposals for rail transit, because of the difficulty of coordination across state boundaries, the reluctance of the Port Authority to assume a leadership role in a financially unattractive venture, and the reluctance of New York City transit officials to permit suburban trains on city subway system tracks. The result was progress in public highway construction without parallel development of transit facilities, leaving a decentralized region by 1965 rather than the 'recentralized' region envisioned by Thomas Adams. The plan also failed to make adequate provision for new or expanded business centres, even though it was quite apparent that a considerably increased and recentralized population would require such centres.

Despite its failures, the regional plan unquestionably had a significant effect on the physical development of the region. And it also brought important changes in the institutional and public bodies concerned with regional development. Numerous planning and zoning agencies were established at the county and municipal level as a direct result of field work by the staff of the regional plan. State policies toward regional development also were augmented. The Port of New York Authority was greatly aided in its uncertain early years by the support of the Regional Plan Committee. Differences occurred, of course, but the influence of the regional plan on the Port Authority was usually salutary and broadening, as in the convincing case made in the plan for providing rapid transit capacity in the design of the George Washington Bridge. The ultimate location of the bridge at 178th Street was also influenced by the regional plan's conception of a 'Metropolitan Loop' of which the bridge was a key element. The loop was intended to move traffic around the centre rather than into it, as would have been the case with the earliest versions of a trans-Hudson bridge proposed for 22nd and, later, 57th Streets.

Robert Moses, in his positions of New York City Park Commissioner and chairman of both the Long Island State Park Commission and the Triborough Bridge Authority, carried many of the plan's highway and park proposals for New York City and Long Island to completion. The regional plan undoubtedly provided Moses with an agenda of public works. But more important, it created for Moses a climate of opinion among the business and political leadership in New York favourable to his programme. Though he would never credit the plan as being of much value, Moses's ascendancy to power was considerably aided by this climate.

Much of the progress made toward the partial though considerable realization of the plan during the Depression years from 1929 to 1941 was due to the vigorous efforts of the Regional Plan Association which pressed local governments to undertake park and transportation projects. In a number of instances, the Association provided detailed design proposals based on the plan which directly resulted in acquisition and development.

The Regional Plan Association, created to carry out the proposals of the regional plan, is perhaps the most enduring and significant institutional heritage of the efforts of the Russell Sage Committee. [. . .]

The plan's impacts on the theory and techniques of urban and regional planning were also important. Population forecasting, economic base theory, and Clarence Perry's neighbourhood theory, were all fostered or spawned by the plan. Edward Bassett, in his work on the regional plan, refined the zoning tool he had helped to create in 1916. As a consequence of the plan, local zoning was adopted earlier and more widely in the New York Region than elsewhere. Bassett also significantly developed the law of public open space and shorefront rights in connection with his work on the plan.

The plan gave early warning of several urban problems which continue to be serious: the pollution of air and water, the inequitable social effects of large-lot zoning, and the destruction of public investment in highways by marginal commercial development. But perhaps its most important contribution was to articulate the connections between the physical framework of the city region and its social and economic conditions. No longer could engineering or architectural works be promoted solely on narrow grounds of individual project efficiency. Relationships between public investments in infrastructure and the distribution of jobs and people would have to be taken into account. The plan and the regional survey persuasively demonstrated the need to consider complex connections among basic regional systems.

The new ideas and principles promoted by the plan spread rapidly through the region and the nation in the decade following its publication in 1929. Former members and consultants of the regional plan staff took up important positions and tasks, not only in the New York area, but in other American cities and univer-sities, carrying with them ideas and concepts evolved in their work for the Regional Plan Committee. [. . .]

When the New York World's Fair opened in Flushing Meadows in 1939 an enormous three-dimensional model of the regional plan was placed on display in the New York Building. In the midst of economic depression the Fair provided millions of visitors a glimpse of the future possibilities of new technologies. The regional plan had anticipated some of those technologies, though by 1939 it must have seemed somewhat backward-looking to fair-goers dazzled by television and robots.

[. . .] After the war was over, low-density suburban development would spread rapidly in the fringe areas made accessible by the parkways and highways promoted in the regional plan. But the region that evolved, with its sprawling suburbs and decaying central cities, resembled neither the humane vision of Lewis Mumford nor the recentralized plan of Thomas Adams.

Notes

1 Regional Plan of New York and Its Environs (1929) *Physical Conditions and Public Services*. Vol. VIII. New York: Committee on Regional Plan of New York

2 Committee on the Regional Plan (1923) *Draft Report of the Advisory Planning Group, November 12, 1923*. New York: The Committee

3 Mumford, Lewis (1930) The booby prizes for 1929. *New Republic*, January 8, pp. 190–91

4 Lewis Mumford to Thomas Adams, January 18, 1930. Regional Plan papers, Cornell University Library

5 Thomas Adams to Lewis Mumford. January 27, 1930. Regional Plan Papers, Cornell University Library

6 Mumford, Lewis (1932) The Plan of New York, *New Republic*, June 15, pp. 121–26; Mumford, Lewis (1932) The Plan of New York: II. *New Republic*, June 22, pp. 124–25

7 Adams, Thomas (1932) A communication in defence of the Regional Plan, *New Republic*, July 6, pp. 207–210

22

TRANSPORT: MAKER AND BREAKER OF CITIES

by Colin Clark

Source: Colin Clark, ''Transport: Maker and Breaker of Cities'', *Town Planning Review* 28 (1957), pp. 237–50

[. . .]

The railways immeasurably cheapened long-distance transport. But once the goods had been unloaded from the train or ship on to a horse-dray their costs of transport were very high. The inevitable consequence of this was that industry was concentrated in compact and densely populated industrial towns, or directly along the waterfront in seaports. Motor transport unfroze this concentration. Costs are incurred inevitably, in getting goods off ship or rail on to a motor truck; but once this has been done, unlike a horse vehicle, it makes little difference whether you carry them two or twenty miles. It was motor transport which provided the economic basis without which the present-day 'sprawl' of industrial towns would have been impossible.

On a railway, the principal element in costs (which may or may not be reflected in the freight rates charged) is incurred in getting the goods on and off the train, getting the train made up and started, and shunting it through junctions. Once this has been done the additional or marginal cost of running the railway wagon an extra hundred miles or so is small. So a railway, by its nature, has high terminal and low marginal costs. A road vehicle has lower terminal and higher marginal costs; although year by year terminal costs tend to increase, as the growing volume of traffic calls for

more elaborate organization. But, in general, it is clearly more advantageous to use road vehicles for short journeys and rail for longer. There is a point, which can be fairly precisely measured, at which the railways' lower marginal costs offset their higher terminal costs. In the 1920's, when road transport was in its infancy, this critical distance was only about 75 miles – any transport for a greater distance than that was more cheaply done by rail. But development has been swift. The lower marginal costs of the railway now only give it the advantage over road transport in journeys over 200 miles in length in British conditions (300 miles in American conditions). In a country as small as Britain, and with most of the population and industry concentrated in a fairly narrow industrial belt, most journeys in fact are below 200 miles in length. Economists therefore should not have been surprised when a complete statistical survey undertaken for the first time in 1954, showed that, apart from mineral traffic, threequarters of all the transport of the country (measured by ton-mileage) is now carried by road.

It seems safe to predict that road transport will go on improving further, and that industry will make still more use of it in the future. (The simple discovery of articulated vehicles, whereby the engine is uncoupled from the body and goes straight to work on another load without waiting for the unloading of the old one, has by itself effected great economies in road transport costs.) Industrialization in the future, relying upon road transport, will be spread out, we may perhaps predict, over fairly wide zones. But we can also predict that remote districts, over 100 miles distant from the principal industrial centres, will, under these conditions, probably continue to deteriorate economically.

The effects of road transport on industry have been so interesting that we tend to ignore their equally interesting effects upon agriculture. In the first place – this consideration does not apply in Britain, where the railway network is unusually dense, but applies very much in some other countries – road transport has made agriculture possible in many regions where it was formerly uneconomic. Much of what is now good wheat-growing land in Canada and the United States of America was, until as late as the 1920's, used for grazing only because it was fifty or more miles from a railway, and the cost of carrying the entire harvest over that distance by horse-power made wheat growing unremunerative.

Apart from grain, wool and the like, much agricultural produce benefits from quick transport, and agriculturists therefore have been even more interested than industrialists in road transport. In any country with good roads, agriculturists now only make a minimum use of the railways for transport of milk, fruit, vegetables, livestock and meat. In the United States, when the goods are perishable, road consignments, often in refrigerated vehicles, for distances over two thousand miles, are quite common.

The result of this development has been that geographical specialization in agriculture has further greatly increased. It seems almost incredible, but it is in fact the case that most of the potatoes for the whole of the United States market are grown in three highly specialized small regions. Though less striking, increased agricultural specialization is visible in Britain. The mixed farm is declining in favour of the specialized dairy or cattle farm in the West and grain or potato farm in the East.

The road transport age which followed the railway age has set the manufacturer and trader free from the close confines of industrial cities, within which they had previously been held. But by the same token it began setting the people free too. With transport more readily available, homes no longer had to be cramped within a narrow circle around the industrial centre, any more than did workshops.

The result has been what planners generally apostrophize as 'sprawl'. The precise extent of sprawl can be seen from the attached diagram of densities in the various zones of London, at different dates from 1801 to 1951 (Fig. 22.1). Within the innermost three miles, where most of the population used to live, densities are now much lower than they were a century ago. Within a radius of 7 miles, densities are lower now than they were in 1921. It is in the more and more distant suburbs that densities continue to increase.

In 1801, only the richest people, who could afford horses or carriages, were able to live more than 3 or 4 miles away from the centre of the city. There were no buses, and the average man had to rely upon walking for his transport, whether proceeding to work or to recreation. By 1841 the population had greatly increased. There were by then a few buses, expensive in relation to wages at the time, and the great majority of Londoners were still dependent on walking for their transport. Under these circumstances, it was hardly possible for the expansion of the city to take it any further out. The line on the density diagram remained at about the same slope as it had been in 1801, with just more people packed into each zone of the city.

It is interesting to find cities in present-day India which, although they may have a few trams or buses, nevertheless price them

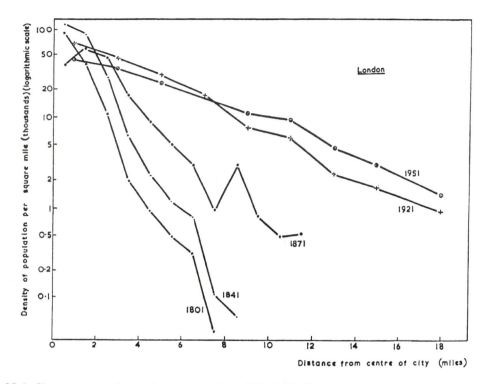

Figure 22.1 Changes in population density in London, 1801–1951. The diagram shows that within the innermost three miles, where most of the population used to live, densities are much lower than they were a century ago. Within a radius of seven miles, densities are lower now than they were in 1921. It is in the more distant suburbs that densities continue to increase.

beyond the means of most of the citizens, who are still dependent upon walking for their transport. Under these circumstances, even where population is very large, we find compact cities of extraordinary density being built up.

In London, we can see the pattern of settlement beginning to loosen by 1871. Some horse-bus and horse-tramway services have been developed by then; but the principal factor had been the provision of reasonably cheap workmen's tickets on the railways. (This concession had been enforced by Parliament, against the wishes of the railway companies.) Moreover, an underground service of steam trains was just coming into operation (for travellers who did not mind soot and fumes). Before the end of the century, this was to be

electrified, and supplemented by a service of electric trams and motor buses. [. . .]

The pattern of 'sprawl' shown for London has been fairly closely followed by other big cities with electric railway (underground or surface) transport – New York, Chicago, Sydney, Manchester, and, less expectedly, Tokyo and Osaka too. But, without electric railways, a similar pattern also develops in modern cities dependent upon private car transport, supplemented by a few buses, such as Los Angeles.

But, the reader may say, must there not be an end to this process? We can understand people living 15 or 20 miles away from their work, perhaps even 30 or 40 miles away if they work in a city as big as New York, and are rich enough to pay for expensive transport.

But that surely must be the limit. Is not 'sprawl' coming to an end?

But – for better or for worse – it is not. There is a defect, a simple defect, in the above argument. A man's work-place need not be in the centre of the city. To an increasing degree now, the work-places are moving out with the people. When there is widespread unemployment, as in the 1930's, employers can expect people to travel considerable distances in search of jobs. In a time of labour shortage, they find it necessary to bring the jobs much nearer to the people, particularly those employers who want to employ women.

So work-places have begun to move out, and also, to an increasing degree, centres of shopping, recreation, education and the like. We can see this process more clearly in the United States and Canada where it has begun earlier and proceeded faster, than here. But there can be no doubt that it is coming. Until virtually every family, as in Canada and the United States, is able to afford a car, we will not get a situation such as is developing in big American cities, where nearly all the new department store development is now in the suburbs, particularly in 'controlled regional shopping centres', where a group of traders combine to provide a shopping centre with acres of car parking. At our present rate of economic progress, we will take some time to reach this objective, especially as our fiscal policy is designed to place heavy taxes on the petrol and cars which people do want, in order to subsidize the building of houses more expensive than people are willing to pay for. But universal car ownership will come sooner or later, and the prudent planner must make his plans accordingly.

By the same token, people living near the centre of a big congested industrial city, be it London or New York, find it hardly worth while running a car, even when they can well afford to do so, because of the lack of parking space and the extremely high costs of garaging.

In an age when all their friends have cars, this will be an additional factor helping to pull population out to the suburbs.

This flight of population to the suburbs is a source of dismay to many – to city councils about the loss of rateable value, to insurance companies and other important property owners who are frightened of a decline in real estate values, to town planners who find the problem of suburban planning beyond them and who would prefer to go back two or three centuries and redesign a nice compact Wren-like city, and, in the Federal Housing Administration in the United States (our own civil servants mostly have more sense) bureaucrats whose actions seem to be based on no rational criteria at all, and who just seem to like tall, and extremely expensive, blocks of flats on congested sites for their own sake.

With us, there are a number of City Councils who have made up their minds (for good reasons or bad) not to export population across the City border if they can help it, and the differential subsidies on expensive high density buildings have enabled them to persist in this policy. Also a good many architects (who have thought little about this problem) and some town-planners (who should have thought about it) have favoured high-density residential development. But even this formidable combination of forces will not be able permanently to withstand the pressure of popular opinion, for the ordinary family really does prefer a house in comparatively spacious surroundings, and certainly resents paying taxation to finance obviously uneconomic blocks of flats.

While in this matter we must move with the current of popular opinion, we should nevertheless realize the destination towards which it is leading us. It is nothing less than the complete disintegration of the city, even of the conurbation, as we know it. Transport has indeed done its work all too well. The final result, if we go on the way we are now going, will be an ugly and planless dispersal of population spreading almost uniformly over a whole industrial zone – perhaps in England an almost continuous built-up area from Liverpool to Dover, with most of the rest of the country impoverished and depopulated. This is undoubtedly one of the examples, perhaps one of the few examples in economics, where the unchecked operation of the free market certainly does not produce the most socially desirable results. For reasons which should be clear, though it is impossible here to state them at length, it is desirable that people should be grouped into communities of manageable size, in which citizens are capable of understanding the affairs of their local community, and taking some part in its government and other activities. Such communities need a certain degree of compactness, so that their citizens do not have to travel too far to reach each other; while at the same time they need to be clearly separated from other communities by an intervening zone of unbuilt agricultural land; for it is

another important social principle that town dwellers should have easy access to open country, just as agriculturists do not want to be too far from towns. For economic, social and political reasons alike[1] we should take (to the best of our present knowledge) communities sized about 150,000 as our objective. A community of this size is large enough to give its inhabitants, and those of the neighbouring rural areas, a full range of economic and administrative services. But we should not allow our community to become any larger if we want its citizens to retain civic pride and interest in their local affairs, if we want its traffic to remain reasonably free from congestion, and if we want to prevent local administration from becoming both expensive and bureaucratic.

Note

1 For more detailed evidence see my article in *Econometrica*, April 1945

23

ORDER IN DIVERSITY: COMMUNITY WITHOUT PROPINQUITY

by Melvin W. Webber

Source: Melvin W. Webber, ''Order in Diversity: Community without Propinquity'', in Lowden Wingo Jr (ed.), *Cities and Space: the future use of urban land*, Baltimore, MD, The Johns Hopkins University Press, 1963, pp. 23–54

The spatial patterns of American urban settlements are going to be considerably more dispersed, varied, and space-consuming than they ever were in the past – whatever metropolitan planners or anyone else may try to do about it. It is quite likely that most of the professional commentators will look upon this development with considerable disfavor, since these patterns will differ so markedly from our ideological precepts. But disparate spatial dispersion seems to be a built-in feature of the future – the complement of the increasing diversity that is coming to mark the processes of the nation's economy, its politics, and its social life. In addition, it seems to be the counterpart of a chain of technological developments that permit spatial separation of closely related people.

At this stage in the development of our thinking, students of the city are still unable to agree even on the nature of the phenomena they are dealing with. But it should surprise no one. For the plain fact of the matter is that, now, when the last rural threads of American society are being woven into the national urban fabric, the idea of city is becoming indistinguishable from the idea of society. If we lack consensus on an organizing conceptual structure of the city, it is mainly because we lack such a structure for society as a whole. The burden, then, rests upon all the arts, the humanities, and the

sciences; and the task grows increasingly difficult as the complexity of contemporary society itself increases and as rapidly accumulating knowledge deprives us of what we had thought to be stable pillars of understanding.

In previous eras, when the goals, the beliefs, the behavior, and the roles of city folk were clearly distinguishable from those of their rural brethren, and when urban settlements were spatially discrete and physically bounded, schoolboy common sense was sufficient to identify the marks of ''urbanness.'' Now all Americans are coming to share very similar cultural traits; the physical boundaries of settlements are disappearing; and the networks of interdependence among various groups are becoming functionally intricate and spatially widespread. With it all, the old symbols of order are giving way to the signs of newly emerging systems of organization that, in turn, are sapping the usefulness of our established concepts of order.

Especially during the last fifteen years, the rapid expansion of the large metropolitan settlements has been paralleled by a rising flood of commentary, reporting and evaluating this remarkable event; and we have developed a new language for dealing with it. Although the scholarly contributions to this new literature tend to be appropriately restrained and the journalistic and polemic contributions characteristically vituperative, the emerging patterns of settlement are typically greeted by both with disapproval if not frantic dismay. By now almost everyone knows that the low-density developments on the growing edge of the metropolis are a form of "cancerous growth," scornfully dubbed with the most denunciatory of our new lexicon's titles, "urban sprawl," "scatteration," "subtopia," and now "slurbs" – a pattern of development that "threatens our national heritage of open space" while "decaying blight rots out the city's heart" and a "demonic addiction to automobiles" threatens to "choke the life out of our cities." Clearly, "our most cherished values" are imperiled by what is synoptically termed "urban chaos." However, such analysis by cliché is likely to be helpful only as incitement to action; and action guided by obsolescent truths is likely to be effective only as reaffirmation of ideology.

We have often erred, I believe, in taking the visual symbols of urbanization to be marks of the important qualities of urban society; we have compared these symbols with our ideological precepts of order and found that they do not conform; and so we have mistaken for "urban chaos" what is more likely to be a newly emerging order whose signal qualities are complexity and diversity.

These changes now taking place in American society may well be compatible with – and perhaps call forth – metropolitan forms that are neither concentrated nor concentric nor contained. Sympathetic acceptance of this proposition might then lead us to new ways of seeing the metropolis, ways that are more sensitive to the environmental qualities that really matter. We might find new criteria for evaluating the changes in metropolitan spatial structure, suggesting that these changes are not as bad as we had thought. In turn, our approach to metropolitan spatial planning would be likely to shift from an ideological campaign to reconstruct the preconceived city forms that

matched the social structures of past eras. Instead, we might see the emergence of a pragmatic, problem-solving approach in which the spatial aspects of the metropolis are viewed as continuous with and defined by the processes of urban society – in which space is distinguished from place, in which human interaction rather than land is seen as the fruitful focus of attention, and in which plans limited to the physical form of the urban settlement are no longer put forth as synoptic statements of our goals.

Metropolitan planning, then, would become the task of mutually accommodating changes in the spatial environment and changes in the social environment. And, because so much of the future is both unknowable and uncontrollable, the orientation of our efforts would shift from the inherently frustrating attempt to build the past in the future to the more realistic strategy of guiding change in desired directions – from a seeking after predesigned end-states to a continuing and much more complex struggle with processes of becoming.

So radical a revision of our thoughtways is not likely to come easily, for we are firmly devoted to the a priori values that we associate with land (especially with open land), with urban centers (especially with the more concentrated and culturally rich centers), and with certain visual attributes of the urban settlement (especially those features that result from the clean boundary line and the physical separation of different types of objects). And, above all, we are devoted to a unitary conception of order that finds expression in the separation of land uses, the classifiable hierarchy of centers, and the visual scene that conforms to classical canons.

So, let us briefly reconsider the idea of city and review some of the current and impending changes to see what their consequences are likely to be for future urbanization in the United States. We can then re-examine the idea of urban space to see how we might allocate it with some greater degree of rationality. [. . .]

The spatial city

[. . .] Although, as some have suggested, there may be certain psychological propensities that induce people to occupy the same place, there seems to be almost universal agreement among urban theorists that population agglomeration is a direct reflection of the specialization of occupations and interests that is at the crux of urbanism and that makes individuals so dependent upon others. Dependency gets expressed as human interaction – whether through direct tactile or visual contact, face-to-face conversation, the transmission of information and ideas via written or electrical means, the exchange of money, or through the exchange of goods or services. In the nature of things, all types of interaction must occur through space, the scale of which depends upon the locations of the parties to the transaction. It is also in the nature of things that there are energy and time costs in moving messages or physical objects through space; and people who interact frequently with certain others seek to reduce the costs of overcoming space by reducing the spatial distances separating them. Population clusterings are the direct expression of this drive to reduce the costs of interaction among people who depend upon, and therefore communicate with, each other.

As the large metropolitan areas in the United States have grown ever larger, they have simultaneously become the places at which the widest varieties of specialists offer the widest varieties of specialized services, thus further increasing their attractiveness to other specialists in self-propelling waves. Here a person is best able to afford the costs of maintaining the web of communications that he relies upon and that, in turn, lies at the heart of complex social systems. Here the individual has an opportunity to engage in diverse kinds of activities, to enjoy the affluence that comes with diversity of specialized offerings; here cultural richness is not withheld simply because it is too costly to get to the place where it can be had.

The spatial city, with its high-density concentrations of people and buildings and its clustering of activity places, appears, then, as the derivative of the communications patterns of the individuals and groups that inhabit it. They have come here to gain accessibility to others and at a cost that they are willing and can afford to pay. The larger the number of people who are accessible to each other, the larger is the likely number of contacts among pairs, and the greater is the opportunity for the individual to accumulate the economic and cultural wealth that he seeks.

Having come to the urban settlement in an effort to lower its costs of communication, the household or the business establishment must then find that location within the settlement which is suitable to it. The competition for space within the settlement results in high land rents near the center, where communication costs are low, and low land rents near the edge of the settlement, where communication costs are high. The individual locator must therefore allocate some portion of his location budget to communication costs and some portion to rents. By choosing an outlying location with its typically larger space he substitutes communication costs (expended in out-of-pocket transportation payments, time, inconvenience, and lost opportunities for communication with others) for rents. And, since rent levels decline slowly as one leaves the built-up portions of the urban settlement and enters the agricultural areas, while communication costs continue to rise as an almost direct function of distance, very few have been wont to move very far out from the center of the urban settlement. The effect has traditionally been a compact settlement pattern, having very high population and employment densities at the center where rents are also highest, and having a fairly sharp boundary at the settlement's margin.

It is this distinctive form of urban settlements throughout history that has led us to equate urbanness with agglomerations of population. Some architects, some city planners, and some geographers would carry it still further, insisting that the essential qualities of the city are population agglomerations and the accompanying building agglomerations themselves; and they argue that the configurations and qualities of spatial forms are themselves objects of value. The city, as artifact or as locational pattern of activity places, has thus become the city planner's specific object of professional attention throughout the world; and certain canons have evolved that are held as guides for designers of spatial cities.

Sensitive to the cultural and economic productivity of populations residing in large and highly centralized urban settlements, some city planners have deduced that the productivity is caused by the spatial form; and plans for future growth of the settlement have therefore been geared to perpetuating or accentuating large, high-density concentrations. Other city planners, alert to a different body of evidence, have viewed the large, high-density city as the locus of filth, depravity, and the range of social pathologies that many of its residents are heir to. With a similar hypothesis of spatial environmental determinism and looking back with envy upon an idealization of the small-town life that predominated in the eighteenth and early nineteenth centuries, this group of planners has proposed that the large settlements be dismantled, that their populations and industries be redistributed to new small towns, and that all future settlements be prevented from growing beyond some predetermined, limited size. [. . .]

Despite some important differences among these proposals however, they all conform to two underlying conceptions from which they stem:

1 The settlement is conceived as a spatial *unit*, almost as though it were an independent artifact – an independent object separable from others of its kind. The unit is spatially delineated by a surrounding band of land which, in contrast to the unit, has foliage but few people or buildings. In some of the schemes subunits are similarly delineated by greenbelts; in others they are defined as subcenters, as subsidiary density peaks of resident and/or employed populations; but the unitary conception holds for all.

2 Whether the desired population size within the unit is to be large or small, whether subunits are to be fostered either as subsettlements within greenbelts or as subcenters within continuously built-up areas, the territorial extent of the "urbanized area" is to be deliberately contained, and a surrounding permanent greenbelt is to be maintained. The doctrine calls for distinct separation of land that is "urbanized" and land that is not.

[. . .]

Behind both ideas are the more fundamental beliefs that urban and rural comprise a dualism that should be clearly expressed in the physical and spatial form of the city, that orderliness depends upon boundedness, and that boundaries are in some way barriers. I have already indicated that the social and economic distinctions between urban and rural are weakening, and it is now appropriate that we examine the spatial counterparts of this blurring nonspatial boundary. I believe that the unitary conceptions of urban places are also fast becoming anachronistic, for the physical boundaries are rapidly collapsing; and, even where they are imposed by legal restraints, social intercourse, which has never respected physical boundaries anyway, is increasingly able to ignore them.

Emerging settlement patterns

It is a striking feature of current, physical urbanization patterns that rapid growth is still occurring at the sites of the largest settlements and that these large settlements are to be found at widely scattered places on the continent. The westward population movement from the Atlantic Seaboard has not been a spatially homogeneous spread, but has leapfrogged over vast spaces to coagulate at such separated spots as the sites of Denver, Houston, Omaha, Los Angeles, San Francisco, and Seattle.

This is a very remarkable event. Los Angeles, San Francisco, San Diego, and Seattle, as examples, have been able to grow to their present proportions very largely as the result of a rapid expansion of industries that are located far from both their raw materials and their customers. The most obvious of these, of course, are the producers of aircraft, missiles, and electronic equipment which use materials manufactured in the East, in Canada, and throughout the world, and then sell most of their product to firms and governments that are also spatially dispersed. They seem to have been attracted to the West by its climate, its natural amenities, and by a regional style of life that their employees seem to find attractive. Once there, they are highly dependent upon good long-distance transportation. And, since successful management of these industries depends upon good access to information about technical processes, about markets, and about finance, they are equally dependent upon good long-distance communication.

It seems clear that the scale of growth there would not have been possible without first the railroad, ocean freighters, and the telegraph and then the telephone, the highways, and the airlines. All of these changes, we must remember, are very recent occurrences in the history of urban man. (The centennial of the Pony Express was celebrated in 1961, and the Panama Canal is scarcely two generations old.) These technological changes have made it possible for individual establishments to operate efficiently thousands of miles away from the national business center at New York, the government center at

Washington, and the industrial belt between Boston and Chicago, to which they are very intimately linked. At least at this territorial scale, it is apparent that economic and social propinquity is not dependent upon spatial propinquity.

These distant metropolitan areas continue to attract a wide variety of specialized firms and individuals, and most of them still prefer to locate *inside* these metropolitan settlements. It is impressive that the television industry, which requires such intricate co-ordination and split-second timing, has chosen to operate primarily out of two metropolitan areas at opposite ends of a continent, yet its establishments are located within the midst of each. Similarly, the financial institutions and administrative offices of corporations which also rely upon quick access to accurate information are attracted to locations within the midst of these settlements. The reasons are apparent.

Just as certain businesses must maintain rapid communications with linked establishments in other metropolitan areas throughout the nation and throughout the world, so too must they maintain easy communication with the vast numbers of local establishments that serve them and that in turn are served by them. The web of communication lines among interdependent establishments within the large urban settlements is extremely strong. Today it is possible to break off large chunks of urban America and place them at considerable distances from the national urban center in the East, but it does not yet seem possible for these chunks to be broken into smaller pieces and distributed over the countryside.

Nevertheless, the events that have marked the growth of widely separated metropolitan settlements force us to ask whether the same kinds of processes that induced their spatial dispersion might not also come to influence the spatial patterns of individual metropolitan settlements as well. A business firm can now move from Philadelphia to Los Angeles and retain close contact with the business world in the East while enjoying the natural amenities of the West; yet it has little choice but to locate within the Los Angeles Basin where it would be readily accessible to a large labor force, to suppliers, and to service establishments. It is attracted to the metropolitan settlement rather than the more pleasant Sierra Nevada foothills because here the costs of overcoming distance to linked establishments are lower. *The unique commodity that the metropolitan settlement has to offer is lower communication costs.* This is the paramount attraction for establishments and, hence, the dominant reason for high-density agglomeration.

The validity of this proposition would be apparent if we were to imagine a mythical world in which people or goods or messages could almost instantaneously be transported between any two establishments – say, in one minute of time and without other costs of any sort. One could then place his home on whichever mountaintop or lakeside he preferred and get to work, school, or shops anywhere in the world. Goods could be distributed to factories or homes without concern for their distances from the point of shipment. Decision-makers in industry and government could have immediate access to any available information and could come into almost immediate face-to-face contact with each other irrespective of where their offices were located, just as friends and relatives could visit in each other's living-rooms, wherever each might live. With transport costs between establishments reduced to nearly zero, few would be willing to suffer the costs of high density and high rent that are associated with high accessibility to the center of the metropolitan settlements. And yet, accessibility to all other establishments would be almost maximized, subject only to the one minute travel time and to restraints of social distance. Under these assumptions, urban agglomerations would nearly disappear. Were it not that the immobility of certain landscape

and climatic features would induce many household and business establishments to seek locations at places of high natural amenity, that some people may have attitudinal preferences for spatial propinquity to others, and that some industrial processes cannot tolerate even one-minute travel times between industrial establishments, we would expect a virtually homogeneous dispersion across the face of the globe.

Of course, zero communication costs are an impossibility, but the history of civilization has been marked by a continuous decline in the effective costs of communication. Time costs and the costs of inconvenience between any given pair of geographic points have declined consistently; and the financial capacity to bear high dollar-costs has tended to counterbalance the high expenses attached to high speed and high comfort. The concomitant effect of very high speeds between distant points and slower speeds between nearby points has been nearly to equate the travel times between pairs of points on the surface of the earth. Certain improvements in transportation equipment that are now becoming possible could gradually reduce differential time costs of travel to nearly zero. The effects of this potential change on the spatial patterns of settlements would be dramatic.

Some potential changes in transportation and communication technology

We are all aware of the fact that, within metropolitan areas in the United States, the widespread use of the automobile has freed the family's residence from the fixed transit lines that had induced the familiar star-like form of settlement. The pattern of residential scatteration at the growing edges of most metropolitan areas would clearly not have happened without the private car; indeed, this pattern was not apparent until the auto induced the suburban developments of the twenties. The telephone, the motor truck, and transportable water, fuels, and electricity have further abetted this lacy settlement boundary. And, of course, all these trends have been further nurtured by a rising level of average family income and by credit arrangements that have made it possible for the average family to choose – and get – one or more autos, telephones, and houses. Similarly the new communication devices, higher corporate incomes, and federal financial encouragement have made it possible for some foot-loose manufacturers and certain types of commercial establishments to locate in relatively outlying portions of metropolitan settlements.

To date, however, very few of these families and business establishments have chosen to locate very far from the metropolitan center, because the costs of maintaining the web of communications that are essential to their cultural and their economic well-being would simply be too high. Even though they might like to locate in a mountain setting, the benefits that would accrue from so pleasant a habitat seem to be far outweighed by the difficulties of maintaining contact with the various specialists they rely upon.

But today a great many of them are much farther away from the metropolitan center, in mileage distance, than they were even fifteen years ago, not to mention the differences that have occurred since the beginning of the century. Even so, a great many have chosen outlying locations without increasing their time distances to the center. Increased mileage distance carries a necessary increase in dollar costs, but the more sensitive component of communications costs in the locator's calculus seems to be the time costs, as the recent traffic studies and the phenomenal rise in long-distance telephone usage indicate.

Increases in travel speeds within most of the metropolitan settlements have been relatively modest as compared to the changing speeds of intermetropolitan travel that the airlines have brought. In part because the potentials of the new freeway systems have been so severely restrained by the countereffects of congestion and in part because the improvements in transit systems have been rare indeed, peak-hour travel speeds have not increased appreciably. But off-peak increases have been great in some places, and some changes are imminent that are likely to cause an emphatic change.

Where the urban freeway systems are uncongested, they have induced at least a doubling in speed and in some places a quadrupling – and the freeways do run freely in off-peak hours. As the urban freeway systems that are now under construction are extended farther out and connected to one another, an unprecedented degree of freedom and flexibility will be open to the traveler for moving among widely separated establishments in conducting his affairs. A network of freeways, such as that planned for the Los Angeles area, will make many points highly accessible, in direct contrast to the single high-access point that resulted from the traditional radial transit net. Even if new or improved high-speed fixed-route transit systems were to be superimposed on freeway networks, the freeway's leveling effect on accessibility would still be felt. And the positive advantage of automobiles over transit systems – affording, at their best, door-to-door, no-wait, no-transfer, private, and flexible-route service – make it inconceivable that they will be abandoned for a great part of intrametropolitan travel or that the expansion of the freeway systems on which they depend will taper off. We would do well, then, to accept the private vehicle as an indispensable medium of metropolitan interaction – more, as an important instrument of personal freedom.

There has been a great deal of speculation about characteristics of the evolutionary successor to the automobile, but it is probably too early to predict the exact form it will take. I would hazard some confident guesses, though, that it will not be a free flight personal vehicle because the air-traffic control problems appear to be insoluble, that it will be automatically guided when on freeways and hence capable of traveling safely at much higher speeds, but that it will continue to be adaptable to use on local streets. If bumper-to-bumper movement at speeds of 150 miles per hour or more were to be attained, as current research-and-development work suggests is possible, greater per lane capacities and greater speeds would be realized than any rapid transit proposals now foresee for traditional train systems. When these on-route operating characteristics are coupled with the door-to-door, no-wait, no-transfer, privacy, and flexible route-end service of the personal vehicle, such a system would appear to be more than competitive with any type of rapid transit service now planned – with two important qualifications. The costs would have to be reasonable, and the land use patterns would have to be compatible with the operating characteristics of the transportation system.

A system that would be capable of moving large numbers of cars into a small area within a short period of time would face the parking dilemma in compounded form. Although unpublished reports of the engineers at The RAND Corporation suggest that it would be mechanically possible and perhaps even economically feasible to build sufficient underground parking facilities on Manhattan to store private cars for all employees and shoppers who arrive there daily, the problem of moving large numbers of cars into and out of the garages during brief periods would call for so elaborate and costly a maze of access ramps as to discourage any serious effort to satisfy a parking demand of such magnitude. Before such an all-out effort is made to accommodate the traditional central business district to the private motor car, the summary effect of thousands of locational decisions by individual entrepreneurs would probably have been to evolve a land use pattern that more readily conforms to the auto's operating characteristics. With further increases in mass auto usage – especially if it could attain bumper-to-bumper, 150 mph movement – we are bound to experience a dispersion of many traditionally central activities to outlying but highly accessible locations. The dispersed developments accompanying the current freeways suggest the type of pattern that seems probable. Here, again, Los Angeles offers the best prototype available.

In what sense is urban space a resource?

I have been suggesting that the quintessence of urbanization is not population density or agglomeration but specialization, the concomitant interdependence, and the human interactions by which interdependencies are satisfied. Viewed from this orientation, the urban settlement is the spatial adaptation to demands of dependent activities and specialists for low communication costs. It is helpful, therefore, to view the spatial city as a communications system, as a vastly complex switchboard through which messages and goods of various sorts are routed.

Information, ideas, and goods are the very stuff of civilization. The degree to which they are distributed to all individuals within a population stands as an important indicator of human welfare levels – as a measure of cultural and economic income. Of course, the distribution of this income is determined predominantly by institutional rather than spatial factors – only the rare Utopian has even suggested that the way to "the good society" is through the redesign of the spatial city. And yet, space intervenes as a friction against all types of communication. Surely, salvation does not lie in the remodeled spatial city; but, just as surely, levels of cultural and economic wealth could be increased if the spatial frictions that now limit the freedom to interact were reduced. This is the important justification for city planning's traditional concern with space.

In the very nature of Euclidean geometry, the space immediately surrounding an urban settlement is limited. Given a transportation–communication technology and its accompanying cost structure, close-in space has greater value than distant space, since nearby inhabitants have greater opportunities to interact with others in the settlement.

But as the transportation–communication technologies change to permit interaction over greater distances at constant or even at falling costs, more and more outlying space is thereby brought into the market, and the relative value of space adjacent to large settlements falls. Urban space, as it has been associated with the economies of localization and agglomeration, is thus a peculiar resource, characterized by increasing supply and by ever-declining value.

These cost-reducing and space-expanding effects of transportation-communication changes are being reinforced by most of the technological and social changes we have recently seen. The patterns of social stratification and of occupations, the organizational structures of businesses and of governments, the goods and the ideas that are being produced, and the average individual's ranges of interests and opportunities are steadily becoming more varied and less tradition-bound. In a similar way, the repercussions of these social changes and the direct impacts of some major technological changes have made for increasing diversity in the spatial structures of urban settlements.

Projections of future change, and especially changes in the technologies of transportation and communication, suggest that much greater variation will be possible in the next few decades. It is becoming difficult to avoid the parallel prediction that totally new spatial forms are in the offing.

To date, very few observers have gone so far as to predict that the nodally concentric form, that has marked every spatial city throughout history, could give way to nearly homogeneous dispersion of the nation's population across the continent; but the hesitancy may stem mainly from the fact that a non-nodal city of this sort would represent such a huge break with the past. Yet, never before in human history has it been so easy to communicate across long distances. Never before have men been able to maintain intimate and continuing contact with others across thousands of miles; never has intimacy been so independent of spatial propinquity. Never before has it seemed possible to build an array of specialized transportation equipment that would permit speed of travel to increase directly with mileage length of trip, thus having the capability of uniting all places within a continent with almost-equal time distance. And never before has it seemed economically feasible for the nodally cohesive spatial form that marks the contemporary large settlement to be replaced by drastically different forms, while the pattern of internal centering itself changes or, perhaps, dissolves.

A number of informed students have read the same evidence and have drawn different conclusions. Observing that the consequences of ongoing technological changes are spatially neutral, they suggest that increased ease of intercourse makes it all the more possible for households and business establishments to locate in the midst of high-density settlements. This was essentially the conclusion that Haig drew when he wrote, "[. . .] Instead of explaining why so large a portion of the population is found in urban areas, one must give reasons why that portion is not even greater. The question is changed from 'Why live in the city?' to 'Why not live in the city?' "[1]

I am quick to agree that many of the recent and the imminent developments are ambiguous with respect to space. They could push urban spatial structure toward greater concentricity, toward greater dispersion, or, what I believe to be most likely, toward a very heterogeneous pattern. Since administrative and executive activities are so sensitive to the availability and immediacy of accurate information – and hence of good communications – they may be the bellwether of future spatial adjustments of other activities as well, and they therefore warrant our special attention.

The new electronic data-processing equipment and the accompanying procedures permit much more intensive use of downtown space than was ever possible with nonautomated office processes; but they can operate quite as effectively from an outlying location, far removed from the executive offices they serve. The sites adjacent to the central telephone exchange may offer competitive advantages over all others, and establishments relying upon computers, that in turn are tied to the long-distance telephone lines, seem to be clustering about the hub of those radial lines in much the manner that they once clustered about the hub of the radial trolley lines. At the same time we can already observe that outlying computer centers are attracting establishments that use their services.

The recent history of office construction in midtown New York, northwest Washington, and in the centers of most large metropolitan areas is frequently cited as clear evidence of the role that face-to-face contacts play in decision-making and of the importance of spatial propinquity in facilitating face-to-face contact. And yet, simultaneously, large numbers of executive offices have followed their produc-tion units to suburban locations, and some have established themselves in outlying spots, spatially separated from their production units and from all other establishments. The predominant movement in the New York area has been to the business center, but the fact that many have been able to move outside the built-up area suggests that a new degree of locational freedom is being added. [. . .]

If most of the social and technological changes I have mentioned were in fact neutral in their spatial impacts, this itself would represent a powerful new factor at work on the spatial organization of cities. Prior dominant modes of transportation and communication, traditional forms of organization of business and government, the older and more rigid patterns of economic and social stratification, and prior educational and occupational levels and opportunities all exerted positive pressures to population agglomeration around dominant high-density business–industrial–residential centers. If these pressures for concentration and concentricity are ebbing, the effects of counter processes will be increasingly manifest. [. . .]

Note

1 Robert M. Haig, ''Toward an Understanding of the Metropolis,'' *New York Regional Survey, Regional Survey of New York and Its Environs*, Vol. 1 (New York: Regional Plan Association, Inc., 1927)

24

SQUARING THE CIRCLE: CAN WE RESOLVE THE CLARKIAN PARADOX?

by Peter Hall

Source: Peter Hall, "Squaring the Circle: Can We Resolve the Clarkian Paradox?", *Environment and Planning B: planning and design* 21 (1994), pp. s79–s94

In 1957 the economist Clark published a paper, destined to become a classic in the urban literature called "Transport: maker and breaker of cities". In it he argued that – at least since the first industrial revolution, two hundred years ago – the growth of cities had been shaped by the development of their transport facilities. But these in turn were dependent on the evolution of transport technologies. For each successive development of the technology, there was a corresponding kind of city. However, the relationship was more complex than that: it was a *mutual* one. The transport system shaped the growth of the city, but on the other hand the previous growth of the city shaped and in particular constrained the transport alternatives that were available. So the pattern of activities and land uses in the city, and the transport system, existed in some kind of symbiotic relationship.

But, Clark stressed, the two could get out of step, and indeed very often did so. That was particularly the case because cities change more slowly than the available technologies change. As a result of this, old cities have trouble in accommodating modern traffic; this is why many of them – in Europe especially – have excluded the automobile from their downtown areas, returning by choice to the pedestrian world of the middle ages.

Cities and transport: four stages, four crises

In consequence, as explained in an earlier paper (Hall, 1992), over the past 150 years cities worldwide have experienced four successive crises of transport technology and urban form, of which the fourth is still in progress.

Down to about 1850, most cities – even the biggest ones – did virtually without what we would today regard as a transport system. Within this pre-public-transport city, personal travel was on foot. Such cities were extraordinarily small and dense. The result, as Clark showed in another classic paper (1951), was an extraordinarily steep density gradient: within the city densities were very high, but they soon descended to rural levels. About 1850, such cities experienced a crisis: they could not grow much further as long as the jobs remained in or near the centre, and as long as there was no personal transport. The answer was horse trams and commuter railways: steel wheels on steel rails reduced the frictional resistance of travel and made it possible to shift not merely bulk goods, but also bulk loads of people, over land; from around 1870, 'streetcar suburbs' (Warner, 1978) developed around North American and European cities, and equivalent suburbs in almost every German city. Cities that developed on this pattern, characterized by London, Manchester, Paris, Berlin, or New York around 1890, can be described as *early public-transport cities*.

But, because the technologies were relatively unsophisticated, a second crisis was in the making. The bigger cities, which had reached populations of over one million by 1890, could not grow much further on the basis of the existing bundle of technologies; there was a social housing crisis. At this point, around 1890–1900, came a new phase: electricity was applied to propelling trams and commuter trains; underground electric railways were constructed in the biggest cities, and soon they were extended above ground to serve new suburban rings, earlier and more aggressively in London and New York than in Paris or Berlin. The result, by the 1920s and 1930s, was a new kind of city, the *late-public-transport city*.

Then came a third crisis, which Los Angeles experienced in the 1920s and other cities only in the 1960s: the clash between the old urban morphology and the car. The Los Angeles answer was typically innovative: it was to let the public transport system decline, and to become an *auto-oriented city*. A few other cities, all of them in the American west – Phoenix, Salt Lake City, Dallas – effectively adopted full motorization with low-density dispersal of homes and jobs.

But other, older, cities kept strong central business districts (CBDs), and in the great redevelopment boom of the 1960s they often became stronger still. They could not adopt full motorization on the Angeleno model. The biggest, like New York, Paris, Tokyo, and London – are strong centre cities, as Thomson (1977) calls them; they have one million and more downtown office workers crowded into an area typically ten square miles or less, and it is this feature that makes them rail dependent: they depend very largely on their early 20th-century rail systems to get people to work, because only rail systems can handle those kinds of flows along corridors: up to 40 000 people per hour on one lane or track.

The next rung of cities, which includes most of the great regional centres of Europe and the United States, and some smaller European capitals – Birmingham and Manchester, Brussels and Copenhagen, Frankfurt and Milan, Atlanta and San Francisco – are what Thomson (1977) calls *weak centre cities*: they typically have between a quarter million and a half million workers in their downtown office cores, and they bring them in by a mixture of buses, trams, and light rail, with a minority using commuter rail.

Strong and weak centre cities are both hybrids: outside their dense cores they are as car dependent as the Los Angeles type of city. Everywhere, even in the strong centre cities, only a minority of the total metropolitan area workforce works downtown, and very commonly that workforce is shrinking. There are suburban clusters of offices and shops, and these are growing, resulting in the phenomenon of the *edge city* (Garreau, 1991; compare Cervero, 1985; 1989). These clusters are highly car dependent, and often have weakly developed public transport systems (though that varies greatly from one city to another, and European cities in particular have better suburban public transport – a point to which I shall return). Further, the people who live in these suburbs use the car almost exclusively in all their nonwork journeys: for shopping, entertainment, recreation, and weekend social life. So, in effect, such urban areas have two transport systems: one used by a minority of commuters and a small group of car-less people, one used by virtually everyone else and indeed by the commuters outside commute hours.

The result is a set of contradictions, which constitute a fourth crisis in the organization of urban transportation.

Inefficiencies and contradictions

The first contradiction, which is fairly well known, is that between private and social cost and benefit: there is a massive social cost which all car drivers impose on all other users at congested times. A second problem, which is really a special case of the first, is the energy-consumption and environmental-impact implications of our increasing dependence on the car as the only available means of moving around. Newman and Kenworthy's well-known work (1989a; 1989b; 1992) indicates that average petrol consumption in US cities was nearly twice as high as in Australian cities, four times higher than in European cities, and ten times higher than in Asian cities. Even within the United States, per capita petrol consumption was as much as 40% higher in "full motorization" cities such as Houston than in "strong centre" cities such as New York or even 'weak centre' cities such as Boston.

Differences in gasoline prices, income, and vehicle efficiency explained only about half of these variations. What was significant was the urban structure: cities with strong concentrations of central jobs, and accordingly a better developed public transport system, had much lower energy use than cities where the jobs were scattered. Cities such as Houston or Phoenix, they conclude, could achieve fuel savings of 20–30% if they became more like Boston or Washington in their urban structure.

Looking at cities worldwide, they also found a strong correlation between energy use and overall density. Toronto records only just over half the fuel consumption of US cities, though petrol prices were actually lower and vehicle efficiency even poorer. Transit use was 0.8% in Detroit, compared with 16.7% in Toronto. However, that does not entirely explain the difference between the per-capita petrol consumption of the two cities. If all of Toronto's transit passengers switched to private cars, gasoline use would increase by 53 gallons, making Toronto's usage still 184 gallons per capita lower than Detroit. The reason, they conclude, is Toronto's more compact land-use patterns, grouped around access to public transport, which are a result of conscious land-use planning – a difference evident even to a first-time visitor flying into Toronto airport from any US city. Toronto is one of the few cities in the world with well-developed policies for transportation energy conservation based on land-use strategies. "Low density", the authors say, "appears to have a multiplicative effect, not only ensuring longer distances for all kinds of travel, but making all nonautomobile modes virtually impossible, since many people live too far from a transit line and walking and biking become impossible" (Newman and Kenworthy, 1989b, page 29).

Comparing European and US cities, they find another important factor: Hamburg has about as many central jobs as Houston, but it makes commuting by rail easier and cheaper than commuting by car, whereas Houston provides plenty of cheap parking. Houston is almost totally automobile oriented, and Hamburg is extremely rail oriented. Overall, there is a strong relationship between energy use and the use of public transport, especially rail, and provision for the car (whether through provision of roadspace and/or parking, or cheap fuel). In European cities, 25% of all passenger travel is by transit and only 44% use a car for the journey to work. The importance of walking or biking in these more compact cities is highlighted by the fact that 21% use these modes for their work trip. In Amsterdam the proportion rises to 28% and in Copenhagen to 32%. Asian cities show even higher proportions.

Newman and Kenworthy (1989a; 1989b) show that a rail-based transit system can compete with the automobile and that in Europe and Asia train speeds are generally faster than the average traffic speed. Central area pedestrianization, which is so extensive and so popular as a means of revitalizing central areas of European cities, is made possible by strong transit operations.

But there is yet a further problem: as already noted, in many large urban areas the central workforce, on which was based the whole traditional radial pattern of public transport, is no longer growing. It is at best static and at worst declining, as homes and jobs both decentralize to evermore far-flung suburbia. In any case, employment is everywhere decentralizing in a relative sense, because the growth of jobs in the suburbs is far faster than that in the CBD. This is true not only of US cities, but - ominously for the Newman–Kenworthy recommendations - in Europe also (Cheshire and Hay, 1989). Far from the United States adopting energy-conserving European patterns, it appears that Europe is going down the profligate US road.

Meanwhile, the new jobs are mainly in the suburbs. That indeed is one reason why public transport systems are showing poor financial results: they are still serving patterns of travel, predominantly radial, that are declining. In the United States, during the 1980s, one of the most remarkable geographical changes has been the appearance of new so-called back office complexes, as along the I-680 corridor in the San Francisco Bay Area, or the Zip Strip around Princeton in New Jersey. There are exactly similar campus office complexes at the periphery of major metropolitan areas elsewhere in the United States and in Europe: Stamford in Connecticut, 43 miles from downtown New York City; Reading, 40 miles west of London; and Parisian new towns such as St Quentin-en-Yvelines. They are part of the edge city phenomenon. However, all these three are different from the earlier examples in having good access to public transport, including feeder buses and commuter rail.

Answers: (1) traffic restraint

In European countries, there is increasing recognition that in the larger cities some form of direct restraint on traffic levels will become necessary as car ownership continues to rise; vehicle–kilometre growth rates are projected at 30–50% over the next decade. It has been estimated that the total economic loss through urban and interurban congestion and detours amounts to 500 billion ECU per year (Jones and Hervik, 1992). However, there is little empirical evidence concerning the likely response of drivers to road charges, and the size of elasticities.

There are two main types of demand-management measures that can be applied to moving traffic: regulatory or fiscal. Regulatory measures aim to restrict the space or time for vehicular movement, for instance by allowing only certain classes of road user access to a link or an area. They include:

(a) bans on vehicles in certain areas (pedestrianization);
(b) traffic calming (Hass-Klau, 1990);
(c) development of a hierarchical road network;
(d) restrictions based on vehicle characteristics (such as licence plate number);
(e) blanket restrictions with some exemptions based on characteristics of the owner (Jones and Hervik, 1992).

Cities such as Athens, Gothenburg, Freiburg, Munich, and Bologna have all pioneered experiments along one or more of these lines (Bendixson and Lombart, 1988). None of the physical restrictions appear by themselves to lead to significant reduction in the use of motor vehicles - though they can be effective locally at redistributing the pattern of movement - and in some instances can actually result in an increase in vehicle miles travelled and in carbon dioxide emissions. Further, unless additional steps are taken, all road vehicles are equally subject to delay (Jones and Hervik, 1992).

Answers: (2) road pricing

Practical experience with congestion pricing has been limited. The oldest congestion pricing experiment is the Area Licensing scheme (ALS) in Singapore, introduced in 1975. Its initial aim was to discourage commuter automobiles from entering the crowded central area of Singapore during the morning peak period by requiring vehicles to display a special licence (Orski, 1992).

In Hong Kong, a large-scale test of road pricing was conducted in 1983–85. Though the experiment was pronounced a technical success, authorities opted not to introduce the system because of adverse public reaction. The scheme had two vocal enemies: the Hong Kong Automobile Association and an urban councillor and chairman of a firm that imports Mercedes-Benz cars (Borins, 1988; Orski, 1992).

In Norway, in-bound toll ring schemes are currently in operation in Bergen (since 1986), Oslo (February 1990), and Trondheim (October 1991); their approach is modeled after that in Singapore. Several other towns (notably Ålesund) have tolled links. Remote Tromsö has introduced a local petrol tax (Jones and Hervik, 1992; Orski, 1992).

In the Oslo system all motorists coming into the inner city, whether they want to stay there or go through, have to pass a toll plaza – eighteen in all, at a distance of about 5 km from the City Hall – and pay a toll equivalent to £1. They can choose, if they wish, a regular subscription for unlimited entry: £22 a month, £220 a year. Trucks pay £2. Any car without a permit can have its registration number automatically photographed, and a £25 fine will result. People living inside the toll ring still travel free – unless they leave the city, as many do at the weekend, when they too must pay on return.

The scheme was introduced not to restrain traffic but to help pay for road improvements, notably a tunnel under the city centre. That reflects Oslo's unusual problem, which is that mountain and water barriers cause much of Norway's traffic to converge on the city, without the possibility of building an outer ring highway; the Oslo tunnel is the equivalent. Because of its astronomic cost, road pricing could be sold to the citizenry in what is called the 'Oslo package'. The scheme provides that, after fifteen years, the tolls will be removed.

As a modification to the scheme, at the end of 1991 Oslo introduced automatic vehicle recording. As an alternative to stopping at the toll plazas to pay, a vehicle can carry a small piece of plastic, clipped behind the driver's mirror. It transmits a unique identification number to antennae at the toll plaza. This technology is used to bill subscribers monthly according to their use of the system, if they prefer it. But it could also be used to vary the tolls by time of day – that is, according to the degree of congestion. The problem here is not technical, but political: the scheme operates under an old law which allows tolls for expensive bridges or tunnels, and any modification would require new legislation.

Sweden plans to introduce the so-called 'Dennis package' for Stockholm in 1996. The product of a painstaking political compromise between the main political parties and between local and national interests, it includes construction of two new toll road systems, one an inner ring around the central and inner areas, the other an outer western tangent, tolling on these roads and on access to the central area, and public transport improvements. It has a double objective: as well as helping to finance the investments, it will restrain traffic in the entire inner city, with a predicted decrease in car traffic of no less than 34% (Söderström, 1992; Tegnér, 1994). [. . .]

Overall, it appears that in Singapore, the ALS and related measures have:

1 restrained car use, reduced rates of car ownership, and substantially reduced congestion;

2 shifted peak-hour travel from car to car pool
   and car to bus;
3 been perceived by the business community
   as beneficial to Singapore; it is clear that
   their removal would result in worsened congestion (Behbehani *et al.*, 1988; Hau, 1992;
   Orski, 1992).

In Norway, traffic within the controlled area of
Bergen has been reduced by 6–7% (Orski,
1992). In Oslo, local media forecast chaos; in
fact, there was none. 240 000 vehicles a day
passed through the system in early months, 100
000 of them regular subscribers. After the toll
ring was introduced, there was a 3–4% drop in
traffic, but after three months, traffic had
returned to its original level. The scheme is
thus collecting plenty of revenue – an estimated £60 million a year – but it has not
deterred motorists. The overall travel impact
of the Norwegian toll schemes appears to be
some 8% reduction for peak, 4% for interpeak,
and 16% for off-peak traffic (Hau, 1992).

Jones (1992) notes an important difference,
between Asian and Western cities, in attitudes
toward the effect of pricing schemes on decentralization. He concludes that where the city
centre is very dominant and well served by
public transport,

> it will retain its attraction; but where the non-car
> alternatives are considered poor, or there are
> major competing centres nearby, a reduction in
> non-work trips to the centre occurs. There is no
> evidence yet in Oslo or Trondheim of significant
> effects on land use patterns, but they would take
> some time to work through (1992, page 111).

It is quite clear from the full-scale operation of
the Norwegian schemes, and from the Hong
Kong pilot project, that electronic road pricing
is technically feasible. The problem is political
acceptability. One of the most important lessons to be learned from experience around the
world is that road pricing in isolation is unlikely
to prove acceptable. It must form but one part
of an overall package of measures designed to

ensure that mobility is not seriously impeded,
while reducing the congestion and environmental impacts of traffic (Richards, 1992).
However, experience in London suggests that
in highly congested cities, road pricing may be
increasing in political acceptability.

If such schemes were implemented, however, the critical question concerns their
impact on activity and land-use patterns in the
wider metropolitan area and surrounding
region. On this, there is really evidence only
from the limited Singapore experiment,
because only this has been in operation for a
sufficiently long time. And conclusions from
Singapore may not necessarily apply to other
countries because of cultural and other differences. The Singapore scheme seems to have
had relatively few major impacts on activity
patterns; the CBD controlled zone has
remained strong and attractive. But this may
reflect the combined influence of other conditions and policies, notably the very limited possibilities of decentralization within the island,
the continuing growth of Singapore as a financial services and tourist centre, and the heavy
investment in top-quality public transport
focusing on the CBD. The tentative conclusion
might be that other cities able to replicate
these conditions might also introduce electronic road pricing without too much trepidation
as to the consequences. But, given the fact that
Singapore has been an almost uniquely successful urban economy over the last twenty years,
there may not be many such comparable
places.

Answers: (3) combining transportation and land-use policies

The basic problem, as set out in the introduction, is that public transport can seldom extend
fast enough or far enough to keep up with the
suburbanization wave, which layers homes and
jobs like a marble cake, all sunk in a dough of
car dependency. To grapple with this problem,

starting in the 1960s, metropolitan areas have created new commuter networks on the model of the San Francisco BART (Bay Area Rapid Transit), the Parisian RER (Regional Express Rail), or the Frankfurt and Munich S-Bahn networks: express systems which link the downtown areas directly with distant suburbs and satellite towns. Especially in Paris, this strategy was linked with the creation of strong subcentres in the satellites, thus distributing and reconcentrating employment, and creating balancing sets of commuter flows: San Quentin-en-Yvelines, earlier mentioned, is one of these. Such commuter networks may extend 70 or 80 km from the urban cores, thus serving the great majority of all demands for radial transport (though in both London and Tokyo, a small minority of commuters are now beginning to use high-speed trains to commute up to 160 km each way each day). The main problem with this approach was still that it cannot cater adequately for the cross-flows, which are left on an overloaded highway system designed for the quite different purpose of carrying intercity or bypass traffic. Frankfurt offers a dramatic example here.

There is thus a tendency to invest very large capital sums in an attempt to solve the problem of urban congestion: build more motorways, develop light rail schemes, extend U-Bahn lines. It is almost certain that such solutions will prove self-defeating. Neither the policy of pouring more concrete to build roads, nor of welding more steel rail to extend underground or build new light rail lines, is likely to solve the suburban gridlock problem. Some of the investments do not even serve the main patterns of demand, and others will run the risk that customers will refuse to use them. There is very little evidence that commuters in low-density suburban areas will abandon their cars for less-convenient public transport, and that will be true even if the quality of the car commute declines quite notably. The reason, first noted with San Francisco's BART system in the 1970s,

is that people care not about the speed on the line haul but on the total door-to-door elapsed journey time, and on this criterion the car has a big built-in advantage. Thus the majority of new light rail lines opened in North America in the 1980s appear to have failed to live up to traffic projections, in some cases by very large margins (Pickrell, 1992).

This is because, as jobs as well as homes deconcentrate, they will seldom if ever reconcentrate in ways that give the same advantage to public transport as the old central areas did. That is clear from Europe, as well as from one or two North American metropolitan centres such as Toronto. A regional public transport authority may work in conjunction with a regional planning agency to group jobs as well as apartment homes around public transport nodes, as Stockholm, Paris, and Toronto have done since the 1950s and 1960s, thus giving public transport the best possible chance to compete with the private car for the commuters, including that very elusive but important breed, the reverse commuter. The Toronto example is particularly interesting, because – as the Newman–Kenworthy evidence proves – it shows that North Americans will readily accept ways of living and working that are radically different from those in the typical North American city.

The other, closely related, problem is that conventional public transport cannot readily scoop up the increasing number of cross-commuters who use corridors that are insufficiently dense to support good public transport. Research in three major metropolitan areas – the San Francisco Bay Area, the Rhine–Main region and the Île-de-France – shows conclusively that in Europe as in North America, the decentralization of homes and jobs is followed by a shift to cross-commuting, mainly by private car (Hall *et al.*, 1993). The only way out of this dilemma would seem to be to provide money to create networks that would cater for these flows. The right way may be to start from that fact, and to consider how to provide a system that will compete with the car on its own terms. It would need to offer door-to-door service in the same or less time, in conditions of equal personal comfort and convenience, and it would need to compete on price. Only some kind of demand-responsive service is likely to achieve that, in the form of dial-a-bus or frequently circulating minibuses (the model used in many cities in developing countries) or van or car pools. These should be encouraged by fiscal incentives as well as by physical priorities such as diamond lanes for high occupancy vehicles (HOVs). These might indeed be physically separated from ordinary freeway traffic and allowed to operate at higher speeds. Experience shows that buses running in this way can even deliver the same numbers of downtown commuters, per hour and per lane, as traditional commuter rail systems. So they may provide a good solution for downtown trips as well.

Very belatedly, some cities have begun to think of adapting their rail systems to give a more even pattern of grid-type access between any subcentre and any other. This was explicit in the original, never-realized plan for the San Francisco BART; the Frankfurt S-Bahn, which is a net connecting the cities of Frankfurt, Mainz, Wiesbaden, and Darmstadt, has some of the same characteristics.

The most spectacular example so far comes from Paris. Although the Parisian transportation system has performed well in serving traditional radial commute patterns, planners in the region Île-de-France have become increasingly aware of the suburb-to-suburb commute problem that has resulted from suburbanization. Because there is little or no transit service to connect suburbs with one another (except when they lie in a direct line along the radial system), as elsewhere most trips of this kind are made by private automobile; congestion and other auto-related externalities have grown to serious proportions.

The first element in the plan, called ORBI-TALE, was unveiled in December 1990 (IAURIF, 1990); it is incorporated into the new regional strategy (IAURIF, 1991) published in definitive form in 1992. ORBITALE (Organisation Régionale dans le Bassin Intérieur des Transports Annulaires Libérés d'Encombrements) is a new transit system to serve the high density inner suburbs.

It is a combination of proven technologies specific to certain sites and axes: conventional tramways, automated light rail and a reserved busway through the southern suburbs. It is currently designed to be some 175 km in length, 148 km of which will be in 'belt' configuration, the remainder in axial configuration. 30 km of the system are currently in the construction phase or approved for construction; the entire network is planned to be completed during the 1990s. It will be integrated into the regional transportation network as it exists and will include approximately fifty points of transfer to the radial transit system. [. . .]

The problem is that, mapped onto metropolitan areas of the size of Paris, such solutions – however ambitious – still leave a very large number of traffic desires inadequately catered for. This suggests an alternative answer based on buses and so-called paratransit vehicles, which are small buses or vans intermediate in size between cars and conventional buses. Most such systems use dedicated rights-of-way, either exclusive busways or so-called HOV lanes on freeways (Richards, 1990). [. . .]

Such schemes are well adapted to suburban, low-density, many-to-many-type journeys from dispersed origins to dispersed destinations. But they can be used, indeed have been used, in radial journeys to city centres also – as in Adelaide, Nancy, and Curitiba. They may provide low-cost alternatives to expensive rail public transport schemes, especially where the future of downtown employment is uncertain. The point is that, once a high-occupancy exclusive right-of-way comes into existence, then it can be utilized by many different kinds of vehicles in a way that an exclusive rail right-of-way cannot. And these vehicles can fan out to serve dispersed suburban origins and destinations in a way that rail vehicles cannot.

All these experiments offer a way of developing public transport systems that specifically cater for the new journey patterns typical of the dispersed metropolis. The problem is that the basic unit, which is the conventional bus, may be too large to cater for such dispersed demand patterns. A system based on small vans, like the ones that are being used so successfully for service to and from US airports, may be the answer. These systems essentially provide a way of moving small numbers of people at a premium fare between their own homes and traffic hubs. They could be as suitable for commuting as for airport access, as the manager of one of the most successful among them suggested in *The Wall Street Journal* (Ferguson, 1990). California has made extensive use of van pools and ride sharing generally, and such schemes have proved highly successful for the daily commuter trip across the bridge from Oakland into San Francisco, which ironically has taken passengers away from the local bus system.

That, of course, is not to say that rail schemes are entirely irrelevant to a solution of the urban transport problem. In Canada, Calgary's light rail system, which uses old rail right-of-way and existing downtown streets, provides an effective radial network to a concentrated CBD that has had a lot of growth in the last decade, and it seems to be carrying good traffic. Washington's Metro seems to be justifying itself, because it similarly serves traffic which is suitable for rail. But in many contexts, road-based solutions seem to be the only practicable answer.

The paradox is that these systems do not use notably new technologies. As the thesis of this paper is that cities change under the influence of their transport systems and that these change under the impact of technological change, one would expect that we might see some dramatic shifts. The odd fact is that we have acquired virtually no new urban transport technology for a century. Electric trains, subways, light rail, even the private car were all around by 1890. It is surely about time for a change.

Such a 21st-century metropolitan transport system would give many-to-many accessibility partly over ordinary streets, partly over an automated guideway system. It would be based on small van-like vehicles rather like the airport shuttles now widely employed in the United States. They would be either electrically powered – a real possibility, given the huge push to the development of electrical vehicles which has come from the recent California air quality regulations – or perhaps dual mode, capable of switching between petrol and electricity. At nodal points they would be coupled together, either physically or electronically, to run as automated trains along special guideways. At yet other nodes they would split again to serve local destinations. Of course, no metropolitan area in the world currently offers anything of the kind. But elements of the system exist: in the guided busways in use in Adelaide and Essen, in the dual-mode vehicles already in service in Essen and Nancy, and in automated public transport systems in Lille, Vancouver, London Docklands, Osaka, and Kobe.

There is an alternative – perhaps only a partial alternative – which is to try to cater for these flows by transport systems management techniques designed to divert demand from the conventional, single-driver no-passenger automobile into more collective or shared modes such as vans and car pools. This is achieved by a variety of carrots and sticks ranging from access to preferential HOV lanes and parking slots, through direct cash incentives, to physical limitation and charging for access to employment nodes. Washington, DC, has gone the whole hog, at the peak hours restricting one freeway to HOVs only. At least one suburban downtown area, Bellevue in the state of Washington, has been equally audacious in restricting parking spaces while exploiting its position as a public transport hub. Road pricing could be employed in the same way not only in congested downtown areas, but also in the suburban nodes. All these experiments deserve intensive study, because they are far more likely to succeed than the expensive rail systems, many of which have proved to be serious failures in terms of ridership and revenue.

This is only a partial alternative, but it could evolve into the first system. There is already a blurring of traditional distinctions between one mode of transport and another: the difference between a rubber-tyred Metro and a guided trolleybus is an extremely subtle one, as is the distinction between a small van and a taxi. And, as more and more information technology is injected into the traditional highway system, it increasingly takes on the characteristics of a public transport operation. So, within the next decade, we could look forward to a convergence of the different modes, wherein cars were electronically locked together on transitways while public transport systems increasingly offered on-demand service from any place to any other place.

The clue to this evolution could be the injection of information technology into the operation of all kinds of transportation. Automated rail and light rail systems are now commonplace and have been operating very successfully, sometimes for a long period; they include heavy rail systems such as the Victoria Line in London and the San Francisco BART system (in both of which manual override is possible) and light rail systems such as those in Lille, Vancouver, and London Docklands. However, the real challenge is the application of such technologies to the highway system, with its vastly more complicated problems of reconciling different traffic movements. Con-

siderable progress has been made in the last five years, in separate initiatives in Europe, the United States, and Japan, in developing sophisticated on-board driver information systems to supplement the on-road systems, which are themselves becoming more complex; experiments are taking place, in Germany, Britain, and the United States, in integrating these into a total information-management system.

However, all these still leave the driver in full control of the car (Hall and Hass-Klau, 1987). The real challenge lies in full or partial automation, which is likely to come first on selected stretches of road, such as specially equipped freeway lanes. Work at the University of California has succeeded in maintaining satisfactory longitudinal clearances, but the problem of lateral clearance is proving more complex. The short-term solution may come in intense management of a single congested lane without passing possibilities (save in emergencies, where manual control might resume). This in itself could permit a great increase in the peak-hour capacity of a highway system.

The impact of such technologies, which are probably still decades from full implementation, might be complex. On the one hand, they would most likely be introduced first on the most congested stretches of the highway system, and so might somewhat perversely provide additional road space on these stretches, thus providing an incentive to turn back to use of the private car for congested inner-urban trips. On the other hand, they might well be associated with introduction of road pricing schemes, which might encourage dispersion. Unless the technology could be almost infinitely extended, which seems unfeasible, it would not be likely to encourage large-scale dispersal. Nevertheless, if it were applied on busy intercity highways, it might encourage longer distance car commuting between distant residential areas and traditional downtowns. In that case, the overall impact might be to encourage dispersal at the regional scale and some degree of concentration at the local scale, around highway interchanges. All this is, however, highly hypothetical.

Overall, in the future as in the past, there is likely to be a close connection between transportation technology and land use. Whatever system evolves to take the place of the conventional single-occupant, driver-controlled car, it is likely to stimulate certain types of land-use response. The most likely speculation is that it will entail some degree of local concentration, because of the investment in infrastructure that may be involved, but that this could be compatible with a considerable degree of regional dispersion. The resulting journeys might be on automated highway systems, or might be provided for by new forms of public transport, for instance high-speed steel-wheel railways [the French TGV (Train à Grande Vitesse) or German ICE (Inter-City Express)] or maglev intercity transportation. If the result were a considerable increase in long-distance transport, that could compromise the achievement of sustainability which is so much currently under discussion. It is therefore relevant to turn to questions of land use.

Transport, land use, and sustainability

The challenge is how to create an environmentally sustainable urban environment which would work effectively for the people living and working in it. Though we know in very general terms what such an environment would be like, we do not know in detail how to measure or achieve it.

At a local level, interesting experiments are now taking place in the Californian cities of Sacramento and San Jose, where architects such as Peter Calthorpe are joining forces with city planning offices to develop areas of single-family homes quite different from the traditional tract development: they are based on maximizing pedestrian access to public

transport spines and on restricting use of the car (Kelbaugh *et al.*, 1989; Calthorpe, 1993). The point is that these designs are not, as sometimes in Western Europe, forced on reluctant public housing tenants who have no choice; they are being sold in the open market, admittedly to a public that is reeling under the impact of some of the highest housing prices in the United States, but constituting a market nonetheless.

The problem is that the larger consequences – at the subregional and regional scale – have not been worked out. Some 16 km south of the California state capitol of Sacramento, the first Calthorpe pedestrian pocket development is nearing completion. The question is how it will relate to the wider Sacramento metropolitan area. It is designed to relate via a planned extension of the new Sacramento Light Rail system, but this is some years away; meanwhile, it will be connected to the Sacramento downtown via an express bus system running on the nearby freeway. The point is that many of the residents may have jobs in other places, not directly connected (indeed, perhaps not connected at all) by public transport. So, despite a design intended to encourage the opposite, they may end up dependent on their cars for the daily commute. This is a more than academic point, because many of the new home buyers in this part of California have jobs in the San Francisco Bay Area and endure very long-distance, car-captive commute journeys.

So the question remains: what would be the ideal subregional urban structure to minimize commuting and above all car-dependent commuting? This has been highlighted by the work of Newman and Kenworthy (1989a; 1989b; 1992), already noted in a previous section. They conclude that "physical planning agencies have a major contribution to make in the conservation of transportation energy in cities" (Newman and Kenworthy, 1989b, page 35). Specifically, Newman and Kenworthy suggest a variety of policies with potential to save fuel:

increasing urban density;
strengthening the city centre;
providing a good transit option;
restraining the provision of automobile infrastructure.

However, Gordon and Richardson (1989) provide a notable counterpoint. They argue that
Newman and Kenworthy's analysis is faulty,
that the problems are wrongly diagnosed, and
that their policy and planning prescriptions are
inappropriate and infeasible. They say that
gasoline savings, by themselves, do not provide
a significant objective. They essentially give the
classic argument that the market itself will
adjust to the level of optimal efficiency and
social welfare.

Newman and Kenworthy, they say, neglect
the considerable suburbanization of employment that has occurred. "The co-location of
firms and households at decentralized locations
has reduced, not lengthened, commuting times
and distances. Decentralization reduces pressures on the CBD, relieves congestion, and
avoids 'gridlock' " (Gordon and Richardson,
1989, page 343). The striking recent change in
US travel behaviour, they point out, is the rise of
nonwork trips. They say the case for rail is very
weak, citing the huge costs involved and its
inefficiency, considering low US densities. Citing Webber's BART study (1976), they say that
rail may not generate the expected land-use
impacts such as densification around stations.
"Since 1976, probably not coincidentally,
downtown San Francisco has suffered from relative decline, and Los Angeles has pulled far
ahead of San Francisco as a corporate headquarters hub" (Gordon and Richardson, 1989, page
344). And Giuliano concluded that within the
Bay Area, "growth in non-BART zones is greater
than in BART zones" (quoted in Gordon and
Richardson, 1989, page 344). [. . .]

It is also interesting that Breheny (1992a)
specifically attacks the European Commission's
emphasis – in their 1990 *Green Paper on the*

Urban Environment (CEC, 1990) – on the
existing high-density city as the basis for future
urban development. He argues that the compact city – which, as he points out, runs counter to all recent trends in EC (and indeed
OECD) countries – is open to objection on a
number of counts. It is not necessarily energy
efficient (on that, the evidence is still not clear,
as is evident from the debate between Newman
and Kenworthy, and Gordon and Richardson);
it is not even feasible, given the amount of
population to be accommodated; it does not
seem to correspond to the life-style that people
want. In any event, it should not be accepted
uncritically without extensive further research.

Any overall informed view, such as comes
from a study of the recent expert symposium
edited by Breheny (1992b), would surely have
to conclude that, from the research so far, we
do not actually know whether it would be
better to concentrate more people within existing cities, or to house them in new communities. We might hope, through quantified
research (for instance, on the patterns of vehicle movement) to reach a reasonably robust
conclusion on this point [. . .] But the complexity of the issues should not be understated.

Conclusions

The conclusions of this paper are somewhat
rich and varied: in addition many of them, based
on research still in progress, are necessarily
tentative. The most important seem to be these:

1 There is definite evidence that, throughout
North America and Western Europe, homes
and jobs are decentralizing to produce a
polycentric 'spread city' pattern, however
much land-use planning seeks to control
and even inhibit the trend.
2 The result is a widespread dependence on
the private car for all trips, including the
daily journey to work. Conventional public
transport is less and less capable of serving

the dispersed travel demands. There is some controversy about the precise relationships between density, centralization, and energy use, but little doubt about the general relationship between land-use-activity patterns and mode choice.

3 Traffic restraint policies have been widely applied to city centres but not elsewhere. Road pricing has been applied only to a few city centres, and to only one (Singapore) for a long period. The indirect effects on activity patterns observed there seem to be minimal but may not be replicated elsewhere.

4 Only one city (Paris) has responded to the fact of decentralization by developing new orbital rail systems. Conditions there are unusual, with very high urban-level densities in the inner suburbs. Other cities - in Europe, North America, Australia, and Japan - are experimenting with novel bus or paratransit systems which may prove more apposite and effective.

5 Experiments are occurring at the local scale in developing pedestrian-friendly local neighbourhoods. It is not clear, however, how to develop effective sustainable urban forms at larger subregional and regional scales; there is a great need for further research, both theoretical and empirical.

6 Experience suggests that there is considerable scope for telecommunications to substitute for the daily work journey through 'telecommuting'. There may, however, be some indirect effects in dispersing residences and jobs, which could reduce some of the benefits (Mokhtarian, 1991; 1992). Here, too, further research will be needed.

Note

[This is a shortened version of a report earlier commissioned by the Wüstenrot Stiftung and published in German. The author wishes to acknowledge the support of the foundation in making this research possible]

References

BEHBEHANI, R., PENDAKUR, V.S. and ARMSTRONG-WRIGHT, A. (1988) "Singapore", in *Cities and Transport: Athens/Gothenburg/Hong Kong/London/Los Angeles/Munich/New York/Osaka/Paris/Singapore*, Paris, OECD, pp. 185-204

BENDIXSON, T. and LOMBART, A. (1988) "Athens", in *Cities and Transport: Athens/Gothenburg/Hong Kong/London/Los Angeles/Munich/New York/Osaka/Paris/Singapore*, Paris, OECD, pp. 9-25

BORINS, S. (1988) "Electronic Road Pricing: An Idea Whose Time May Never Come", *Transportation Research* 22A, pp. 37-44

BREHENY, M.J. (1992a) "The Contradictions of the Compact City: A Review", in BREHENY, M.J. (ed.), *Sustainable Development and Urban Form*, London, Pion, pp. 138-59

BREHENY, M.J. (ed.) (1992b) *Sustainable Development and Urban Form*, London, Pion

CALTHORPE, P. (1993) *The Next American Metropolis: Ecology, Community, and the American Dream*, Princeton, Princeton Architectural Press

CEC (1990) *Green Paper on the Urban Environment* EUR 17902, Brussels, Commission of the European Communities

CERVERO, R. (1985) *Suburban Gridlock*, New Brunswick, NJ, Center for Urban Policy Studies, Rutgers University

—— (1989) *America's Suburban Centers: The Land Use-Transportation Link*, Boston, MA, Unwin Hyman

CHESHIRE, P.C. and HAY, D.G. (1989) *Urban Problems in Western Europe: An Economic Analysis*, London, Unwin Hyman

CLARK, C. (1951) "Urban Population Densities", *Journal of the Royal Statistical Society A* 114, pp. 490-6

—— (1957) "Transport: Maker and Breaker of Cities", *Town Planning Review* 28, 237-50

FERGUSON, T.W. (1990) "The Way to the Airport Might be a Shortcut to Work", *Wall Street Journal*, 27 February, p. A25

GARREAU, J. (1991) *Edge City: Life on the New Frontier*, New York, Doubleday

GORDON, P. and RICHARDSON, H.W. (1989) "Gasoline Consumption and Cities: A Reply", *Journal of the American Planning Association* 55, pp. 342-6

HALL, P. (1992) "Transport: Maker and Breaker of Cities", in Sussex, John Wiley, MANNION, A.M. and BOWLBY, S.R. (eds), *Environmental Issues in the 1990s*, Chichester, pp. 265-76

HALL, P. and HASS-KLAU C. (1987) "Urban Transport: Time for a Fresh Look", in HARRISON, A. and RETTON, J. (eds), *Transport UK 1987: An Economic, Social and Policy Audit*, Newbury, Berks, Policy Journals, pp. 83-6

HALL, P., SANDS, B. and STREETER, W. (1993) "Managing the Suburban Commute: A Cross-national Comparison of Three Metropolitan Areas", Berkeley, CA, WP-596, Institute of Urban and Regional Development, University of California at Berkeley

HASS-KLAU C. (1990) *The Pedestrian and City Traffic*, London, Belhaven Press

HAU, T.D. (1992) *Congestion Charging Mechanisms for Road: An Evaluation of Current Practice*, Washington, DC, World Bank

IAURIF (1990) *ORBITALE: Un Réseau de Transports en Commun de Rocade en Zone Centrale*, Paris, Institut d'Aménagement et d'Urbanisme de la Région d'Île-de-France, and Conseil Régional Île-de-France

IAURIF (1991) *La Charte de l'Île-de-France: Project Présenté par l'Exécutif Régional. Cahiers de l'Institut d'Aménagement et d'Urbanisme de la Région d'Île-de-France, 97-98*, Paris, Institut d'Aménagement et d'Urbanisme de la Région d'Île-de-France, and Conseil Régional Île-de-France

JONES, P. (1992) *Review of Available Evidence on Public Reactions to Road Pricing*, London, Transport Studies Group, University of Westminister for London Transportation Unit, Department of Transport

JONES, P. and HERVIK, A. (1992) "Restraining Car Traffic in European Cities: An Emerging Role for Road Pricing", *Transportation Research* 26A, pp. 133-45

KELBAUGH, D. *et al.* (1989) *The Pedestrian Pocket Book: A New Suburban Design Strategy*, New York, Princeton Architectural Press in association with the University of Washington

MOKHTARIAN, P.L. (1991) "Telecommuting and Travel: State of the Practice, State of the Art", *Transportation* 18, pp. 319-42

—— (1992) "Telecommuting in the United States: Letting Our Fingers do the Commuting," *TR News* 158, pp. 2-7

NEWMAN, P.W.G. and KENWORTHY, J.R. (1989a) *Cities and Automobile Dependence: A Sourcebook*, Brookfield, VT, Gower

—— (1989b) "Gasoline Consumption and Cities: A Comparison of U.S. Cities with a Global Survey", *Journal of the American Planning Association* 55, pp. 24-37

—— 1992, "Is There a Role for Physical Planners?", *Journal of the American Planning Association* 58, pp. 353-62

ORSKI, C.K (1992) "Congestion Pricing: Promise and Limitations", *Transportation Quarterly* 46, pp. 157-67

PICKRELL, D. (1992) "A Desire Named Streetcar: Fantasy and Fact in Rail Transit Planning", *Journal of the American Planning Association* 58, pp. 158-76

RICHARDS, B. (1990) *Transport in Cities*, London, Architecture Design and Technology Press

RICHARDS, M.G. (1992) "Road Pricing: International Experience", preliminary draft, prepared for the US Department of Transportation Congestion Pricing Symposium, Washington, DC

SÖDERSTRÖM, J. (1992) "The Dennis Agreement: A 15-year Program for Construction and Financing of Roads and Public Transportation in Stockholm", paper presented at the conference on Mobility and Territory in Major Cities: The Role of a Road Network, Madrid, November; Stockholm, Swedish Association of Local Authorities (mimeo)

TEGNÉR, G., 1994, "The 'Dennis Traffic Agreement': A Coherent Transport Strategy for a Better Environment in the Stockholm Metropolitan Region", paper presented at the STOA International Workshop, Brussels, April; Solna, TRANSEK AB (mimeo)

THOMSON, J.M (1977) *Great Cities and their Traffic*, London, Gollancz

WARNER, S.B. (1978) *Streetcar Suburbs: The Process of Growth in Boston, 1870-1900*, Cambridge, MA, Harvard University Press

WEBBER, M.M. (1976) "The BART Experience – What Have We Learned?", *The Public Interest* 35, pp. 79-108

25

NEW HIGHWAYS

by James Martin

Source: James Martin, *Telematic Society: a challenge for tomorrow*, Englewood Cliffs, NJ, Prentice-Hall, 1981, pp. 5–13

Imagine a city ten or twenty years in the future, with parks and flowers and lakes, where the air is crystal clear and most cars are kept in large parking lots on the outskirts. The high-rise buildings are not too close, so they all have good views and everyone living in the city can walk through the gardens or rainfree pedestrian malls to shops, restaurants, or pubs. The city has cabling under the streets and new forms of radio that provide all manner of communication facilities. The television sets, which can pick up many more channels than today's television, can also be used in conjunction with small keyboards to provide a multitude of communication services. The more affluent citizens have 7-foot television screens, or even larger.

There is less need for physical travel than in an earlier era. Banking can be done from the home, and so can as much shopping as is desired. There is good delivery service. Working at home is encouraged and is made easy for some by the workscreens and videophones that transmit pictures and documents as well as speech. Meetings and symposia can be held with the participants in distant locations.

Some homes have machines that receive transmitted documents. With these machines one can obtain business paperwork, news items selected to match one's interests, financial or stock market reports, mail, bank statements, airline schedules, and so on. Many of these items, however, are best viewed on the home screens rather than in printed form.

There is almost no street robbery, because most persons carry little cash. Restaurants and stores all accept bank cards, which are read by machines and can be used only by their owners. When these cards are used to make payments, funds may be automatically transferred between the requisite bank accounts by telecommunications. Citizens can wear radio devices for automatically calling police or ambulances if they wish. Homes have burglar and fire alarms connected to the police and fire stations.

Pocket terminals mushroom in sales and drop in cost as fast as pocket calculators did a decade earlier. Most people who carried a pocket calculator now have a pocket terminal. The pocket terminal, however, has an almost endless range of applications. It can access many computers and data banks via the public data networks. The pocket terminal becomes a consumer product (as opposed to a product for businesspeople) on sale at supermarkets, with human factoring that is simple and often amusing. The public regard it as a companion which enables them to find good restaurants, display jokes on any subject, book airline and theater seats, contact medical programs, check what their stockbroker computer has to say, send messages, and access their electronic mailbox.

Public data networks are ubiquitous and

cheap, and accessible from every telephone. Their cost is independent of distance within most countries.

Industry is to a major extent run by machines. Automated production lines and industrial robots carry out much of the physical work, and data processing systems carry out much of the administrative work. There is almost no machine tool that does not contain a miniature computer. Even mechanical products have been completely redesigned because of the use of cheap microchips and fabrication by robots with senses and skills totally different from ours. To avoid unemployment, long weekends have become normal and are demanded by the labor unions. Paperwork is largely avoided by having computers send orders and invoices directly to other computers and by making most payments, including salary payments, by automatic transmission of funds into the appropriate bank accounts. What was once called the office-of-the-future became as conventional as coke machines, and with telecommunications and plasma screens in briefcase lids, people took their office-of-the-future home with them.

Inventing and producing ways to fill the increased leisure time is a major growth industry. It cannot be done by driving aimlessly around in an automobile, because petroleum has quadrupled in price again since the start of the 1980s. There are sailboats, sailplanes, windjammer cruises, high-fidelity television, and elaborate hobbies. The home screens can he used for playing games with distant opponents, for transmission of sports events, for requesting new movies, for talking face to face, and for remote attendance at conferences.

Above all, there is superlative education. History can be learned with programs as gripping and informative as Alistair Cooke's "America." University courses modeled on England's "Open University" use television and remote computers; degrees can be obtained via televi-

sion. Computer-assisted instruction, which was usually crude and unappealing in its early days, has now become highly effective. The student watches color film sequences or reads still frames and is asked to respond periodically on a keyboard. His or her response determines what will be shown next. The computer reinforces the material until it is learned. To prepare such programs there has grown up an industry as large as Hollywood and just as professional. Program production is expensive, but one program is often used by hundreds of thousands of students. You can learn hobbies, languages, mathematics, cooking – all manner of academic and leisure subjects. The world supply of such programs is growing rapidly, and most can be obtained on request on the home communications facilities. The automated education leaves teachers free to concentrate on the more human and creative aspects of teaching.

Information retrieval facilities give access to sports and financial information, weather forecasts, encyclopedias, and vast stores of reports and documents. However, far from acting as a subsitute for books, the screens allow citizens to explore the contents of excellent libraries from which books, magazines, or discs can be delivered.

The communication channels provide excellent medical facilities, some computerized and some via videophones and large television screens. Remote diagnostic studios are used, employing powerful television lenses and many medical instruments. With the help of a nurse in the studio, a distant doctor or specialist can examine a patient as though he or she were in the doctor's office. The patient can see and talk to the doctor and the doctor's large color screen can be filled with the pupil of a patient's eye, or tongue, or skin rash. The doctor can listen to a distant stethoscope and see multiple instrument readings and computer analyses of them. Automatic monitoring of chronically sick patients is done with radio devices, sometimes with automatic administering of drugs. Patients can be monitored during normal daily activities by means of miniature instrumentation (as astronauts were in the 1970s). Patients in remote areas have "telemedicine" access to highly qualified and specialized doctors and facilities when they need them. Most hospitals have telecommunications access to the world's specialists and specialized computers.

Many other aspects of city life are automated. Machines keep track of theater bookings, dentist appointments, health surveillance, and so on. The systems are entirely flexible: When the city dwellers in their wanton way break appointments, fail to turn up, or change their minds at the last minute, the computers reschedule as well as possible. Many persons have developed an arrogant attitude toward this, considering it their prerogative to change their mind whenever they wish and expecting the computers to adjust accordingly. Many of the mechanisms of the city, then, have become part of an enormous programmed complex of machines. Far from being ruled by the computer, as some people feared in earlier decades, the citizen now expects to go his own sweet way and have the computers serve him slave-like. When they fail and he is kept waiting in a station or cannot obtain the movie he wants on his home screen, he soon protests that the level of programming or facility planning must be improved.

To a large extent it is a paperless society. Paperwork is regarded as an atrocious human time-waster to be avoided whenever possible. Bank records and other household paperwork are available on the home television screen. Many businesspeople carry a briefcase with a thin flat keyboard and a plasma screen in the lid. This enables them to access their office-of-the-future facilities from any location. They can process their electronic in-basket, move work to other people's in-baskets, deal with customer requests, appointments, and so on.

Technical innovation has changed the news media. Citizens can watch their political representatives in action and can register approval or protest while they talk. Often this is statistically summarized on the screen. The electorate are both better informed and better able to make their views or protests known.

Local television enables all groups to make their own programs and express their own views. Most of these locally originating programs are of poor quality and are seen by few people, but some achieve popularity and can be promoted to the next higher level in the television networking. The best, or most pertinent, are promoted through several levels until they reach nationwide networking. Television has evolved from being a totally centralized medium with broadcasts *from* the television organization *to* the people, into a medium in which the people can participate to some extent with interactive keysets and local studio facilities.

In the city the cabling that provides these facilities becomes as important as its water pipes and electric supply. On a global scale another telecommunications revolution has occurred. Communications satellites have increased in power so that the images on the home screens, the signals that control industry, the electronic mail, and messages can pass around the world at low cost. The news of Lincoln's assassination took twelve days to reach London, but these new signals pass around the world in a fraction of a second. An executive in Chicago can instantly move items from his electronic in-basket to that of a subordinate in Rio or Peking.

If new telecommunications have changed the city, they have changed the rural districts even more. For a century there had been movement of people from the country to the city – leaving the farms as farm mechanization spread, looking for work in the cities, seeking their fortune, sending their children away to where there was better education.

Many people would have preferred the countryside if only it had the facilities of the city – good education, excellent medical facilities, superb entertainment, and, most important, highly paid jobs. *Now* telecommunications provide these. Many country villages have a satellite antenna. People can have their own garden or farmstead and can walk in the fields and woods; they eat fresh vegetables and bread from the local bakery: but they are no longer cut off from the world. Offices in the village street are part of big insurance companies or global corporations. They no longer need to be in big cities. The big city offices have become distributed. Their television screens, meeting studios, electronic mail, and secretarial services connect them to the world. They can pay the same salaries as in the cities.

The cost of physical transport has risen appallingly with the worsening petroleum shortage. There is an increasing tendency to consume local produce rather than transport it from distant places. Commuting into cities, an unpleasant and time-wasting occupation, has declined. There is a growing trend to small communities which are self-dependent except for their use of the new telecommunications highways.

Country villages have electronic access to the world's medical specialists and computers. Large numbers of television channels are available via the satellite and cable systems, and the excellent education channels reach town and country alike.

Some of the most brilliant individuals are content to live in the remotest villages, where they find peace and beautiful scenery, because they can commune with distant colleagues in many locations at someone else's expense and can use distant machines to conduct research, make video programs, design corporate facilities, or otherwise earn a living. Not all such people want to be remote. Some need the bustling, turbulent, human pressure cookers of the cities and campuses. In a telecommunications society they have a choice. Some spend part of their year in each environment.

There has been a massive increase in "soft" industries – those that do not require a large immobile plant: for example, software creation, chip design, design of games, videodisc programming, and creating material for the now vast electronic teaching industry. The percentage of blue-collar workers in the labor force of advanced countries steadily drops and is replaced by white-collar workers, many of whom are highly mobile because of the telecommunications infrastructure. A high proportion of the wealth-generating people in society become able to choose where they live. [. . .]

As physical transportation costs rise, communities tend to be more self-contained, growing their own food and providing for people's needs locally. Along with this, government becomes increasingly localized in some countries. People become free to choose their government to a greater extent. It is, perhaps, only this element of free-marketplace competition between governments that places real constraints on the excesses of bureaucracy.

Countries that provide tax havens flourish as the information industries spread. Many more tax havens come into existence and link themselves to the world's data networks and telecommunications. A characteristic of the information entrepreneur is that he can often make a fortune fast if one of his products becomes a best-seller. An "offshore" software and information industry grows to large proportions.

The new village culture plays a part in saving energy. The houses use solar energy and wood stoves, backed up by conventional power. The villagers walk to work and often shop by telecommunications. Not much energy is consumed by the microminiature electronics and communications links. The villages are more self-sufficient than in an earlier age, and more ecologically balanced.

Small is beautiful if the pieces are in communication.

The images on the home screens can come from all over the world. Many television programs are dubbed in multiple languages, and the viewer can select the language he wants. In developing nations, hundreds of millions of people, unable to read and isolated until recently from most communications, join the club of world television. This immensely powerful medium enlightens, educates, entertains, spreads literacy, and spreads better farming methods, but it can also misinform, spread unfulfillable aspirations, foster greed, provoke antagonisms and discontent, and spread themes of chaos. Only occasionally does it spread the best of human culture.

Medical assistance can be brought by telecommunications to parts of the world where doctors or specialists are in short supply. The satellite channels also make available the services to other experts: crop disease experts, consulting engineers, world authorities. The cost of air travel has increased greatly, but telecommunications make the world a small place.

Multinational corporations are laced together with worldwide networks for telephones, instant mail, and links between computers. Video conference rooms and computerized information systems increase the degree to which head-office executives guide corporate operations in other countries. Computers schedule fleets and optimize the use of resources on a worldwide basis. Money can be moved electronically from one country to another and switched to different currencies. There is worldwide management of capital inventory control, product design, bulk purchasing, computer software, and so on. Local problem situations can trigger the instant attention of head-office staff. [. . .]

Many of the new uses of telecommunications [. . .] are in conflict with the established order. They will encounter fierce opposition from vested interests. However, most of the changes are changes for the better: better education, better news media, improvements in the political process, better forms of human communication, better entertainment, better medical resources, less pollution, less human drudgery, less use of petroleum, more efficient industry, and a better informed society with a rich texture of information sources.

Most persons, and most politicians, are simply not aware of the possibilities, and hence the vested interests may succeed in preserving older methods. It is desirable that young people especially, who will inherit the world we are creating, should understand the new opportunities and make their voices heard.

26

TELECOMMUNICATIONS AND THE CHANGING GEOGRAPHIES OF KNOWLEDGE TRANSMISSION IN THE LATE TWENTIETH CENTURY

by Barney Warf

Source: Barney Warf, ''Telecommunications and the Changing Geographies of Knowledge Transmission in the Late Twentieth Century'', *Urban Studies* 32 (1995), pp. 361–78

[. . .]

The late 20th century has witnessed an explosion of producer services on an historic scale, which forms a fundamental part of the much-heralded transition from Fordism to post-Fordism (Coffey and Bailly, 1991; Wood, 1991). Central to this transformation has been a wave of growth in financial and business services linked at the global level by telecommunications. The emergence of a global service economy has profoundly altered markets for, and flows of, information and capital, simultaneously initiating new experiences of space and time, generating a new round of what Harvey (1989, 1990) calls time–space convergence. More epistemologically, Poster (1990) notes that electronic systems change not only what we know, but how we know it.

The rapid escalation in the supply and demand of information services has been propelled by a convergence of several factors, including dramatic cost declines in information-processing technologies induced by the microelectronics revolution, national and worldwide deregulation of many service industries, including the Uruguay Round of GATT negotiations (which put services on the agenda for the first time), and the persistent vertical disintegration that constitutes a fundamental part of the emergence of post-Fordist production regimes around the world (Goddard and Gillespie, 1986; Garnham, 1990; Hepworth, 1990). The growth of traditional financial and business services, and the emergence of new ones, has ushered in a profound – indeed, an historic – transformation of the ways in which information is collected, processed and circulates, forming what Castells (1989) labels the 'informational mode of production'.

This paper constitutes an ambitious overview of the development, spatial dynamics and economic consequences of international telecommunications in the late 20th century as they arise from and contribute to the expansion of a global service economy. It opens with a broad perspective of recent changes in trade in producer services, particularly international finance, as the propelling force behind a large and rapidly expanding telecommunications infrastructure. Secondly, it explores the political economy of one of the largest and most renowned electronic systems, the Internet. Thirdly, it dwells upon the spatial dimensions of the mode of information, including the flowering of a select group of global cities, off-shore banking centres, and the globalisation of clerical functions. Fourthly, it traces the emergence of what is called here 'new informational spaces', nations and regions reliant upon information services at the core of their economic development strategies. The conclusion summarises several themes that arise persistently in this discussion.

The global service economy and telecommunications infrastructure

There can be little doubt that trade in services has expanded rapidly on an international basis (Kakabadse, 1987), comprising roughly one-quarter of total international trade. Internationally, the US is a net exporter of services (but runs major trade deficits in manufactured goods), which is one reason why services employment has expanded domestically. Indeed, it could be said that as the US has lost its comparative advantage in manufacturing, it has gained a new one in financial and business services (Noyelle and Dutka, 1988; Walter, 1988). The data on global services trade are poor, but some estimates are that services comprise roughly one-third of total US exports, including tourism, fees and royalties, sales of business services and profits from bank loans.

From the perspective of contemporary social theory, services may be viewed within the context of the enormous series of changes undergone by late 20th-century capitalism. In retrospect, the signs of this transformation are not difficult to see: the collapse of the Bretton Woods agreement in 1971 and the subsequent shift to floating exchange rates; the oil crises of 1974 and 1979, which unleashed $375bn of petrodollars between 1974 and 1981 (Wachtel, 1987), and the resulting recession and stagflation in the West; the explosive growth of Third World debt, including a secondary debt market and debt–equity swaps (Corbridge, 1984), the growth of Japan as the world's premier centre

of financial capital (Vogel, 1986); the explosion of the Euromarket (Pecchioli, 1983; Walter, 1988); the steady deterioration in the competitive position of industrial nations, particularly the US and the UK, and the concomitant rise of Japan, Germany and the newly industrialising nations, particularly in east Asia; the transformation of the US under the Reagan administration into the world's largest debtor, the emergence of flexible production technologies (e.g. just-in-time inventory systems) and computerisation of the workplace; the steady growth of multinational corporations and their ability to shift vast resources across national boundaries; the global wave of deregulation and privatisation that lay at the heart of Thatcherite and Reaganite post-Keynesian policy; and finally, the integration of national financial markets through telecommunications systems. In the 1990s, one might add the collapse of the Soviet bloc and the steady integration of those nations into the world economy. This series of changes has been variably labelled an 'accumulation crisis' in the transition from state monopoly to global capitalism (Graham *et al.*, 1988), or the end of one Kondratieff long wave and the beginning of another (Marshall, 1987). What is abundantly clear from these observations is the emergence of a new global division of labour, in which services play a fundamental role.

The increasing reliance of financial and business services as well as numerous multinational manufacturing firms upon telecommunications to relay massive volumes of information through international networks has made electronic data collection and transmission capabilities a fundamental part of regional and national attempts to generate a comparative advantage (Gillespie and Williams, 1988). The rapid deployment of such technologies reflects a conjunction of factors, including: the increasingly information-intensive nature of commodity production in general (necessitating ever larger volumes of technical data and related inputs on financing, design and engineering, marketing and so forth); the spatial separation of production activities in different nations through globalised sub-contracting networks; decreases in price and the elastic demand for communications; the birth of new electronic information services (e.g. on-line databases, teletext and electronic mail); and the high levels of uncertainty that accompany the international markets of the late 20th century, to which the analysis of large volumes of data is a strategic response (Moss, 1987b; Akwule, 1992). The computer networks that have made such systems technologically and commercially feasible offer users scale and scope economies, allowing spatially isolated establishments to share centralised information resources such as research, marketing and advertising, and management (Hepworth, 1986, 1990). Inevitably, such systems have profound spatial repercussions, reducing uncertainty for firms and lowering the marginal cost of existing plants, especially when they are separated from one another and their headquarters over long distances, as is increasingly the case.

Central to the explosion of information services has been the deployment of new telecommunications systems and their merger with computerised database management (Nicol, 1985). This phenomenon can be seen in no small part as an aftershock of the microelectronics revolution and the concomitant switch from analogue to digital information formats: the digital format suffers less degradation over time and space, is much more compatible with the binary constraints of computers, and allows greater privacy (Akwule, 1992). As data have been converted from analogue to digital forms, computer services have merged with telecommunications. When the cost of computing capacity dropped rapidly, communications became the largest bottleneck for information-intensive firms such as banks, securities brokers and insurance companies. Numerous corporations, especially in financial services,

invested in new communications technologies such as microwave and fibre optics. To meet the growing demand for high-volume telecommunications, telephone companies upgraded their copper-cable systems to include fibre-optics lines, which allow large quantities of data to be transmitted rapidly, securely and virtually error-free. By the early 1990s, the US fibre-optic network was already well in place (Fig. 26.1). In response to the growing demand for international digital data flows beginning in the 1970s, the United Nations International Telecommunications Union introduced Integrated Service Digital Network (ISDN) to harmonise technological constraints to data flow among its members (Akwule, 1992). ISDN has since become the standard model of telecommunications in Europe, North America and elsewhere.

The international expansion of telecommunications networks has raised several predicaments for state policy at the global and local levels. This topic is particularly important because, as we shall see, state policy both affects and is affected by the telecommunications industry. At the international level, issues of transborder data flow, intellectual property rights, copyright laws, etc., which have remained beyond the purview of traditional trade agreements, have become central to GATT and its successor, the International Trade Organization. At the national level, the lifting of state controls in telecommunications had significant impacts on the profitability, industrial organisation and spatial structure of information services. In the US, for example, telecommunications underwent a profound reorganisation following the dissolution of ATT's monopoly in 1984, leading to secular declines in the price of long-distance telephone calls. Likewise, the Thatcher government privatised British Telecom, and even the Japanese began the deregulation of Nippon Telegraph and Telephone.

Telecommunications allowed not only new

Figure 26.1 The US fibre-optics network, 1992.
Source: Office of Technology Assessment (1993)

volumes of inter-regional trade in data services, but also in capital services. Banks and securities firms have been at the forefront of the construction of extensive leased telephone networks, giving rise to electronic funds transfer systems that have come to form the nerve centre of the international financial economy, allowing banks to move capital around [at] a moment's notice, arbitraging interest rate differentials, taking advantage of favourable exchange rates, and avoiding political unrest (Langdale, 1985, 1989; Warf, 1989). Citicorp, for example, erected its Global Telecommunications Network to allow it to trade $200bn daily in foreign exchange markets around the world. Such networks give banks an ability to move money – by some estimates, more than $1.5 trillion daily (*Insight*, 1988) – around the

globe at stupendous rates. Subject to the process of digitisation, information and capital become two sides of the same coin. In the securities markets, global telecommunications systems have also facilitated the emergence of the 24-hour trading day, linking stock markets through the computerised trading of stocks. Reuters and the Chicago Mercantile Exchange announced the formation of Globex, an automated commodities trading system, while in 1993 the New York stock exchange began the move to a 24-hour day automated trading system. As Fig. 26.2 indicates, the world's major financial centres are easily connected even with an 8-hour trading day. The volatility of stock markets has increased markedly as hair-trigger computer trading programmes allow fortunes to be made (and lost) by staying

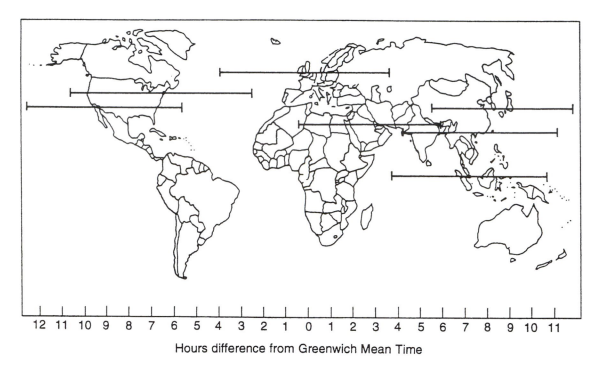

Figure 26.2 Trading hours of major world financial centres.

microseconds ahead of (or behind) other markets, as exemplified by the famous crashes of 19 October 1987. It is vital to note that heightened volatility, or the ability to switch vast quantities of funds over enormous distances, is fundamental to these capital markets: speculation is no fun when there are no wild swings in prices (Strange, 1986).

Within the context of an expanding and ever more integrated global communications network, a central role in the formation of local competitive advantage has been attained by teleports, which are essentially office parks equipped with satellite earth stations and usually linked to local fibre-optics lines (Lipman *et al.*, 1986; Hanneman, 1987a, 1987b and 1987c). The World Teleport Association defines a teleport as:

> An access facility to a satellite or other long-haul telecommunications medium, incorporating a distribution network serving the greater regional community and associated with, including, or within a comprehensive real estate or other economic development. (Hanneman, 1987a, p. 15)

Just as ports facilitate the transshipment of cargo and airports are necessary for the movement of people, so too do teleports serve as vital information transmission facilities in the age of global capital. Because telecommunications exhibit high fixed costs and low marginal costs, teleports offer significant economies of scale to small users unable to afford private systems (Stephens, 1987; Burstyn, 1986). Teleports apparently offer a continually declining average cost curve for the provision of telecommunications services. Such a cost curve raises important issues of pricing and regulation, including the tendency of industries with such cost structures to form natural monopolies. Government regulation is thus necessary to minimise inefficiencies, and the pricing of telecommunications services becomes complex (i.e. marginal revenues do not equal marginal costs, as in non-monopolistic, non-regulated sectors) (Rohlfs, 1974; Saunders *et al.*, 1983; Guldmann, 1990).

In the late 1980s there were 54 teleports in the world, including 36 in the US (Hanneman, 1987a). Most of these are concentrated in the industrialised world, particularly in cities in which data-intensive financial and business services play a major economic role. In Europe, London's new teleport in the Docklands will ensure that city's status as the centre of the Euromarket for the near future; Hamburg, Cologne, Amsterdam and Rotterdam are extending telematic control across Europe.

Tokyo is currently building the world's largest teleport. In the 1980s, the Japanese government initiated a series of high-technology 'technopolises' that form part of a long-term 'teletopia' plan to encourage decentralisation of firms out of the Tokyo region to other parts of the nation (Rimmer, 1991). In 1993 the city initiated the Tokyo Teleport on 98 ha of reclaimed land in Tokyo harbour (Tokyo Metropolitan Government Planning Department, 1993). The teleport's 'intelligent buildings' (those designed to accommodate fibre optics and advanced computational capacity), particularly its Telecom Centre, are designed to accommodate ISDN requirements. Wide Area Networks (WANs) provide local telecommunications services via microwave channels, as do Value Added Networks on fibre-optic routes. The site was originally projected to expand to 340 ha, including office, waterfront and recreational functions, and employ 100 000 people, but may be scaled back in the light of the recent recessionary climate there.

The world's first teleport is named, simply, The Teleport, located on Staten Island, New York, a project jointly operated by Merrill Lynch and the Port Authority of New York and New Jersey. Built in 1981, The Teleport consists of an 11-acre office site and 16 satellite earth stations, and is connected to 170 miles of fibre-optic cables throughout the New York region, which are, in turn, connected to the expanding national fibre-optic network. Japanese firms have taken a particularly strong interest in The Teleport, comprising 18 of its 21 tenants. For example, Recruit USA, a financial services firm, uses it to sell excess computer capacity between New York and Tokyo, taking advantage of differential day and night rates for supercomputers in each city by transmitting data via satellite and retrieving the results almost instantaneously (Warf, 1989).

In addition to the US, European and Japanese teleports, some Third World nations have invested in them in order to secure a niche in the global information services economy. Jamaica, for example, built one at Montego Bay to attract American 'back office' functions there (Wilson, 1991). Other examples include Hong Kong, Singapore, Bahrain and Lagos, Nigeria (Warf, 1989).

The Internet: political economy and spatiality of the information highway

Of all the telecommunications systems that have emerged since the 1970s, none has received more public adulation than the Internet. The unfortunate tendency in the popular media to engage in technocratic utopianism, including hyperbole about the birth of cyberspace and virtual reality, has obscured the very real effects of the Internet. The Internet is the largest electronic network on [the] planet, connecting an estimated 20m people in 40 countries (Broad, 1993). Further, the Internet has grown at rapid rates, doubling in networks and users every year (Fig. 26.3); by mid 1992, it connected more than 12 000 individual networks worldwide. Originating as a series of public networks, it now includes a variety of private systems of access, in the US including services such as Prodigy, CompuServe or America On-Line (Lewis, 1994), which allow any individual with a microcomputer and modem

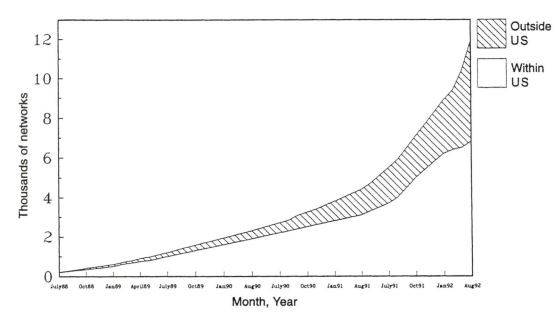

Figure 26.3 Growth of the Internet.
Source: Broad (1993)

to 'plug in', generating a variety of 'virtual communities'. By 1994, such services connected almost 5m people in the US alone (Lewis, 1994).

The origins of the Internet can be traced back to 1969, when the US Department of Defense founded ARPANET, a series of electronically connected computers whose transmission lines were designed to withstand a nuclear onslaught (Schiller, 1993). Indeed, the very durability and high quality of much of today's network owes its existence to its military origins. In 1984, ARPANET was expanded and opened to the scientific community when it was taken over by the National Science Foundation, becoming NSFNET, which linked five supercomputers around the US (Fig. 26.4). The Internet, which emerged upon a global scale via its integration with existing telephone, fibre-optic and satellite systems, was made possible by the technological innovation of packet switching, in which individual messages may be decomposed, the constituent

parts transmitted by various channels (i.e. fibre optics, telephone lines, satellite), and then reassembled, seamlessly and instantaneously, at the destination. In the 1990s such systems have received new scrutiny as central elements in the Clinton administration's emphasis on 'information superhighways'.

The Internet has become the world's single most important mechanism for the transmission of scientific and academic knowledge. Roughly one-half of all of its traffic is electronic mail, while the remainder consists of scientific documents, data, bibliographies, electronic journals and bulletin boards (Broad, 1993). Newer additions include electronic versions of newspapers, such as the *Chicago Tribune* and *San Jose Mercury News*, as well as an electronic library, the World Wide Web. In contrast to the relatively slow and bureaucratically monitored systems of knowledge production and transmission found in most of the world, the Internet and related systems permit a thoroughly unfiltered, non-hierarchical flow

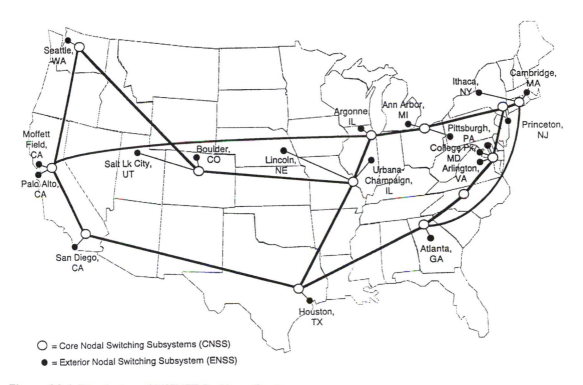

Figure 26.4 Distribution of NSFNET Backbone Service.
Source: Office of Technology Assessment (1993)

of information best noted for its lack of over-lords. Indeed, the Internet has spawned its own unregulated counterculture of 'hackers' (Mungo and Clough, 1993). However, the system finds itself facing the continuous threat of commercialisation as cyberspace is progressively encroached upon by corporations, giving rise, for example, to new forms of electronic shopping and 'junk mail' (Weis, 1992). The combination of popular, scientific and commercial uses has led to an enormous surge in demand for Internet capacities, so much so that they frequently generate 'traffic jams on the information highway' as the transmission circuits become overloaded (Markoff, 1993).

Despite the mythology of equal access for everyone, there are also vast discrepancies in access to the Internet at the global level (Schiller, 1993; Cooke and Lehrer, 1993). As

measured by the number of access nodes in each country, it is evident that the greatest Internet access remains in the most economically developed parts of the world, notably North America, Europe and Japan (Fig. 26.5). The hegemony of the US is particularly notable given that 90 per cent of Internet traffic is destined for or originates in that nation. Most of Africa, the Middle East and Asia (with the exceptions of India, Thailand and Malaysia), in contrast, have little or no access. There is, clearly, a reflection here of the long-standing bifurcation between the First and Third Worlds. To this extent, it is apparent that the geography of the Internet reflects previous rounds of capital accumulation – i.e. it exhibits a spatiality largely preconditioned by the legacy of colonialism.

There remains a further dimension to be

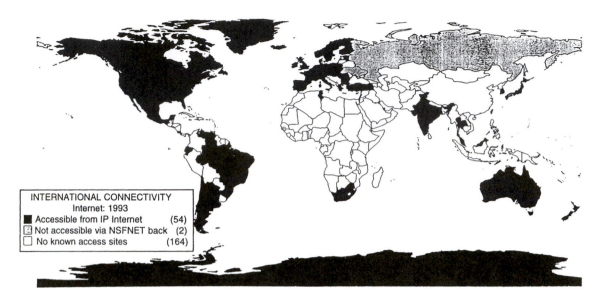

Figure 26.5 The geography of access to the Internet, 1993.
Source: Unpublished data from the Internet Supervisory Oversight Committee

explored here, however, the bifurcation between the superpowers following World War II. As Buchner (1988) noted, Marxist regimes favoured investments in television rather than telephone systems: televisions, allowing only a one-way flow of information (i.e. government propaganda), are far more conducive to centralised control than are telephones, which allow multiple parties to circumvent government lines of communication. Because access to the Internet relies heavily upon existing telephone networks, this policy has hampered the emerging post-Soviet 'Glasnet'. Superimposed on top of the landscapes of colonialism, therefore, is the landscape of the Cold War. [. . .]

Geographical consequences of the mode of information

As might be expected, the emergence of a global economy hinging upon producer services and telecommunications systems has led to new rounds of uneven development and spatial inequality. Three aspects of this phenomenon are worth noting here, including the growth of world cities, the expansion of offshore banking centres and the globalisation of back offices.

World cities

The most readily evident geographical repercussions of this process have been the growth of 'world cities', notably London, New York and Tokyo (Moss, 1987a; Sassen, 1991), each of which seems to be more closely attuned to the rhythms of the international economy than the nation-state in which it is located. In each metropolitan area, a large agglomeration of banks and ancillary firms generates pools of well-paying administrative and white-collar professional jobs; in each, the incomes of a wealthy stratum of traders and professionals have sent real estate prices soaring, unleashing rounds of gentrification and a corresponding impoverishment for disadvantaged populations. While such predicaments are not new historically – Amsterdam was the Wall

Street of the 17th century (Rodriguez and Feagin, 1986) – the magnitude and rapidity of change that global telecommunications have unleashed in such cities is without precedent.

London, for example, boomed under the impetus of the Euromarket in the 1980s, and has become detached from the rest of Britain (Thrift, 1987; Budd and Whimster, 1992). Long the centre of banking for the British Empire, and more recently the capital of the unregulated Euromarket, London seems to have severed its moorings to the rest of the UK and drifted off into the hyperspaces of global finance. State regulation in the City – always loose when compared to New York or Tokyo – was further diminished by the 'Big Bang' of 1986. Accordingly, the City's landscape has been reshaped by the growth of offices, most notably Canary Wharf and the Docklands. Still the premier financial centre of Europe, and one of the world's major centres of foreign banking, publishing and advertising, London finds its status challenged by the growth of Continental financial centres such as Amsterdam, Paris and Frankfurt.

Similarly, New York rebounded from the crisis of the mid 1970s with a massive influx of petrodollars and new investment funds (i.e. pension and mutual funds) that sustained a prolonged bull market on Wall Street in the 1980s (Scanlon, 1989; Mollenkopf and Castells, 1992; Shefter, 1993). Today, 20 per cent of New York's banking employment is in foreign-owned firms, notably Japanese giants such as Dai Ichi Kangyo. Driven by the entrance of foreign firms and increasing international linkages, trade on the New York stock exchange exploded from 12m shares per day in the 1970s to 150m in the early 1990s (Warf, 1991). New York also boasts of being the communications centre of the world, including one-half million jobs that involve the collection, production, processing, transmission or consumption of information in one capacity or another (Warf, 1991). This complex, including 60 of the largest advertising and legal services firms in the US, is fuelled by more word-processing systems than in all of Europe combined. The demand for space in such a context has driven an enormous surge of office construction, housing 60 headquarters of US Fortune 500 firms. Currently, 20 per cent of New York's office space is foreign-owned, testimony to the need of large foreign financial firms to establish a presence there.

Tokyo, the epicentre of the gargantuan Japanese financial market, is likely the world's largest centre of capital accumulation, with one-third of the world's stocks by volume and 12 of its largest banks by assets (Masai, 1989). The Tokyo region accounts for 25 per cent of Japan's population, but a disproportionate share of its economic activity, including 60 per cent of the nation's headquarters, 65 per cent of its stock transactions, 89 per cent of its foreign corporations, and 65 per cent of its foreign banks (Cybriwsky, 1991). Tokyo's growth is clearly tied to its international linkages to the world economy, particularly in finance, a reflection of Japan's growth as a major world economic power (Masai, 1989; Cybriwsky, 1991). In the 1980s, Japan's status in the global financial markets was unparalleled as the world's largest creditor nation (Vogel, 1986; *Far Eastern Economic Review*, 1987). Tokyo's role as a centre of information-intensive activities includes a state-of-the-art telecommunications infrastructure, including the CAPTAIN (Character and Pattern Telephone Access Information Network) system (Nakamura and White, 1988).

Offshore banking

A second geographical manifestation of the new, hypermobile capital markets has been the growth of offshore banking, financial services outside the regulation of their national authorities. Traditionally, 'offshore' was synonymous with the Euromarket, which arose in the 1960s as trade in US dollars outside the US. Given the collapse of Bretton Woods and the instability of world financial markets, the Euromarket has since expanded to include other currencies as well as other parts of the world. The recent growth of offshore banking centres reflects the broader shift from traditional banking services (loans and deposits) to lucrative, fee-based non-traditional functions, including debt repackaging, foreign exchange transactions and cash management (Walter, 1989).

Today, the growth of offshore banking has occurred in response to favourable tax laws in hitherto marginal places that have attempted to take advantage of the world's uneven topography of regulation. As the technological barriers to capital have declined, the importance of political ones has thus risen concomitantly. Several distinct clusters of offshore banking may be noted, including, in the Caribbean, the Bahamas and Cayman Islands; in Europe, Switzerland, Luxembourg and Liechtenstein; in the Middle East, Cyprus and Bahrain; in southeast Asia, Singapore and Hong Kong; and in the Pacific Ocean, Vanuatu, Nauru and Western Samoa. Roberts (1994, p. 92) notes that such places "are all part of a worldwide network of essentially marginal places which have come to assume a crucial position in the global circuits of fungible, fast-moving, furtive money and fictitious capital." Given the extreme mobility of finance capital and its increasing separation from the geography of employment, offshore banking can be expected to yield relatively little for the nations in which it occurs; Roberts (1994), for example, illustrates the case of the Cayman Islands, now the world's fifth-largest banking centre in terms of gross assets, where 538 foreign banks employ only 1000 people (less than two apiece). She also notes that such centres are often places in which 'hot money' from illegal drug sales or undeclared businesses may be laundered.

Offshore markets have also penetrated the global stock market, where telecommunica-tions may threaten the agglomerative advantages of world cities even as they reinforce them. For example, the National Association of Securities Dealers Automated Quotation System (NASDAQ) has emerged as the world's fourth-largest stock market; unlike the New York, London, or Tokyo exchanges, NASDAQ lacks a trading floor, connecting half a million traders worldwide through telephone and fibre-optic lines. Similarly, Paris, Belgium, Spain, Vancouver and Toronto all recently abolished their trading floors in favour of screen-based trading.

Global back offices

A third manifestation of telecommunications in the world service economy concerns the globalisation of clerical services, in particular back offices. Back offices perform many routinised clerical functions such as data entry of office records, telephone books or library catalogues, stock transfers, processing of payroll or billing information, bank cheques, insurance claims, magazine subscriptions and airline frequent-flyer coupons. These tasks involve unskilled or semi-skilled labour, primarily women, and frequently operate on a 24-hour-per-day basis (Moss and Dunau, 1986). By the mid 1980s, with the conversion of office systems from analogue to digital form largely complete, many firms began to integrate their computer systems with telecommunications.

Historically, back offices have located adjacent to headquarters activities in downtown areas to ensure close management supervision and rapid turnaround of information. However, under the impetus of rising central-city rents and shortages of sufficiently qualified (i.e. computer-literate) labour, many service firms began to uncouple their headquarters and back office functions, moving the latter out of the downtown to cheaper locations on the urban periphery. Most back office relocations, therefore, have been to suburbs (Moss and Dunau, 1986; Nelson, 1986). Recently, given the increasing locational flexibility afforded by satellites and a growing web of inter-urban fibre-optics systems, back offices have begun to relocate on a much broader, continental scale. Under the impetus of new telecommunications systems, many clerical tasks have become increasingly footloose and susceptible to spatial variations in production costs. For example, several firms fled New York City in the 1980s: American Express moved its back offices to Salt Lake City, UT, and Phoenix, AZ; Citicorp shifted its Mastercard and Visa divisions to Tampa, FL, and Sioux Falls, SD, and moved its data processing functions to Las Vegas, NV, Buffalo, NY, Hagerstown, MD, and Santa Monica, CA; Citibank moved its cash management services to New Castle, DE; Chase Manhattan housed its credit card operations in Wilmington; Hertz relocated its data entry division to Oklahoma City; Avis went to Tulsa. Dean Witter moved its data processing facilities to Dallas, TX; Metropolitan Life repositioned its back offices to Greenville, SC, Scranton, PA, and Wichita, KS; Deloitte Haskins Sells relocated its back offices to Nashville, TN; and Eastern Airlines chose Miami, FL.

Internationally, this trend has taken the form of the offshore office (Wilson, 1991). The primary motivation for offshore relocation is low labour costs, although other considerations include worker productivity, skills, turnover and benefits. Offshore offices are established not to serve foreign markets, but to generate cost savings for US firms by tapping cheap Third World labour pools. Notably, many firms with offshore back offices are in industries facing strong competitive pressures to enhance productivity, including insurance, publishing and airlines. Offshore back office operations remained insignificant until the 1980s, when advances in telecommunications such as trans-oceanic fibre-optics lines made possible greater locational flexibility just when the demand for clerical and information processing services grew rapidly (Warf, 1993). Several New York-based life insurance companies, for example, have erected back office facilities in Ireland, with the active encouragement of the Irish government (Lohr, 1988). Often situated near Shannon Airport, they move documents in by Federal Express and the final product back via satellite or the TAT-8 fibre-optics line that connected New York and London in 1989 (Fig. 26.6). Despite the fact that back offices have been there only a few years, Irish development officials already fret, with good reason, about potential competition from Greece and Portugal. Likewise, the Caribbean has become a particularly important locus for American back offices, partly due to the Caribbean Basin Initiative instituted by the Reagan administration and the guaranteed access to the US market that it provides. Most back offices in the Caribbean have chosen Anglophonic nations, particularly Jamaica and Barbados. American Airlines has paved the way in the Caribbean through its subsidiary Caribbean Data Services (CDS), which began when a data processing centre moved from Tulsa to Barbados in 1981. In 1987, CDS opened a second office near Santo Domingo, Dominican Republic, where wages are one-half as high as Barbados (Warf, forthcoming). Thus, the same flexibility that allowed back offices to move out of the US can be used against the nations to which they relocate.

New information spaces

The emergence of global digital networks has generated growth in a number of unanticipated places. These are definitely not the new industrial spaces celebrated in the literature on post-Fordist production complexes (Scott, 1988), but constitute new 'information spaces' reflective of the related, yet distinct, mode of information. Three examples – Singapore, Hungary and the Dominican Republic – illustrate the ways in which contemporary telecommunications generate repercussions in the least expected of places.

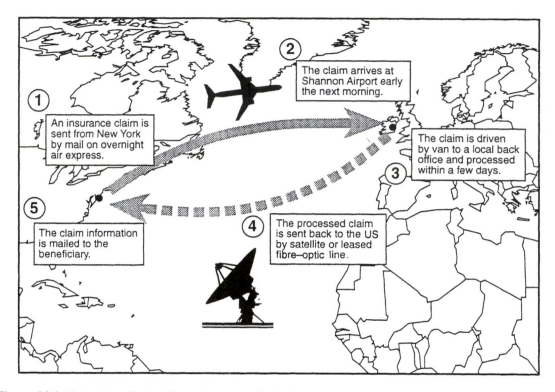

Figure 26.6 Mechanics of back office relocation to Ireland.

Singapore

Known best perhaps as a member of the East Asian newly industrialised countries (NICs), Singapore today illustrates what may be the most advanced telecommunications infrastructure in the world, creating an 'intelligent island' with high-speed leased circuits, a dense telephone and fibre-optic network, household teleboxes for electronic mail and ubiquitous remote computer access (Dicken, 1987; Corey, 1991). Singapore's government has led the way in this programme through its National Computer Board and Telecommunications Authority. This transformation has occurred as part of a sustained shift in the island's role from unskilled, low-wage assembly functions to exporter of high value-added business services and as the financial hub of south-east Asia, a process hastened by the flight of capital from Hong Kong (Jussawalla and Cheah, 1983). Exports of services have now become Singapore's largest industry in terms of employment and foreign revenues. Reuters, for example, uses Singapore as its news hub in south-east Asia. In part, this transformation reflects the island's relatively high wages and fears of competition from its larger neighbours. Today, more than one-third of Singapore's labour force is engaged in skilled, white-collar employment. In addition, Singapore uses its telecommunications network for advanced Electronic Data Interchange (EDI) services to facilitate maritime shipping, in congruence with its status as the world's largest port.

Hungary

Before the collapse of the Soviet Union, Hungary suffered many of the same telecommunications problems as other underdeveloped nations: outdated technology, unsatisfied demand and few advanced services. Today, largely due to deregulation and foreign investment, the Hungarian telecommunications system is the most advanced in the former Soviet bloc, subsuming 10 per cent of the nation's total investment capital. The leader in this process has been the postal service, Magyar Posta, and its successor, the Hungarian Telecommunications Company (Matav), which introduced innovative pricing based on market, not political criteria, fees for telephone connections, time-differentiated and distance-sensitive pricing and bond financing. Concomitantly, an administrative reorganisation decentralised control of the firm, breaking the inefficient stranglehold of the bureaucratic, Communist *apparatchik* (Whitlock and Nyevrikel, 1992). The birth of the new Hungarian telecommunications network was invaluable to the nation's emerging financial system, centred in Budapest, which has expanded beyond simple loans and stocks to include database management and stock transfers (Tardos, 1991). Thus, in this respect, Hungary serves as a model for other nations making the transition from state socialism to market economies.

The Dominican Republic

In the 1980s, the Dominican government introduced a policy designed to develop non-traditional exports, particularly tourism and information services, as part of a strategy to reduce the country's reliance upon agricultural exports. For a small, relatively impoverished nation, the nation possesses a well-endowed information services infrastructure (Warf, forthcoming). The national telephone company Codetel (Compania Dominicana de Telecommunicaciones), for example, has provided the Dominican Republic with near-universal telephone access, high-speed data transmission services on fibre-optics lines, digital switching equipment, cellular telephones and microwave service to all neighbouring nations except Cuba. Codetel also sells a variety of high value-added services such as electronic mail and databases, telex, remote terminals, facsimile services, Spanish–English translations and leased lines. This infrastructure has made the Dominican Republic the most advanced nation in telecommunications in Latin America and has attracted numerous foreign firms. IBM-Santo Domingo, for example, engages in a complex, worldwide system of sub-contracting with its subsidiaries, purchasing, for example, printers from Argentina, disk drives from Brazil, CPUs from the US or Brazil and software, written in Canada, the US and Denmark, through its distributor in Mexico. A similar firm is Infotel, which performs a variety of computer-related functions for both domestic and international clients, including compilation of telephone directories, photocomposition, data conversion, computerised, on-line sale of advertising images, desktop publishing and map digitising. Infotel serves a variety of domestic and foreign clients, including Dominican utilities and municipal governments, the GTE telephone-operating companies, the US Geological Survey and the Spanish telephone network. Another service attracted to the Dominican Republic is back offices. American Airlines, and its subsidiary Caribbean Data Services, processes medical and dental insurance claims, credit card applications, retail sales inventories, market surveys and name and address listings at a Free Trade Zone near the capital.

Concluding comments

What lessons can be drawn from these observations about the emergence of a globalised service economy and the telecommunications networks that underpin it? As part of the broad sea-change from Fordist production regimes to the globalised world of flexible accumulation, about which so much has already been said, it is clear that capital – as data or cash, electrons or investments – in the context of global services has acquired a qualitatively increased level of fluidity, a mobility enhanced by the worldwide wave of deregulation unleashed in the 1980s and the introduction of telecommunication networks. Such systems give banks, securities, insurance firms and back offices markedly greater freedom over their locational choices. In dramatically reducing the circulation time of capital, telecommunications have linked far-flung places together through networks in which billions of dollars move instantaneously across the globe, creating a geography without transport costs. There can be no doubt that this process has real consequences for places, as attested by the current status of cities such as London, New York, Tokyo and Singapore and the Cayman Islands. Generally, such processes tend to concentrate skilled, high value-added services, e.g. in global cities, while dispersing unskilled, low value-added services such as back offices to Third World locations.

In short, it is vital to note that, contrary to early, simplistic expectations that telecommunications would 'eliminate space', rendering geography meaningless through the effortless conquest of distance, such systems in fact produce new rounds of unevenness, forming new geographies that are imposed upon the relics of the past. Telecommunications simultaneously reflect and transform the topologies of capitalism, creating and rapidly recreating nested hierarchies of spaces technically articulated in the architecture of computer networks. Indeed, far from eliminating variations among places, such systems permit the exploitation of differences between areas with renewed ferocity. As Swyngedouw (1989) noted, the emergence of hyperspaces does not entail the obliteration of local uniqueness, only its reconfiguration. That the geography engendered by

this process was unforeseen a decade ago hardly needs restating; that the future will hold an equally unexpected, even bizarre, set of outcomes is equally likely.

References

AKWULE, R. (1992) *Global Telecommunications: The Technology, Administration, and Policies*, Boston, Focal Press

BROAD, W. (1993) "Doing Science on the Network: A Long Way From Gutenberg", *New York Times*, 18 May, B5

BUCHNER, B. (1988) "Social Control and the Diffusion of Modern Telecommunications Technologies: A Cross-National Study", *American Sociological Review* 53, pp. 446-53

BUDD, L. and WHIMSTER, S. (eds) (1992) *Global Finance and Urban Living: A Study of Metropolitan Change*, London, Pergamon

BURSTYN, H. (1986) "Teleports: At the Crossroads", *High Technology* 6(5), pp. 28-31

CASTELLS, M. (1989) *The Informational City*, Oxford, Basil Blackwell

COFFEY, W. and BAILLY, A. (1991) "Producer Services and Flexible Production: An Exploratory Analysis", *Growth and Change* 22, pp. 95-117

COOKE, K. and LEHRER, D. (1993) "The Internet: The Whole World is Talking", *The Nation*, 257, pp. 60-3

CORBRIDGE, S. (1984) "Crisis, What Crisis? Monetarism, Brandt II and the Geopolitics of Debt", *Political Geography Quarterly* 3, pp. 331-45

COREY, K. (1991) "The Role of Information Technology in the Planning and Development of Singapore", in BRUNN, B. and LEINBACH, T. (eds) *Collapsing Space and Time*, London, HarperCollins, pp. 217-31

CYBRIWSKY, R. (1991) *Tokyo: The Changing Profile of an Urban Giant*, Boston, G.K. Hall & Co

DICKEN, P. (1987) "A Tale of Two NICs: Hong Kong and Singapore at the Crossroads", *Geoforum* 18, pp. 151-64

Far Eastern Economic Review (1987) "Japan Banking and Finance", 9 April, pp. 47-110

GARNHAM, N. (1990) *Capitalism and Communication: Global Culture and the Economics of Information*, Beverly Hills, Sage

GILLESPIE, A. and WILLIAMS, H. (1988) "Telecommunications and the Reconstruction of Comparative Advantage", *Environment and Planning A*, 20, pp. 1311-21

GODDARD, J. and GILLESPIE, A. (1986) *Advanced Telecommunications and Regional Development*, Newcastle-upon-Tyne, Centre for Urban and Regional Development Studies

GRAHAM, J. *et al.* (1988) "Restructuring in U.S. Manufacturing: The Decline of Monopoly Capitalism", *Annals of the Association of American Geographers* 78, pp. 473-90

GULDMANN, J. (1990) "Economies of Scale and Density in Local Telephone Networks", *Regional Science and Urban Economics* 20, pp. 521–33

HANNEMAN, G. (1987a) "The Development of Teleports", *Satellite Communications*, March, pp. 14–22

—— (1987b) "Teleport Business", *Satellite Communications*, April, pp. 23–6

—— (1987c) "Teleports: The Global Outlook", *Satellite Communications*, May, pp. 29–33

HARVEY, D. (1989) *The Condition of Post-modernity*, Oxford, Basil Blackwell

—— (1990) "Between Space and Time: Reflections on the Geographical Imagination", *Annals of the Association of American Geographers*, 80, pp. 418–34

HEPWORTH, M. (1986) "The Geography of Technological Change in the Information Economy", *Regional Studies* 20, pp. 407–24

—— (1990) *Geography of the Information Economy*, London, Guildford Press

Insight (1988) "Juggling Trillions on a Wire: Is Electronic Money Safe?" 15 February, pp. 38–40

JUSSAWALLA, M. and CHEAH, C. (1983) "Towards an Information Economy: The Case of Singapore", *Information Economics and Policy* 1, pp. 161–76

KAKABADSE, M. (1987) *International Trade in Services: Prospects for Liberalisation in the 1990s*, London, Croom Helm

LANGDALE, J. (1985) "Electronic Funds Transfer and the Internationalisation of the Banking and Finance Industry", *Geoforum* 16, pp. 1–13

—— (1989) "The Geography of International Business Telecommunications: The Role of Leased Networks", *Annals of the Association of American Geographers* 79, pp. 501–22

LEWIS, P. (1994) "A Boom for On-line Services", *New York Times*, 12 July, C1

LIPMAN, A., SUGARMAN, A. and CUSHMAN, R. (1986) *Teleports and the Intelligent City*, Homewood, IL, Dow Jones

LOHR, S. (1988) "The Growth of the Global Office", *New York Times*, 18 October, D1

MARKOFF, J. (1993) "Traffic Jams Already on the Information Highway", *New York Times*, 3 November, p. 1, C7

MARSHALL, M. (1987) *Long Waves of Regional Development*, London, Macmillan

MASAI, Y. (1989) "Greater Tokyo as a Global City", in KNIGHT, R. and GAPPERT, G. (eds), *Cities in a Global Society*, Newbury Park, CA, Sage

MOLLENKOPF, J. and CASTELLS, M. (eds) (1992) *Dual City: Restructuring New York*, New York, Russell Sage Foundation

MOSS, M. (1987a) "Telecommunications, World Cities and Urban Policy", *Urban Studies* 24, pp. 534–46

—— (1987b) "Telecommunications, and International Financial Centres", in BROTCHIE, J., HALL, P. and

NEWTON, P. (eds), *The Spatial Impact of Technological Change*, London, Croom Helm

MOSS, M. and DUNAU, A. (1986) "Offices, Information Technology, and Locational Trends", in BLACK, J., ROARK, K. and SCHWARTZ, L. (eds), *The Changing Office Workplace*, Washington, DC, Urban Land Institute, pp. 171–82

MUNGO, P. and CLOUGH, B. (1993) *Approaching Zero: The Extraordinary World of Hackers, Phreakers, Virus Writers and Keyboard Criminals*, New York, Random House

NAKAMURA, H. and WHITE, J. (1988) "Tokyo", in DOGAN, M. and KASARDA, J. (eds), *The Metropolitan Era*, Vol. 2: *Mega-Cities*, Newbury Park, CA, Sage

NELSON, K. (1986) "Labor Demand, Labor Supply and the Suburbanization of Low-Wage Office Work", in SCOTT, A. and STORPER, M. (eds), *Production, Work, Territory*, Boston, Allen Unwin

NICOL, L. (1985) "Communications Technology: Economic and Spatial Impacts", in CASTELLS, M. (ed.), *High Technology, Space, and Society*, Beverly Hills, CA, Sage, pp. 191–209

NOYELLE, T. and DUTKA, A. (1988) *International Trade in Business Services*, Cambridge, MA, Ballinger

OFFICE of TECHNOLOGY ASSESSMENT (1993) *Automation of America's Offices*, Washington, DC, US Government Printing Office

PECCHIOLI, R. (1983) *The Internationalization of Banking: The Policy Issues*, Paris, OECD

POSTER, M. (1990) *The Mode of Information: Post-structuralism and Social Context*, Chicago, University of Chicago Press

RIMMER, P. (1991) "Exporting Cities to the Western Pacific Rim: The Art of the Japanese Package", in BROTCHIE, J. *et al.* (eds), *Cities of the 21st Century*, Melbourne, Longman Cheshire

ROBERTS, S. (1994) "Fictitious Capital, Fictitious Spaces: The Geography of Offshore Financial Flows", in CORBRIDGE, S., MARTIN, R. and THRIFT, N. (eds), *Money, Power and Space*, Oxford, Basil Blackwell

RODRIGUEZ, N. and FEAGIN, J. (1986) "Urban Specialization in the World-System", *Urban Affairs Quarterly* 22, pp. 187–219

ROHLFS, J. (1974) "A Theory of Interdependent Demand for a Communications Service", *Bell Journal of Economics and Management Science* 5, pp. 13–37

SASSEN, S. (1991) *The Global City: New York, London, Tokyo*, Princeton, NJ, Princeton University Press

SAUNDERS, R., WARFORD, J. and WELLENIUS, B. (1983) *Telecommunications and Economic Development*, Baltimore, The Johns Hopkins University Press

SCANLON, R. (1989) "New York City as Global Capital in the 1980s", in KNIGHT, R. and GAPPERT, G. (eds) *Cities in a Global Society*, Newbury Park, CA, Sage

SCHILLER, H. (1993) "The Information Highway: Public Way or Private Road?", *The Nation*, 257, pp. 64–5

SCOTT, A.J. (1988) *New Industrial Spaces*, London, Pion

SHEFTER, M. (1993) *Capital of the American Century: The National and International Influence of New York City*, New York, Russell Sage Foundation

STEPHENS, G. (1987) "What Can Business Get from Teleports?", *Satellite Communications*, March, pp. 18–19

STRANGE, S. (1986) *Casino Capitalism*, Oxford, Basil Blackwell

SWYNGEDOUW, E. (1989) "The Heart of the Place: The Resurrection of Locality in an Age of Hyperspace", *Geografiska Annaler* 71, pp. 31–42

TARDOS, A. (1991) "Problems of the Financial Information System in Hungary", *Acta Oeconomica* 43, pp. 149–66

THRIFT, N. (1987) "The Fixers: The Urban Geography of International Commercial Capital", in HENDERSON, J. and CASTELLS, M. (eds), *Global Restructuring and Territorial Development*, Beverly Hills, Sage Publications

TOKYO METROPOLITAN GOVERNMENT PLANNING DEPARTMENT (1993) *Tokyo Teleport*, Tokyo, Tokyo Metropolitan Government Information Centre

VOGEL, E. (1986) "Pax Nipponica?", *Foreign Affairs* 64, pp. 752–67

WACHTEL, H. (1987) "Currency Without a Country: The Global Funny Money Game", *The Nation*, 26 December, 245, pp. 784–90

WALKER, R. (1985) "Is There a Service Economy? The Changing Capitalist Division of Labor", *Science and Society*, Spring, pp. 42–83

WALTER, I. (1988) *Global Competition in Financial Services: Market Structure, Protection, and Trade Liberalization*, Cambridge, MA, Ballinger

—— (1989) *Secret Money*, London, Unwin Hyman

WARF, B. (1989) "Telecommunications and the Globalization of Financial Services", *Professional Geographer* 41, pp. 257–71

—— (1991) "The Internationalization of New York Services", in DANIELS, P. (ed.), *Services and Metropolitan Development: International Perspectives*, London, Routledge, pp. 245–64

—— (1993) "Back Office Dispersal: Implications for Urban Development", *Economic Development Commentary* 16, pp. 11–16

—— (forthcoming) "Information Services in the Dominican Republic", *Yearbook of the Association of Latin American Geographers*

WEIS, A. (1992) "Commercialization of the Internet", *Electronic Networking* 2(3), pp. 7–16

WHITLOCK, E. and NYEVRIKEL, E. (1992) "The Evolution of Hungarian Telecommunications", *Telecommunications Policy*, pp. 249–58

WILSON, M. (1991) "Offshore Relocation of Producer Services: The Irish Back Office", Paper presented at the Annual Meeting of the Association of American Geographers, Miami

WOOD, P. (1991) "Flexible Accumulation and the Rise of Business Services", *Transactions of the Institute of British Geographers* 16, pp. 160–72

27

CITIES AND THEIR AIRPORTS: POLICY FORMATION, 1926–1952

by Paul Barrett

Source: Paul Barrett, "Cities and their Airports: Policy Formation, 1926-1952", *Journal of Urban History* 14 (1987), pp. 112–37

The past two decades have witnessed a proliferation of literature on the role of urban infrastructure in altering and responding to changes in urban spatial patterns. The reciprocally shaping functions of technology, engineering, planning, and business and political institutions and their goals have likewise been studied from a variety of perspectives, as have the cultural, policy, commercial, and technological bases of suburbanization. This new literature is vast and, of course, reflects differences of ideology and of perspective. Still contested, for instance, are the roles of trolley and factory in fixing the social and spatial city of the northeast. Today, however, no one doubts the importance of the commercial airport [in] shaping urban transportation and land use patterns as well as the job prospects and comfort levels of travelers and metropolitan residents.

In the 1970s, for example, Chicago's O'Hare (the nation's busiest transfer airport) provided 33,000 jobs in an essentially suburban setting. Built and improved at a cost of over $6,000,000,000, O'Hare in the mid-1980s was an overcrowded center of traffic jams and noise pollution as well as convention centers and corporate headquarters. City and suburban officials and airline representatives argued bitterly over whether a third airport, still further from the central business district (CBD), constituted the best means of integrating air transport with urban and suburban life.

Airports were fulfilling the role described in 1974 by urban planner Richard L. Meier. "Hotels, conference rooms, communications facilities, computation capabilities and recreational services," Meier wrote, "are accumulating [around airports] to handle the interfaces between producers and suppliers, parent organizations and their subsidiaries, markets and their distributors, consulting groups and their clients." Meier went on to call for careful, cooperative planning of such new suburban centers.[1] Although planners and general observers alike recognize the signal influence of the airport on its city, few airports were built with that city in mind.

Why did the United States develop a policy tradition that treated major commercial airports as transportation facilities and nothing more? [. . .] What are the roots of this powerful tradition? Where and how did it intersect with highway, mass transit, and urban renewal policy? What was the role of aviation technology in determining the size and location of major airports?

These questions have real importance for urban life. Unlike other suburban centers, airports were the subjects of city and federal as well as suburban and corporate decision making. They presented, as planner George B. Ford observed in 1929, "the first chance that history has offered to apply the modern principle of prevention to a new art and science in a big way."[2] The fact that major airports were not effectively integrated into metropolitan planning in the formative period of commercial aviation still affects inner-city job seekers and noise-ridden suburban home-owners alike.

Planning for the coming air age: 1926–1952

Between 1926, when airmail contracts began to be awarded to commercial airlines, and 1952, when federal agencies first gave significant attention to the need to plan for airport-area land use, academic planners and architects gave the airport serious, if sometimes fanciful, treatment. "Aerodromes," of course, were important in Le Corbusier's "radial cities." [. . .] Architect Thomas Mitchell suggested in 1919 that city centers be rebuilt around sixty-story skyscrapers supporting 3,000-foot landing strips, with glass surfaces allowing the entry of light into streets below. This suggestion was unusual only in that it required the complete rebuilding of the CBD. At the other extreme, Richard Nuetra called in 1930 for the placement of shopping centers and hotels around the ideal outlying airport.

Early on, planners such as George B. Ford and John Nolan, as well as academics at the Harvard School of City Planning, gave serious consideration to the location of commercial airports and showed awareness of their potential impact on land use. Nolan warned in 1928 that the tendency of railroad termini to gather urban functions around themselves demonstrated the need to keep airports away from the already cluttered CBD. While these discussions went on, however, the number of urban area airports skyrocketed. The ruminations of academic planners appear to have had little to do with site choice for real airports or with the way land around them was developed. In short,

it was not planners, but the leaders of corporations and chambers of commerce along with politicians, city engineers, and federal bureaucrats whose decisions determined the relationship between the airport and the city.

American airports, 1926–1952

In 1926, the United States had 823 airports of all sorts including military and emergency landing strips. By 1930 municipal airports and those owned by airlines, flying schools, and the like numbered 1,037. Privately owned airports constituted over half of this total. By 1938 municipal fields alone numbered 1,833, and private investment had virtually ceased.[3] More than $340,000,000 had been invested in municipal airports but the Civil Aeronautics Authority argued that an additional $435,000,000 was needed to produce a really adequate system.[4] The postwar era brought continued rapid change, and by 1952 Lyndon Baines Johnson was moved to observe that "[airlines are] doing such a job promoting traffic that even down in the big, fine and perfect state that Texas is [. . .] they are a thorn in our side."[5] In this atmosphere of ballyhoo and free spending, American cities built their first (and often second or third) municipal airports. Like boosters [optimistic promoters] in nineteenth-century cities seeking railroad connections, civic leaders in our period seemed preoccupied with commercial growth, real estate values, and symbols of urban greatness. Scant attention was paid to how the land surrounding these icons and profit generators was developed.

Through the early 1930s, when most airports were profit-seeking ventures, their owners sought to attract paying crowds with swimming pools, dodg'em cars, miniature golf courses, roof-top dance floors, and a wide range of similar enticements. Chambers of commerce linked municipal airport development with attempts to attract capital and prestige. In the later 1920s the Chicago Chamber of Commerce engaged in a vigorous and unsuccessful effort to make its city the "Detroit of aviation."[6]

Only during World War II, when the full potential of the 1930s revolution in aviation technology became apparent, did government officials begin to speak out on the city-shaping power of airports. William A. Burden, Special Aviation Assistant to the Secretary of Commerce, observed in 1943 that "airports can act as a means for healthy decentralization of congested centers of population, for the development of new focal points of surface transport systems, for the development of new communities along existing and projected trade routes of the air."[7] By the war's end, views of this sort were common enough that a critic could refer to the idea that "new cities will spring up around an airport" as a "current facile conception."[8] Yet this new awareness did not bring new forms of action.

Decision makers were becoming aware of the potential impact of airports on the urban fringe, and some planners and geographers thought they saw quite clearly what needed to be done. Why, then, was the impact of air terminals on developing suburbs left unguided? An answer may lie in the fact that, rhetoric aside, policy was created by short-term political, legal, bureaucratic, and economic considerations and by changes in aviation technology that appeared more nearly autonomous than they were.

Actual policy was shaped by several often unrelated factors. Cities viewed airports as adjuncts of the CBD, and this fact diverted attention to highway building and other panaceas that did not entail substantive land use planning. Changing aviation technology created pressures of its own. The Civil Aeronautics Administration (CAA) and its predecessor bodies increasingly monopolized legitimized expertise on airport issues and used this power primarily to promote commercial aviation, not rational land use. Airlines preferred single-passenger terminals as opposed to multiple airports in one city, and this in turn affected airport size and location. The use of major air terminals by private planes also affected their size and location. Finally, zoning power was limited, both within cities and among cities and suburbs. Airport-area land use planning was thereby rendered more difficult.

Airports and the central business district

Air terminals were planned throughout our period as adjuncts of urban CBDs. The fact that businessmen working from downtown offices were among the principal airport boosters helped foster a conception of air service favoring the CBD. One consequence was that far more attention was paid to linking the airport with the old business district than to conditioning its impact on land use.

Early air travelers were overwhelmingly CBD-bound business-people. United Air Lines reported in 1932 that corporate officers, sales managers, engineers, and other executives made up eighty percent of its passengers. In 1954, sixty percent of 194,100 patrons of New York City's airports were traveling on business. Sixty-seven percent were professionals; a mere two percent held blue-collar jobs. Information for the intervening periods suggests that this pattern was constant. Until the later 1950s, air fares remained out of reach for most Americans. This meant that the "need" to serve businessmen (always thought to be exceedingly busy) conditioned the drive for highways linking airports with city centers.

Indeed, the building of highways between airports and city centers quickly became an important aspect of airport "planning." In 1928 John Nolan contended that "in large cities the airplane calls for improved motorways of the superhighway type." Improved facilities for the automobile and the motorbus, he continued, should be one result of aviation. Studies of the "Airport problem" between 1928 and 1941 backed this view, and the idea that distance should be measured in time, not miles became something of a catch phrase.[9] One engineering marvel – the superhighway – was to reconcile another technological triumph – the airliner – with the traditional city.

During World War II the drives for postwar superhighways and for airport expansion grew together. The National Interregional Highway Committee recommended that new airports be located with the projected national highway network in mind. Planning consultant Julian Whittlesey was more explicit in 1945: "You should have a fast radial road from the [air]port to the center; for other and future ports you need suburban or belt roads to interconnect them and feed them from outlying districts and other towns. Such a road system is essential to any modern city and is long overdue." As historian Mark Rose has argued, land clearance and reassembly as part of a program of expressway construction were often viewed as panaceas for a wide range of urban ills during this period.[10]

Highway interest happily joined the call for city-to-airport roads. Many had participated in the drive for municipal airports since the 1920s, and the American Roadbuilders' Association (ARBA) now came forth with widely cited statistics on the time and "timevalue" lost by air travelers through the absence of expressways. Walter Macatee, the ARBA representative who gathered these statistics, later became highway transportation specialist for the CAA. This collaboration produced a CAA publication advocating "city to airport highways," which [. . .] stressed city and state responsibility for highways that "benefit all highway users," whether airport-bound or not.[11]

City officials moved rapidly to connect existing airports to their CBDs during the 1945–1952 period. Reinforcing this tendency, a 1952 presidential commission warned that "the inadequacy of present road networks [. . .] [is] one of the greatest detriments to the future development of transport aviation."[12] Before 1952, no federal, state, or local

government or agency save one appears to have produced or commissioned a study that dealt seriously with the land use changes brought by highways to airports.

Metropolitan area aviation planning, when it occurred, also served to distract attention from the issue of land use. The New York Regional Plan of 1926, like aviation planning per se, treated aviation in isolation. This path was followed by regional aviation planning for other areas. Much attention was necessarily given to keeping aircraft from neighboring airports apart. More important, through the late 1940s such plans were primarily concerned with an anticipated boom in private aviation: a distracting factor that receives more attention below.

Finally, land use experts contributed to the apathy toward planning for airport areas. Preoccupied with suburban *residential* land values, they periodically announced that airports reduced the value of the land around them. This was sometimes a platitude rather than a fact: The text of a 1930 study by the Harvard School of City Planning reported that airports tended to decrease the value of developed land; however, summarizing the results of a 62-city survey suggested that most early airports were built in underdeveloped areas where surrounding land increased in value more often than otherwise. Nonetheless the assumption that airports lowered land values remained an enduring and misleading myth throughout the 1926–1952 period.

Actually, the relationship between airports and the land surrounding them was more complex. Service facilities – from hotels to hot dog stands – congregated near airports in the early 1930s. A survey of nineteen cities, published in 1942, showed that residential values did decline slightly near airports, but that the value of land used for certain kinds of business tended to rise. Business meant jobs, and so housing – albeit not middle-class suburban development as desired by real estate promoters.

By the early 1950s, studies comparing airport areas with similar districts not near airports found little difference in overall land values. Airports did not stand out on the geographers' maps as potential urban subcenters, nor did local officials treat them as special. Chicago's Municipal Airport had a public school at the end of one of its main runways in 1945, and the Los Angeles Municipal Airport (not then a major airline terminal) leased some ''spare'' land for construction of a hospital. The special role of air terminals in metropolitan development had not been defined, and air travel still seemed an adjunct of the CBD.

Urban leaders, then, could reasonably view airports as remote CBD commuter depots, built for a limited but economically essential elite: people whose plants and offices were and always would be located downtown. Contemporary beliefs about airports and land values, and about the ease of linking one part of the metropolis with another via superhighways, simply reinforced this elite vision of the airport.

Aircraft technology and airport size and location

The relationship between aircraft technology and airport location is complex and paradoxical. It is true, but insufficient, to say that during our period larger planes required longer runways and airports ever further removed from the CBD. As always, technological change was guided and interpreted by contemporaries: It did not follow a straightforward, autonomous path.

Through 1935 an airport with 3,000-foot runways was considered large indeed. New planes and (just as important) CAA requirements that runways be long enough to allow an aircraft to come to a safe stop if an engine failed at take-off, changed this figure rapidly after 1937. The DC-3 needed 1,700 feet of runway to reach an altitude of fifty feet, but 1944 CAA regulations banned it from airports with runways of under

3,500 feet. The DC-4 and Lockheed Constellation, introduced during World War II as military transports, had CAA runway requirements of 5,200 and 5,300 feet, respectively. Planes that needed mile-long runways seemed huge, yet by 1957 the first jet transports called for runways of nearly two miles' length.

During our period, predictions that runway lengths had reached their zenith alternated with warnings of ever larger planes to come. [. . .] Some urban centers (Oklahoma City was one) scrambled to become air transport hubs by planning the largest possible airports and, after 1936, most aviation writers warned that airports must be planned with space for expansion so that ever larger numbers of ever larger airplanes could be lured in.

This metropolitan version of the cargo cult was made possible in most places by federal funding, first in the form of emergency depression projects, then wartime emergency improvements, and finally (1946) direct federal assistance for runway construction on an ongoing basis. After 1946 the Federal Government also ''gave'' cities ''war surplus'' military aviation fields, thus creating another wild card in the airport location game.

Yet cities did not simply accept the runway requirements of the largest possible airliners as defining the size of their airports. Neither did technological innovation work *only* to assure that airports would be larger and farther from the CBD. Airport planners were not, as one Los Angeles Airport engineer complained in 1952, ''at the mercy of the plane designers.''[13]

As early as 1937 officials of smaller cities complained vociferously that planes were becoming unnecessarily large. The American Society of Planning Officials and the American Municipal Association agreed, adding that ''consideration must be given to establishing maximum limits in the construction of airplanes and airports.'' In 1950 Eastern Airlines' chief engineer complained that ''airport size cannot continue to increase indefinitely.''

Aircraft designers were preoccupied with speed "due primarily to the military design philosophy [. . .] in which speed is supreme." Airspeed achieved at the price of longer runways and airports far from the passenger's real destination added at the beginning and end of the journey much of the time that was saved in flight.[14]

Indeed there *was* a limit to the size of viable commercial aircraft in the postwar era. Douglas Aircraft's chief engineer assured airport managers in 1952 that "the manufacturers are certainly going to have to design airplanes so they don't require airports of ten thousand feet and up, because if we do, we'll never sell any." Boeing engineers, at the time, held similar views.[15]

Boeing's Stratocruiser was the only airliner approaching "superliner" status before the jet age. Able to carry up to 100 passengers, this two-decked craft required 5,400 feet of runway for actual takeoff as against 3,720 feet for the period's most popular transport, Douglas' DC-6. A viable engineering design, the Stratocruiser did not fit existing airports. For this and other reasons, only forty-eight were flying in 1958, while 1,315 DC-6s, Lockheed Constellations, and related craft handled most long-range airline traffic. In the early 1950s Boeing engineers were planning jet transports with *smaller* passenger capacities than the propeller-driven Stratocruiser, precisely because of resistance to giganticism in airport planning. "This [jet] design requirement," said Boeing engineer Paul Buck in 1950, "is due to *available* landing field lengths and *not* from the choice of the designers."[16]

Just as the dimensions of existing airports, along with airline perceptions of the air travel market, influenced aircraft design, so engineering and architectural innovation combined after World War II to help accommodate more passengers and larger planes at fields of limited size. The 1945–1952 period saw the introduction of bilevel terminal buildings,

which segregated passengers and visitors from cargo and luggage handling. Houston, Boston, Baltimore, and Philadelphia all had two-deck terminals planned or under construction by 1950, and Chicago and St. Louis soon followed suit. "Nose-in" berthing (wherein planes are "parked" perpendicular to terminal fingers and are towed in and out by tractors) was gaining acceptance by the early 1950s. Predecessors of the "jetway" also made their appearance by 1952. Combined, these innovations meant more planes and passengers could be served through a given amount of terminal space.

One more set of technological innovations helped limit the necessary size of airports. "Tricycle" landing gear (which render a plane "tail high" when standing on the ground as opposed to the "nose high" posture of the DC-3 and earlier craft), combined with improved flight control systems to enable postwar liners to land in crosswinds of up to forty miles per hour. This innovation eliminated the need for criss-cross or tangential runway patterns and left more space on the airfield for terminal, repair, and cargo facilities.

Thus while runways did become much longer between 1935 and 1952, the postwar airport was shaped by a mixture of innovations, some of which worked against the trend toward giganticism. Airlines and some cities also resisted the dictation of airport size by the requirements of the largest possible planes. The "need" for bigger airports *did* send civic leaders scurrying after new sites and new sources of revenue, thus reducing the likelihood of serious planning for airport areas. At the same time, commercial aviation technology during this period clearly reflected the political economy within which its creators worked.

Expertise and the definition of the airport problem

Into the early 1930s, professionals in a variety of fields could successfully assert their expertise in airport planning, location, and design. Architects and planners studied and rendered verdicts on such matters. Companies sprang up that specialized in airport construction – the Austin Airport Company was most prominent. Cement manufacturers and highway interests took an active part, most visibly in the Lehigh Portland Cement Company Airport Design Competition of 1930. Ownership and administration also varied: Park boards sometimes ran airports (since planes at first seemed primarily a form of recreation). Private developers built airports as commercial ventures, as did aviation manufacturers (most notably Curtiss-Wright) and airlines themselves.

By 1930 civil engineers engaged in airport design had mounted a campaign to establish their field as a distinct profession. Up to that point, engineers whose chief experience lay in the design of athletic fields felt free to comment with authority on airport issues. Planners, architects, and state aeronautical agencies also gave the "airport engineers" lively competition. Most important was the federal government, whose agencies (such as the Department of Commerce's Aeronautics Branch) made the growth of commercial aviation the primary aim of their increasingly influential pronouncements on airport design.

In 1930 the Aeronautics Branch was offering cities free advice on airport planning; advice that, airport engineers complained, disposed cities to leave airport design to city engineers instead of consulting "real" experts. In 1932, the Aeronautics Branch issued a document on "airport design and construction," which dealt with the location of airports as well as with the engineering aspects of their design. The federal government had entered the sphere of the planner as well as that of the engineer.

Federal aid for airports that formed a part of New Deal relief efforts could sometimes shape city airport policy. More important in the long run, however, was the role of the Civil Aeronautics Administration, which came into existence in 1938. The CAA acted in part as a guarantor of airline safety – Boeing, for example, took delight in the credibility offered by CAA certification tests. Testimony of the Air Transport Association of America's representative at the hearings that led to the CAA's creation suggests that established airlines also hoped the CAA would regulate airports and their use in ways that would structure the industry.

The role of the CAA in commercial aviation history is complex, but clearly it did set standards for airport construction, enforceable by means of its power to determine what kind of planes could use what sorts of airports. Consulting firms continued to advise cities on airport location and design. Airport engineering *was* a profession, one whose members might work in consulting firms or even for airlines. But the CAA was the final authority. It vitiated existing state aeronautical commissions and, through 1951, legitimized the treatment of airports and highways leading to them as transportation facilities only. In brief, the CAA critically narrowed the meaning of expertise in airport planning.

Other factors affecting airport size and location

Private flying also helped assure that passenger airports would become ever larger and thus be located at the urban fringe. Private aviation played a dual role in this giganticization of urban airports. First, the expectation that personal aircraft would play a Flivver-like role[17] caused government agencies and planners to divert much energy and attention to the solution of problems that never emerged. Rather typically, two urban researchers reported in 1942 that "a new urban dislocation as disruptive as that which accompanied the coming of the automobile," would follow in the wake of postwar private aviation.[18] Equally important, private aviation made demands on municipal airports that conflicted with those of commercial carriers, thus contributing to increasing airport size.

Neither the type nor the timing of airport use was effectively planned during our period. The need to remove business and recreational craft from municipal fields was accepted in theory by 1945. Yet private planes tied up runway time at commercial airports long after 1952. A presidential commission typified the federal view in 1948, calling for equal access to commercial airports for planes of all categories. When air travel became a means of mass transportation, airliners still had to compete for public landing space, just as buses and streetcars competed with private autos for street space. For aviation, bigger airports seemed the only answer.

Airlines contributed their share to the problem by scheduling as many flights as possible at peak hours. Airlines thus used more gate, runway, and approach space than was necessary to handle the number of people moved. Excess facilities made necessary by ''peaking'' denoted another resemblance between air travel and mass transit of the more traditional sort.

Zoning and planning power and airport-area development

Most airports were built in open country. In 1929 one zoning expert warned that such sites might some day be hemmed in but that zoning was not the answer, since courts would demand that it be based on ''the health, safety, morals and general welfare of the community.'' This was only the most obvious of the problems that emerged over the next generation as cities and suburbs tried to control airport-area land use.[19]

The relationship between aviation and existing common law produced many ironies. State courts differed on the rights of aviators: Noise that disturbed a farmer's cattle caused the closing of a Pennsylvania airport in 1944, but in another state it was evidently not illegal for flyers to frighten the residents of a fox farm

into aborting their young. Airport neighbors who erected poles or grew tall trees to harass low-flying pilots were ordered to remove the obstructions, but several airports, built in undeveloped areas, were ordered closed as public nuisances when housing was built near their borders. Despite protracted efforts to mesh common law with the needs of commercial aviation, zoning for air safety did not provide a fully reliable tool. City officials complicated the matter in 1944 by working through the US Conference of mayors to defeat federal legislation that would have allowed the CAA to impose zoning rules around airports.

During a period of six weeks in 1952, events around Newark, New Jersey, finally brought the safety aspect of the airport–community relationship to the fore. Three planes departing from or landing at Newark Airport crashed within one square mile of residential Elizabeth, New Jersey. The final crash, in which a plane rammed a building, killed six persons inside the residence as well as the craft's passengers and crew.

Newark Airport was closed after the third disaster. Aviation industry spokesmen rushed to reassure the public and formed an effective public relations committee. Airlines began a drive for airport improvements perhaps, as *Aviation Week* observed, because they hoped to avert federal action restricting their own operations. Some land use planners used the occasion to argue that airports should be placed at least ten miles from populated areas. Most important, the disasters led to the new kind of federal response to the airport ''problem'': one that recognized that people *would* settle near airports, wherever they were built.

Following the 1952 crashes, a presidential commission, headed by World War II hero Jimmy Doolittle, was appointed to look into airport safety. After interviewing city officials and aviation experts, it issued a report, *The Airport and Its Neighbors*, which called for positive planning. While it reiterated much of the conventional wisdom of the 1940s, the commission did take a new step in recommending large, fan-shaped cleared zones at the ends of runways. It also presented six pages of drawings illustrating suggested uses of these zones and insisted that "airports must be planned as a part of the total system of metropolitan transportation and land use."[20]

In July, 1952, the CAA issued a new edition of its manual on airport planning. Whereas earlier editions sometimes pictured airports surrounded by virtually empty land, the CAA now called for special planning boards to design commercial and industrial centers for airport environs. The examples it presented were idealized but the proclamation that "airport planning should be concerned with much more than the . . . development of the landing area proper" marked a genuine shift of focus.[21]

This recognition that airports had contexts did not mark a new era. Real recognition of the city-shaping power of airports awaited the jet age, and the means of dealing with that power have yet to be discovered. The fact that many airports do, de facto become urban subcenters raises one final question: Was the idea of major airports adjoining traditional CBDs really as absurd as it has long appeared to be?

Downtown airports?

In retrospect, early proposals for downtown airports were a form of technological utopianism. Rooftop landing strips were among the first such proposals: Hotels in New York City and San Francisco proposed such accommodations for wealthy guests in 1920 and 1921. Far more grandiose suggestions along these lines proliferated through the 1920s and into the early 1930s, but unpredictable winds around tall buildings and the increasing weight of planes were insuperable problems.

The desire for downtown airports persisted. In 1938, a meeting of planning and city officials asserted that airports should be "as close to the

center of commercial activity as land area, land costs and safety will permit." Serious plans for airports atop railyards, public buildings, piers, and the like continued at least through 1946.[22]

Another means of obtaining close-in airports was the "two-birds-with-one-stone" approach of urban renewal. As early as 1930 the Harvard School of City Planning suggested that "in some cases a blighted district might be so low in value and so much of a public menace . . . that public money might legitimately be used to acquire it as part of a municipal airport." By 1945, CAA officials were urging cities to buy up "slum" land for commercial or general aviation airfields.[23]

One reason for the failure of such schemes was reflected in a study of Chicago airport problems conducted in 1946. "The possibility of transferring the occupants of the most heavily blighted area . . . to [outlying] areas with an average density of only 4,800 is admittedly attractive," the General Airport Company reported to the city fathers, "but the means to implement such a move are not now known."[24] Even had the legal and financial means for such a massive relocation of population been available, the political will required to move tens of thousands of inner-city blacks to the city's largely middle-class white rim would almost certainly have been lacking.

Waterfront or island airports were the remaining CBD options. Here no technological obstacles or threats to the existing social order supervened. Instead, economic, ecological, and political considerations ruled out close-in waterfront or island airports for most cities.

In Washington, Oakland, Boston, and New York, close-in waterfront airports became an enduring fact of life, despite persistent problems. For other waterfront cities, airports on natural or man-made islands seemed the most plausible means of bringing planes (and so, business) downtown. Officials in San Francisco, Cleveland, and Chicago all gave serious consideration to such plans before 1945, and one competent author suggested that foggy Seattle could accommodate a major island airport. Chicago's drive for an island airport was the longest lasting, and it reveals much about the reasons for the failure of such plans.

Basing its request on plans dating from the 1920s, Chicago in 1934 submitted to the Public Works Administration (PWA) an $8,600,000 design for a triangular 200-acre airport to be built 1½ miles south of the city's center and 1,600 feet offshore in Lake Michigan. Airline officials and US Department of Commerce officials gave the plan full backing, and concerns over the effects of lake fogs and shifting winds do not appear to have determined the PWA's decision.

Crucial to the plan's failure was Harold Ickes's animosity toward Chicago's political machine. Ickes used fear on the part of some members of the city's civic elite that an airport would commercialize the lakefront to help drum up resistance. Important too was the badly mistaken belief of spokesmen for some west side landowners that a superhighway (an inevitable by-product of an outlying airport) would raise land values in the sections of the city through which it passed. Despite backing from the Chicago Plan Commission and the city's chamber of commerce, the PWA turned down the island airport plan, officially because materials (as opposed to labor) constituted too high a proportion of the proposed airport's cost.

This was not the end of the island plan. Mayor Edward Kelly revived it in 1937, complete with recreational facilities. In 1941, a report sponsored by local and regional planning bodies and the Chicago Chamber of Commerce considered an island airport on landfill protected by dikes. A politically connected engineering firm presented an elaboration of this plan; a triangular airport with 8,000-foot runways to be built one mile offshore just north of the CBD.

In 1946 Chicago abandoned this plan in favor of the conversion of a "free" military surplus field (O'Hare). The General Airport Company's study ratified this decision (which was made before the company's report was filed). General Airport cited fog, wind, and ice hazards, the difficulty of expanding the proposed island, and its effect on CBD traffic. Above all, however, it stressed the plan's $30,000,000 price tag.

The island airport idea surfaced again in 1955 and in 1967–1969 plans for a full-scale jetport in the lake were actively pushed by Chicago's mayor. Although this idea was abandoned in 1973 (partly due to environmentalist pressure), New York City, Los Angeles, and Cleveland had by then undertaken studies for multibillion dollar offshore airports. These plans, ironically, were in part a *response* to environmentalist concerns over the noise and air pollution effects of onshore airports. As environmentalism receded and the nation's economy declined, island jetports again seemed nothing but an expensive dream.

Conclusion

Airport location, and the lack of effective planning for land use around airports, can be understood in part as functions of a policy tradition, developed side by side with the highway planning tradition, which treated airports as transportation facilities only. Aviation technology set the parameters within which this tradition worked, but its role must be understood in the context of public, business, and government attitudes toward transportation and land use planning.

During our period, only elite environmentalism, the demand for the continued isolation of the urban poor, or simple politics could challenge the assumption that airports were to be judged solely according to their role in intercity transportation. This made planning for airport-centered development a remote side issue, and inner-city airports an absurd proposition.

At this writing suburbs remain class segregated, and public transportation to suburban job sites is usually poor and expensive. Inner-city unemployment remains high, while occasional complaints are heard that low-skill suburban jobs cannot be filled. Airport location and land use planning are but a small part of this problem. They are, of course, a major factor lowering the quality of life in some suburbs. It would appear that America's "first chance . . . to apply the modern principle of prevention to a new art and science in a big way" was muffed. Distraction by the transportation "necessities" of the time, together with a tradition of treating transportation issues in isolation, help to account for yet another incremental disaster in American urban life.

Notes

Author's Note: *I wish to offer special thanks to Mark H. Rose, for his excellent editorial comments, and to Mark S. Foster, for the inspiration upon which this study is based*

1 Richard L. Meier, *Planning for an Urban World: The Design of Resource Conserving Cities* (Cambridge, MA, 1974), 289-290, 299

2 George B. Ford, "The Airport and the City Plan," *American City* 40 (March, 1929), 129

3 [US] Civil Aeronautics Administration, *Airport Survey* (Washington, DC, 1939), XV, XVIII; *American City* 43 (August, 1930), 164; US Department of Commerce, Bureau of Air Commerce, *Proceedings of the National Airport Conference, December 6, 7, 1937* (Washington, DC, 1938), 11

4 US Department of Commerce, Bureau of Air Commerce, *Proceedings of the National Airport Conference, December, 6, 7, 1937* (Washington, DC, 1938), 11; CAA, *Airport Survey*, XVIII

5 US Department of Commerce, Civil Aeronautics Administration, *Legislative History of the Federal Airport Act* (Washington, DC, 1947), 109; *Air Transportation* 6 (March, 1945), 35; United States Senate, [82nd Congress, 2nd Session], *Hearings Before the Preparedness Sub-Committee of the Committee on Armed Services . . .*

on Portland International Airport, July 1, 1952 (Washington, DC, 1952), 10

6 *City Planning* 6 (January, 1930), 43; *Airports* 3 (January, 1930), 32-33; *American City* 43 (August, 1930), 164; 51 (March, 1936), 11; Chicago Association of Commerce, *Chicago Facts* (Chicago, 1927), 18; *Chicago Commerce* 25 (April 18, 1927), 9-10

7 On changing aviation technology see Richard K. Smith. "The Intercontinental Airliner and the Essence of Airplane Performance, 1929-1939," *Technology and Culture* 24 (1983), 428-449

8 *American City* 58 (December, 1943), 75; 59 (August, 1944), 51

9 John Nolan, *Airports and Airways and Their Relation to City and Regional Planning* (New York, 1928), 12-13

10 National Interregional Highway Committee, *Inter-Regional Highways, Outlining and Recommending a National System of Inter-Regional Highways* (mimeograph, 1944), 43-44; Julian Whittlesey, "City Problems in the Air Age," *American City* 60 (May, 1945), 84; Mark H. Rose, *Interstate: Express Highway Politics, 1941-1956* (Lawrence, KS, 1979), x, 20-21, 27-28, 60-61

11 *American City* 41 (September, 1929), 5; Lehigh Portland Cement Company, *American Airport Designs, Containing 44 Prize Winning and Other Drawings from the Lehigh Airports Competition* (New York and Chicago, 1930). On the ARBA and Macatee see *Aviation Week* 50 (March 7, 1949), 19; 21; *Journal of the American Institute of Planners* 19 (Summer, 1953), 127; US Department of Commerce, Civil Aeronautics Administration, *City to Airport Highways* (Washington, DC, 1953), 9

12 General Airport Company, *Comprehensive Study Relating to Aeronautical Facilities for Metropolitan Area of Chicago Projected to 1970* (Stanford, CT, 1946), 72-73, 143; US President's Airport Commission, *The Airport and Its Neighbors* (Washington, DC, 1952), 17; *Architectural Record* 122 (July, 1952), 38

13 *Public Works* 83 (May, 1952), 60

14 American Society of Planning Officials, *The Airport Dilemma* (Chicago, 1938), 132; US Department of Commerce, *National Airport Conference*, 14, 88-94; Charles Froesch, "Air Cargo and Passenger Trends and Their Influence on Airport Development," New York State Airport Conference, *Minutes* (Syracuse, 1950), 108, 110

15 Airport Operators' Council, *Papers Presented and a Selective Discussion Report Covering the Fifth Annual Meeting* (Washington, DC, 1952), 18; Airport Operators' Council. *Papers Presented Sixth Annual Meeting* (Washington, DC, 1953), 77

16 Airport Operators' Council *Papers Presented Third Annual Meeting* (Washington, DC, 1950), 49, 51

17 [Flivver is an affectionate nickname for the Model-T.]

18 Katherine Perring and Melville C. Branch, Jr., "Urban

Land Use and the New Mobility," *American City* 57 (May, 1942), 45

19 Edward M. Bassett, "Zoning Roundtable," *City Planning* 5 (July, 1929), 194

20 United States President's Airport Commission, *The Airport*, 7-9, 81-89

21 US Department of Commerce, Civil Aeronautics Administration, *Airport Planning* (Washington, DC, 1952), 39-54

22 *Airport Dilemma*, 42; *Air Transportation* 7 (January, 1946), 26-30

23 Henry V. Hubbard, Miller McClintock, and Frank B. Williams, *Airports: Their Locations, Administration, and Legal Basis* (Cambridge, MA, 1930), 21-22; *American City* 60 (May, 1945), 88; (April, 1945), 119

24 General Airport Company, *Comprehensive Study Chicago*, 145

28

THE GROWTH OF THE CITY

by Ernest W. Burgess

Source: Ernest W. Burgess, "The Growth of the City: An Introduction to a Research Project", in Robert E. Park, Ernest W. Burgess and Roderick D. McKenzie (eds), *The City*, with an introduction by Morris Janowitz, Chicago, IL, The University of Chicago Press, 1967 [1925], pp. 47–62

The outstanding fact of modern society is the growth of great cities. Nowhere else have the enormous changes which the machine industry has made in our social life registered themselves with such obviousness as in the cities. In the United States the transition from a rural to an urban civilization, though beginning later than in Europe, has taken place, if not more rapidly and completely, at any rate more logically in its most characteristic forms.

All the manifestations of modern life which are peculiarly urban – the skyscraper, the subway, the department store, the daily newspaper, and social work – are characteristically American. The more subtle changes in our social life, which in their cruder manifestations are termed "social problems," problems that alarm and bewilder us, as divorce, delinquency, and social unrest, are to be found in their most acute forms in our largest American cities. The profound and "subversive" forces which have wrought these changes are measured in the physical growth and expansion of cities. That is the significance of the comparative statistics of Weber, Bücher, and other students.

These statistical studies, although dealing mainly with the effects of urban growth, brought out into clear relief certain distinctive characteristics of urban as compared with rural populations. The larger proportion of women to men in the cities than in the open country,

the greater percentage of youth and middle-aged, the higher ratio of the foreign-born, the increased heterogeneity of occupation increase with the growth of the city and profoundly alter its social structure. These variations in the composition of population are indicative of all the changes going on in the social organization of the community. In fact, these changes are a part of the growth of the city and suggest the nature of the processes of growth.

The only aspect of growth adequately described by Bücher and Weber was the rather obvious process of the *aggregation* of urban population. Almost as overt a process, that of *expansion*, has been investigated from a different and very practical point of view by groups interested in city planning, zoning, and regional surveys. Even more significant than the increasing density of urban population is its correlative tendency to overflow, and so to extend over wider areas, and to incorporate these areas into a larger communal life. This paper, therefore, will treat first of the expansion of the city, and then of the less-known processes of urban metabolism and mobility which are closely related to expansion.

Expansion as physical growth

The expansion of the city from the standpoint of the city plan, zoning, and regional surveys is thought of almost wholly in terms of its physical growth. Traction studies have dealt with the development of transportation in its relation to the distribution of population throughout the city. The surveys made by the Bell Telephone Company and other public utilities have attempted to forecast the direction and the rate of growth of the city in order to anticipate the future demands for the extension of their services. In the city plan the location of parks and boulevards, the widening of traffic streets, the provision for a civic center, are all in the interest of the future control of the physical development of the city.

This expansion in area of our largest cities is now being brought forcibly to our attention by the Plan for the Study of New York and Its Environs, and by the formation of the Chicago Regional Planning Association, which extends the metropolitan district of the city to a radius of 50 miles, embracing 4,000 square miles of territory. Both are attempting to measure expansion in order to deal with the changes that accompany city growth. In England, where more than one-half of the inhabitants live in cities having a population of 100,000 and over, the lively appreciation of the bearing of urban expansion on social organization is thus expressed by C.B. Fawcett:

> One of the most important and striking developments in the growth of the urban populations of the more advanced peoples of the world during the last few decades has been the appearance of a number of vast urban aggregates, or conurbations, far larger and more numerous than the great cities of any preceding age. These have usually been formed by the simultaneous expansion of a number of neighboring towns, which have grown out toward each other until they have reached a practical coalescence in one continuous urban area. Each such conurbation still has within it many nuclei of denser town growth, most of which represent the central areas of the various towns from which it has grown, and these nuclear patches are connected by the less densely urbanized areas which began as suburbs of these towns. The latter are still usually rather less continuously occupied by buildings, and often have many open spaces.
>
> These great aggregates of town dwellers are a new feature in the distribution of man over the earth. At the present day there are from thirty to forty of them, each containing more than a million people, whereas only a hundred years ago there were, outside the great centers of population on the waterways of China, not more than two or three.[1] [. . .]

In Europe and America the tendency of the great city to expand has been recognized in the term "the metropolitan area of the city," which far overruns its political limits, and in the case of New York and Chicago, even state lines. The metropolitan area may be taken to include urban territory that is physically contiguous, but it is coming to be defined by that facility of transportation that enables a business man to live in a suburb of Chicago and to work in the loop, and his wife to shop at Marshall Field's and attend grand opera in the Auditorium.

Expansion as a process

No study of expansion as a process has yet been made, although the materials for such a study and intimations of different aspects of the process are contained in city planning, zoning, and regional surveys. The typical processes of the expansion of the city can best be illustrated, perhaps, by a series of concentric circles, which may be numbered to designate both the successive zones of urban extension and the types of areas differentiated in the process of expansion.

The chart [in Figure 28.1] represents an ideal construction of the tendencies of any town or city to expand radially from its central business district – on the map "The Loop" (I). Encircling the downtown area there is normally an area in transition, which is being invaded by business and light manufacture (II). A third area (III) is inhabited by the workers in industries who have escaped from the area of deterioration

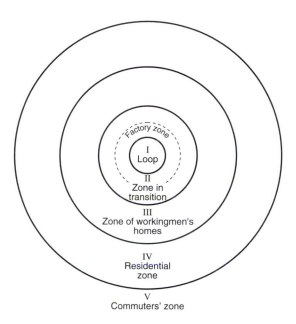

Figure 28.1 The growth of the city.

(II) but who desire to live within easy access of their work. Beyond this zone is the "residential area" (IV) of high-class apartment buildings or of exclusive "restricted" districts of single family dwellings. Still farther, out beyond the city limits, is the commuters' zone – suburban areas, or satellite cities – within a thirty- to sixty-minute ride of the central business district.

This chart brings out clearly the main fact of expansion, namely, the tendency of each inner zone to extend its area by the invasion of the next outer zone. This aspect of expansion may be called succession, a process which has been studied in detail in plant ecology. If this chart is applied to Chicago, all four of these zones were in its early history included in the circumference of the inner zone, the present business district. The present boundaries of the area of deterioration were not many years ago those of the zone now inhabited by independent wage-earners, and within the memories of thousands of Chicagoans contained the residences of the

"best families." It hardly needs to be added that neither Chicago nor any other city fits perfectly into this ideal scheme. Complications are introduced by the lake front, the Chicago River, railroad lines, historical factors in the location of industry, the relative degree of the resistance of communities to invasion, etc.

Besides extension and succession, the general process of expansion in urban growth involves the antagonistic and yet complementary processes of concentration and decentralization. In all cities there is the natural tendency for local and outside transportation to converge in the central business district. In the downtown section of every large city we expect to find the department stores, the skyscraper office buildings, the railroad stations, the great hotels, the theaters, the art museum, and the city hall. Quite naturally, almost inevitably, the economic, cultural, and political life centers here. The relation of centralization to the other processes of city life may be roughly gauged by the fact that over half a million people daily enter and leave Chicago's "loop." More recently sub-business centers have grown up in outlying zones. These "satellite loops" do not, it seems, represent the "hoped for" revival of the neighborhood, but rather a telescoping of several local communities into a larger economic unity. The Chicago of yesterday, an agglomeration of country towns and immigrant colonies, is undergoing a process of reorganization into a centralized decentralized system of local communities coalescing into sub-business areas visibly or invisibly dominated by the central business district. The actual processes of what may be called centralized decentralization are now being studied in the development of the chain store, which is only one illustration of the change in the basis of the urban organization.

Expansion, as we have seen, deals with the physical growth of the city, and with the extension of the technical services that have made city life not only livable, but comfortable, even luxurious. Certain of these basic necessities of urban life are possible only through a tremendous development of communal existence. Three millions of people in Chicago are dependent upon one unified water system, one giant gas company, and one huge electric light plant. Yet, like most of the other aspects of our communal urban life, this economic co-operation is an example of co-operation without a shred of what the "spirit of co-operation" is commonly thought to signify. The great public utilities are a part of the mechanization of life in great cities, and have little or no other meaning for social organization. [. . .]

Mobility as the pulse of the community

Movement, per se, is not an evidence of change or of growth. In fact, movement may be a fixed and unchanging order of motion, designed to control a constant situation, as in routine movement. Movement that is significant for growth implies a change of movement in response to a new stimulus or situation. Change of movement of this type is called *mobility*. Movement of the nature of routine finds its typical expression in work. Change of movement, or mobility, is characteristically expressed in adventure. The great city, with its "bright lights," its emporiums of novelties and bargains, its palaces of amusement, its underworld of vice and crime, its risks of life and property from accident, robbery, and homicide, has become the region of the most intense degree of adventure and danger, excitement and thrill.

Mobility, it is evident, involves change, new experience, stimulation. [. . .]

The elements entering into mobility may be classified under two main heads: (1) the state of mutability of the person, and (2) the number and kind of contacts or stimulations in his environment. The mutability of city popula-

tions varies with sex and age composition, the degree of detachment of the person from the family and from other groups. All these factors may be expressed numerically. The new stimulations to which a population responds can be measured in terms of change of movement or of increasing contacts. Statistics on the movement of urban population may only measure routine, but an increase at a higher ratio than the increase of population measures mobility. In 1860 the horse-car lines of New York City carried about 50,000,000 passengers; in 1890 the trolley-cars (and a few surviving horse-cars) transported about 500,000,000; in 1921, the elevated, subway, surface, and electric and steam suburban lines carried a total of more than 2,500,000,000 passengers.[2] In Chicago the total annual rides per capita on the surface and elevated lines were 164 in 1890; 215 in 1900; 320 in 1910; and 338 in 1921. In addition, the rides per capita on steam and electric suburban lines almost doubled between 1916 (23) and 1921 (41), and the increasing use of the automobile must not be overlooked.[3] For example, the number of automobiles in Illinois increased from 131,140 in 1915 to 833,920 in 1923.[4]

Mobility may be measured not only by these changes of movement, but also by increase of contacts. While the increase of population of Chicago in 1912–22 was less than 25 per cent (23.6 per cent), the increase of letters delivered to Chicagoans was double that (49.6 per cent) – (from 693,084,196 to 1,038,007,854).[5] In 1912 New York had 8.8 telephones; in 1922, 16.9 per 100 inhabitants. Boston had, in 1912, 10.1 telephones; ten years later, 19.5 telephones per 100 inhabitants. In the same decade the figures for Chicago increased from 12.3 to 21.6 per 100 population.[6] But increase of the use of the telephone is probably more significant than increase in the number of telephones. The number of telephone calls in Chicago increased from 606,131,928 in 1914 to 944,010,586 in 1922,[7] an increase of 55.7 per cent, while the population increased only 13.4 per cent.

Land values, since they reflect movement, afford one of the most sensitive indexes of mobility. The highest land values in Chicago are at the point of greatest mobility in the city, at the corner of State and Madison streets, in the Loop. A traffic count showed that at the rush period 31,000 people an hour, or 210,000 men and women in sixteen and one-half hours, passed the southwest corner. For over ten years land values in the Loop have been stationary, but in the same time they have doubled, quadrupled, and even sextupled in the strategic corners of the "satellite loops," an accurate index of the changes which have occurred. Our investigations so far seem to indicate that variations in land values, especially where correlated with differences in rents, offer perhaps the best single measure of mobility, and so of all the changes taking place in the expansion and growth of the city. [. . .]

Notes

1 "British Conurbations in 1921," *Sociological Review*, XIV (April, 1922), 111–12

2 Adapted from W.B. Monro, *Municipal Government and Administration*, II, 377

3 Report of the Chicago Subway and Traction Commission, p. 81, and the *Report on a Physical Plan for a Unified Transportation System*, p. 391

4 Data compiled by automobile industries

5 Statistics of mailing division, Chicago Post-office

6 Determined from *Census Estimates for Intercensual Years*

7 From statistics furnished by Mr. R. Johnson, traffic supervisor, Illinois Bell Telephone Company

29

BOSTON'S HIGHWAY 128: HIGH-TECHNOLOGY REINDUSTRIALIZATION

by Manuel Castells and Peter Hall

Source: Manuel Castells and Peter Hall, *Technopoles of the World: the making of twenty-first century industrial complexes*, London, Routledge, 1994, pp. 29–38

If there is a success story of an old industrial region regenerating its economy from the ashes of its industrial past to become a leading high-technology complex, it must surely be Massachusetts – at least, until the late 1980s. In fact, Greater Boston has gone through at least two waves of reindustrialization in the last 40 years, after losing its traditional industries, mainly in textiles and apparel, during the 1930s and 1940s. The first round took place in the 1950s and early 1960s, and was directly linked to military and space programs that concentrated research and manufacturing in the state, mainly in precision instruments, avionics, missiles, and electrical machinery. But the post-Vietnam War reduction in military spending sent the area into another recession, producing an 11 percent unemployment rate by 1975.

Then, in the 1975–85 decade, there occurred one of the most remarkable processes of reindustrialization in American economic history: while in the 1968–75 period Greater Boston lost 252,000 manufacturing jobs, during the years 1975–80 it gained 225,000 new manufacturing jobs, most of them in high-technology industries. Most of the new firms located along Highway 128, the original suburban beltway of Boston, completed in 1951, which links 20 towns, most of them hubs of manufacturing and service activities from the old industrial era. As growth continued unabated for most of the 1980s, new industrial location took place towards the west and northwest of the state, around Highway 495, Boston's new outer beltway, joining the pioneer location of Wang Laboratories at Lowell, and spilling over the state line into New Hampshire (Figure 29.1). In 1980 there were in the Boston area about 900 high-technology manufacturing establishments, employing over 250,000 workers, to which should be added another 700 firms in consulting and technological services, making Boston the third-largest high-technology complex in the United States, after the Southern California multinuclear technopole, and Silicon Valley.

The renaissance of the town of Lowell, the oldest textile industrial city in New England, well illustrates this remarkable turnaround of the Massachusetts economy. A rural area in 1826, Lowell became a textile manufacturing powerhouse from 1850 onwards, reaching a population of 112,000 in 1920, with 40 percent of its workers employed in the textile industry. But quite suddenly, in the mid-1920s, New England's textile industry shifted to the American South or went bankrupt. Between 1924 and 1932, Lowell lost 50 percent

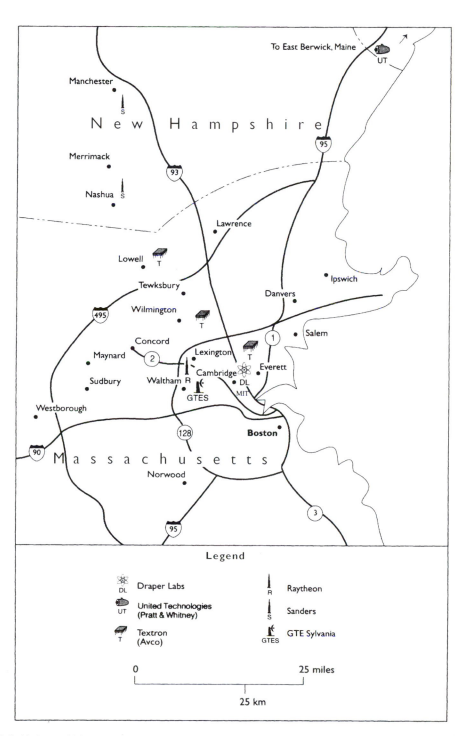

Figure 29.1 Highway 128: general location.
Source: Markusen *et al.* (1991)

of its manufacturing jobs; it then stagnated until the 1960s.

Then, in the 1960s, a new start-up company – originally founded in Boston in 1951 by a Chinese Harvard computer sciences graduate, An Wang – established its headquarters in Lowell to take advantage of empty factory buildings, local incentives, and an experienced industrial labor force ready to settle for relatively low wages: Lowell became the center of what was for some time one of the most successful computer companies in the world, Wang Laboratories. Other companies followed the example, and during the 1970s Lowell's computer manufacturing employment skyrocketed. Electrical and electronic equipment grew at 9 percent per year in the period 1970–6, and the computer industry simply exploded in the late 1970s, growing at an annual 43 percent rate from 1976 to 1982. The computer industry was responsible for 70 percent of the growth in high-technology manufacturing, and for 45 percent of total new jobs growth during the years 1976–82. In 1980, the Lowell area had become again an industrial city, with a population of 227,000 people, and with manufacturing jobs accounting for 39 percent of total employment, in comparison with 21 percent for the entire United States; Lowell had successfully reindustrialized.

The core of Greater Boston's new industrial base is the computer industry. And the growth of this industry in Boston was linked, in the 1970s, to the growth of one particular segment of the industry: mini-computers. Some of the largest and most innovative companies in the world were born in the Boston area in the 1950s, 1960s and 1970s, and remained there: Digital Equipment Corporation (the second producer of mini-computers in the world, immediately after IBM), Data General, Wang Laboratories, Prime Computer, Computervision.

Most of these firms were new, start-up companies, created by engineers and scientists who were graduates or faculty members of MIT (less often, Harvard) or from other electronics firms, in a process of spin-off that parallels the Silicon Valley story. Indeed, a study showed that during the 1960s 175 new Massachusetts firms were created by former employees of MIT's research laboratories. The same study showed that another 39 firms spun off from an older major electronics firm, Raytheon, itself an early spin-off from MIT in the 1930s. Thus, the presence of outstanding research universities, and especially of MIT – to be precise, in Cambridge, on the banks of the Charles River opposite Boston – was certainly a decisive factor in the ability of Massachusetts to reindustrialize.

Furthermore, location surveys conducted on the area's high-technology companies show that the critical factor for maintaining their location here is direct access to one of the largest concentrations of academic, scientific, and engineering talent in the world. Indeed, there are 65 universities and colleges in the Greater Boston area, some of them of top international quality, such as MIT, Harvard, Amherst, Tufts, and Brandeis, and others of respectable quality, such as Boston University, the University of Massachusetts, and Northeastern University, in addition to several elite undergraduate liberal colleges. The whole of New England contained over 800,000 undergraduate and graduate college students by the late 1980s. This factor is critical for high-technology industries, since college graduates represent about 33 percent of all employees in the computer industry. Engineers and technical workers accounted in 1980 for 9.1 percent of employed workers in Massachusetts, in contrast to 7.3 percent in New York, 8.0 percent in Illinois, or 8.9 percent in California.

Thus, apparently, Massachusetts' successful reindustrialization resulted from a unique combination of an entrepreneurial–industrial tradition and the excellence of a university system that provided the necessary new raw material, scientific knowledge and technological skills, which could generate a structural shift from smokestack industries into a high-technology and advanced services economy, in a process largely independent of government policy. Yet, as usual, the story is more complex; it demands careful examination of the historical origins of the Massachusetts scientific–industrial complex.

MIT, the war machine, and the new entrepreneurs

Clearly, at the heart of Greater Boston's high-technology development lies a strong foundation of scientific and technological excellence. Other factors – good communications and tele-communications, cultural amenities, historic heritage, beautiful landscape, and the like – are certainly not unique to Boston. The presence of venture capital firms, accounting for the fourth-largest concentration of such firms in America, after New York, San Francisco, and Chicago, seems to be a consequence of high technology development rather than a causal factor. The formation and subsequent growth of this scientific–technological complex certainly owe a great deal to its strong tradition in higher education and academic research. But the shaping of this academic basis to the service of leading-edge industrial technologies, and the considerable financial and human resources that were concentrated in such an effort, are mainly linked to one institution, MIT, and to a particular process: the technological shift in warfare, first during World War Two, later during the Cold War. We need to trace back the story from its origins.[1]

In the 1930s, MIT had two features that, together, made it different from all other elite academic institutions on the East Coast: it had the oldest and most distinguished electrical engineering department in America, founded as early as 1882; and, because it had much less money than Ivy League universities such as Harvard or Princeton, it was much more

open to conduct contract research with the Government or with private firms. Indeed, such contractual cooperation was (and still is) an established policy at MIT. It also spear-headed the movement towards industrial spin-offs: in 1920, an associate professor in electrical engineering at MIT, Vannevar Bush, created a company, Raytheon, to produce thermostatic controls and vacuum tubes.

Both Bush and his company would prove decisive for the future of high-technology in Massachusetts, with the help of World War Two and of the Defense Department. In 1940, Bush became director of the Federal National Research and Defense Committee, and in this position was fully conscious of the decisive military potential of radar, just invented in Britain. To prevent the capture of such critical knowledge by the Germans in the dark hours of Britain's heroic resistance to the Nazis, Bush convinced both the American and British Governments to bring the British radar research team to the United States to continue secret work on the device, together with Professor Bowles of MIT. After laboratory space was arranged at MIT at a few hours' notice, the team went on to create MIT's Radiation Laboratory, so named to mislead the Germans about the nature of the research being conducted there. Radar and its applications came out of the Radiation Laboratory, later transformed into the Electronics Research Laboratory, one of the most distinguished institutions in the field.

Bowles also worked with Raytheon, still a modest company, and he obtained for his company the contract to manufacture the necessary equipment, sending Raytheon into explosive growth: by 1945 Raytheon, located in Waltham, Massachusetts, had increased its labor force 40 times, to become a large company, producing 80 percent of the world's magnetrons. In the 1950s, Raytheon became a major industrial force in the field of rockets and missiles.

Other advanced laboratories were founded at MIT in the 1930s and 1940s, on the basis of major military research contracts, among them: the Instrumentation Laboratory, today an independent organization; Draper Labs; and the Lincoln Laboratory, established after the war to contract research for the Air Force on radar and computer technology. Harvard also played a role, albeit more limited, in the establishment of military-oriented, electronics research programs, particularly on the basis of the Harvard Countermeasures Laboratory, established in parallel to the Radiation Laboratory.

But perhaps the most direct connection between the MIT-induced wartime programs, and today's high-technology industry in Boston, was the formation and development of a computer science capability at MIT, following the work of Jay W. Forrester, who arrived at MIT from his native Nebraska in 1939, and worked during the war at MIT's Servomechanisms Laboratory. To solve the complex simulation problems he had to resolve to help the demands of aviation, he decided that he needed something that then did not exist: a computer. He visited John von Neumann at the University of Pennsylvania, who was building the machine that would become the first computer, the ENIAC, which became operational in February 1946 – a remote application of the ideas published in England in 1937 by the mysterious genius at the origin of this scientific adventure, Alan Turing.

Forrester was disappointed with Neumann's machine: it was too slow and too unreliable. He started his own project, the Whirlwind Computer Project, and sold it to the Pentagon as the core of a system to build an "electronic radar fence," the "Star Wars" of the 1950s: the Semi-Automatic Ground Environment, or SAGE project. Forrester selected for the work a company that – unlike Raytheon, Sylvania, and Remington Rand, the earliest major electronics firms – was not yet established in electronics. This

new company was IBM, and it was chosen by pure chance. In June 1952, when Forrester was looking for a commercial manufacturer to build his iron core memory for the SAGE project, he met at the Second Joint Computer Conference an IBM engineer, John McPherson, who seized the opportunity, convincing IBM's President Tom Watson to meet President Truman to offer the services of IBM. Having secured the contract, IBM started moving from the punch-card business to computer manufacture. On the basis of the SAGE contract, IBM hired 8,000 engineers and workers, signaling its real head-start in computer manufacturing with its Model 650, delivered in December 1954.

IBM stayed aloof in its northern New York semi-rural estate, mirroring the isolation of the other technological giant, AT&T, in its New Jersey suburb of Murray Hills. But the seeds of innovation, which were sown in the Whirlwind program and in the SAGE project, were germinated in Massachusetts. Jay Forrester, who did not entirely trust IBM, sent one of his graduate students, Kenneth Olson, to supervise the company's work at Poughkeepsie. Olson deeply disliked IBM's corporate style. So, once his work was finished, he decided to start his own computer company in an abandoned factory in Maynard, Massachusetts (near the later Route 495) in 1957. The name of the company was Digital Equipment Corporation. A new industry grew out of new knowledge. But this new knowledge was nurtured by Pentagon financial and institutional support.

In fact, the continuing support of the defense establishment, both for MIT and Harvard work and for MIT-originated companies, was based on the ties of trust, both personal and scientific, that linked the MIT–Harvard network with the commanding heights of the US Government for three decades. At the end of the 1930s, President Roosevelt appointed Vannevar Bush as his science adviser. In the 1940s, Truman's science advisers were James Killian from MIT and George Kistiakowsky from Harvard. In 1960, Kennedy's science adviser was Jerome Wiesner. Johnson made a small variation: he appointed Don Hornig from Princeton, but Hornig was a Harvard college graduate and a Harvard Ph.D. Nixon appointed as science adviser Lee Du Bridge, former director of MIT's Radiation Laboratory. Ford appointed Guyford Stever from MIT, and Carter named Frank Carter, also from MIT. Thus, a long string of science presidential appointees from MIT, and to a lesser extent from Harvard, marks the critical importance of the connection between the centers of decision of the Federal Government, particularly in technology-based defense programs, and the technological potential of MIT. It was thus no surprise when in the 1980s the "Star Wars" Program reactivated the direct linkage between defense and advanced research, MIT again received a substantial proportion of the new contracts, in spite of the established dominance of California as the main high-technology defense-oriented complex.

To sum up, the causal chain of events underlying the development of Greater Boston's high-technology complex operated thus: MIT (and to a lesser extent Harvard) became the hub of advanced electronics research in the 1940s and 1950s, with considerable support through funds and orders from the Defense Department. MIT's Faculty and graduates used their advanced knowledge in new technologies, as well as their excellent personal contacts with the military establishment, to start companies that prospered rapidly because of lack of technological competition in the new industrial complex. These companies reproduced the spin-off process, giving birth to dozens of new companies that clustered in an industrial–technological milieu, developing agglomeration economies, and supported by the high quality of labor in the region, which arose from its educational basis and its industrial tradition in skilled manufacturing.

Yet, really to understand the formation of the Boston high-technology complex, we must add another layer of complexity, by analyzing the evolution of its industrial composition at different stages of its development. This new round of analysis will in fact prove critical for our ability to judge the experience and thus evaluate the region's future prospects.

The fourth industrial wave: the tunnel at the end of the light

While these elements are all key ingredients of high-technology growth in Massachusetts, their relative importance differed quite markedly at various stages of the process.

In the 1940s and 1950s the beginnings of an electronics-based defense industry took place around the old industrial establishments created in the 1920s, Raytheon and Sylvania, and the much older establishment of aircraft engine manufacturing plants by General Electric and Pratt & Whitney in Lynn, an old industrial town north of Boston. But, in spite of the scientific support of MIT, this older defense-oriented complex could hardly survive the decline in defense spending at the end of the 1950s, after the Korean War, and in the early 1970s, when the Vietnam buildup was brought under control.

In particular, the oldest companies were unable to adapt from vacuum-tube technology to transistor technology during the critical evolution of the electronics industry in the late 1950s. Indeed, Raytheon rejected Shockley's[2] offer to cooperate with the company, prompting Shockley to go back to his native Palo Alto to found Shockley Semiconductors, the ancestor of most of Silicon Valley's semiconductors companies. The failure of the old electronics companies to enter the microelectronics field explains why, in spite of its early start and superior scientific base, Massachusetts lost its leadership to California in electronics manufacturing. Thus, the first reindustrialization in the 1950s, tightly linked to the defense industry both in electronics and engines, did not pull the region out of its structural decline, because of the rapid bureaucratization of large companies after their innovative stage, and because of excessive reliance on cozy arrangements with the defense establishment.

The third wave of industrialization in Massachusetts, based on computers from 1975 onwards, was thus quite independent of the old 1950s electronics base.[3] It was started by new companies, the great majority of which were created after 1960, with the addition of the most important of all, DEC, founded in 1957. The computer industry did have an original linkage with military programs, because most of the new entrepreneurs obtained their knowledge in research projects linked to MIT or to other firms generally funded by defense spending. Yet the products of the new companies diversified quite rapidly; by the boom decade, 1975–85, they were mainly aimed at the civilian market. It was the invention of the minicomputer and the introduction of computers into the office (with the workstation concept developed by Wang) that propelled the new high-technology complex in Massachusetts in the late 1970s.

However, the Reagan administration buildup in the 1980s again redirected Massachusetts' high-technology industry toward military programs: New England became the region that received most defense spending per capita in the entire United States during the 1980s. The emphasis of ''Star Wars'' on software and artificial intelligence, one of the strongest scientific fields at MIT as well as in the technological services firms of Massachusetts, created huge, instant, highly-profitable markets for these firms, precisely in the critical 1984–6 period when the world's computer industry went into a downturn. Seizing the opportunity, Route 128's computer companies became increasingly dependent on military markets, reversing the trend that had begun the 1970s.

Thus, what appears at first as a continuing process of high-technology development linked to defense markets, in fact conceals three stages: the World War Two and Cold War buildup of an industrial base, around large firms operating as defense contractors; the spin-off of entrepreneurial firms in the 1960s and 1970s, creating a new, civilian computer industry out of the military-oriented research programs; and the remilitarization of the computer industry, particularly of its technologically advanced components, during the 1980s, under the combined pressures of a slowdown in the world computer market and stepped-up demand from the new Pentagon technological frontier.

The vulnerability of such excessive dependence on military markets became evident during the 1988–91 period, when the Massachusetts miracle came to an abrupt halt: with the reversal of defense spending, because of budgetary constraints and the end of the Cold War, high-technology industries laid off some 60,000 workers in three years, and unemployment surged in the state. But the relative weakness of the Massachusetts high-technology complex may have deeper roots than the standard explanation of excessive reliance on military markets.

One reason may be technological. The region is much less diversified in its industrial high-technology structure than Silicon Valley or other high-technology areas. Computer making accounts for the bulk of the activity. And within computers, one particular kind of computers, minicomputers, dominates the picture. Technological change in microelectronics in the late 1980s dramatically increased the power of microcomputers, and advances in telecommunications, in software, and in the protocols to link up computers, have made possible the shift from minicomputers to networks of micro-computers working together as a unit. Thus, companies heavily specializing in office-oriented minicomputers, such as Wang, were substantially hurt by the new developments, as was Digital, in spite of its greater diversification. The industrial structure of the region must thus adapt to the new conditions of technology and to the growing competition in the world market, as Japan, Europe, and the Asian newly industrializing countries (NICs) enter the computer market.

Given that fact, a second potential reason for the new industrial crisis in Massachusetts appears to stem from the relative incapacity of its industrial structure to adapt to the new technologies and to the new market environment. In her latest work, AnnaLee Saxenian emphasizes the lack of flexibility of the Route 128 firms, in spite of their entrepreneurial origin. Three elements seem to explain such industrial rigidity: first, many of the firms were in fact spin-offs of older industrial firms, such as Raytheon, and their founders brought with them their old-fashioned corporate culture; second, the industry is ruled by professional associations, such as the Massachusetts High Technology Council, that enforce an industrial discipline so as to exercise their lobby power, both in the State Government and in Washington; finally, the renewed reliance on defense contracts in the 1980s diminished the entrepreneurial skills of the firms, removing them from the fast-changing networks of the world's computer markets. So the future of the Greater Boston high-technology complex looks gloomier than that of other similar areas in the United States. Ultimately, its future must depend on the ability of its firms to make the critical organizational and technological transition to the peace economy and to the world market.

Thus, reindustrialization is a magic word with little meaning unless it takes place within a particular historical, technological, and institutional context. The Greater Boston area has undergone not one, not two, but three reindustrialization processes in the last 50 years, each one of them characterized by a particular mix of relationships between research institutions, government markets, and industrial structure. What remains a constant factor, throughout the process of repeated recovery from the depth of crisis, is the state's commitment to excellence in education and research, which it inherited from its founding fathers. It remains to be seen whether that commitment will bounce Massachusetts back for yet a fourth time.

Notes

1 This account relies heavily on the standard sources: Lampe, 1984; Loria, 1984; Saxenian, 1985; and Wildes and Lindgren, 1985

2 [William Shockley, co-inventor of the transistor in 1947 with a team at Bell Laboratories in New Jersey]

3 On the first and second waves, see especially Dorfman, 1982, 1983; Fishman, 1981; Lampe, 1988; and Saxenian, 1985

References

DORFMAN, N.S. (1982) *Massachusetts' High Technology Boom in Perspective: An Investigation of its Dimensions, Causes and of the Role of New Firms*, Cambridge, MA, MIT, Centre for Policy Alternatives (CPA 82-2)

—— (1983) "Route 128: The Development of a Regional High Technology Economy", *Research Policy* 12: 299-316

FISHMAN, K.D. (1981) *The Computer Establishment*, New York, McGraw-Hill

LAMPE, D.R. (1984) "Das M.I.T. und die Entwicklung der Region Boston", in SCHWARZ, K. (ed.), *Die Zukunft der Metropolen: Paris, London, New York, Berlin: Aufsätze*, Berlin, Technische Universität, pp. 554-9

—— (ed.) (1988) *The Massachusetts Miracle: High Technology and Economic Revitalization*, Cambridge, MA, MIT Press

LORIA, J. (1984) "Das Massachusetts Institute of Technology und die Entwicklung der Region Boston", in SCHWARZ, K. (ed.), *Die Zukunft der Metropolen: Paris, London, New York, Berlin: Katlog zur Ausstellung*, Berlin, Technische Universität, pp. 128-46

MARKUSEN, A., HALL, P., CAMPBELL, S. and DIETRICK, S. (1991) *The Rise of the Gunbelt: The Military Remapping of Industrial America*, New York, Oxford University Press

SAXENIAN, A.L. (1985) "Silicon Valley and Route 128: Regional Prototypes or Historic Exceptions?", in CASTELLS, M. (ed.), *High Technology, Space and Society*, Beverly Hills and London, Sage, pp. 81-105

WILDES, K.L. and LINDGREN, N.A. (1985) *A Century of Electrical Engineering and Computer Science at M.I.T., 1882-1982*, Cambridge, MA, MIT Press

30

THE ROLE OF INFORMATION TECHNOLOGY IN THE PLANNING AND DEVELOPMENT OF SINGAPORE

by Kenneth E. Corey

Source: Kenneth E. Corey, ''The Role of Information Technology in the Planning and Development of Singapore'', in Stanley D. Brunn and Thomas R. Leinbach (eds), *Collapsing Space and Time: geographic aspects of communications and information*, London, HarperCollins Academic, 1991, pp. 217–31

[. . .]

Singapore

Singapore is both a country and a metropolis. This metropolitan area has developed on a small island off the southern tip of the Malaysian peninsula. Today the population of Singapore is nearly 2.6 million people.

For the student of modern urban development Singapore is an extraordinary laboratory that is rich with learnings. [. . .]

Singapore's reliance on intervention and planning

The role of the state is fundamental to Singapore's development (Dicken, 1987; Grice and Drakakis-Smith, 1985). Singapore became independent in 1965. Since then it has been governed by the People's Action Party under the leadership of Prime Minister Lee Kuan Yew. The government of Singapore has taken up the task of development in an activist manner that has used intervention and planning in pursuit of the ultimate goal of realizing a high quality of life for Singaporeans.

Nearly 200 years ago Singapore began as a port and it developed a predominantly entrepôt economy. In 1963, just two years before independence, a program of industrialization was conceived. While manufacturing has been and will continue to be of major importance to Singapore's economy, by the late 1960s and the early 1970s the government also had begun to consider complementary development policies that importantly involved scientific and technical knowledge. In 1972, the government had formulated its Development Plan for the 70s which included planned programs for an important future sector of high-technology and brain services (Hon, 1972).

1979 saw government intervention in support of its continued objective of restructuring the economy toward higher productivity and higher-skilled industries and services. This strategic position was sustained by the Singapore government in 1981, when it announced its Economic Development Plan for the 1980s; the principal objective of which was 'to develop Singapore into a modern industrial economy based on science, technology, skills and knowledge' (Goh, 1981).

The 1980 through 1984 period was assessed by the government as having 'been years of exceptional prosperity for Singapore and Singaporeans. Singapore's gross national product (GNP) grew at an average rate of 8.5 per cent per annum in real terms' during these five years (Tan, 1985). However, in 1985, after twenty years of double-digit average annual economic growth, the Singapore economy, as measured by the gross domestic product (GDP), declined by 1.8 percent.

Related to the worldwide recession, this downturn catalyzed Singapore policy makers and planners into an immediate reassessment of their approach to economic development. This resulted in the early 1986 publication of the *New Directions* report (Economic Committee, 1986). Basically, this evaluation reaffirmed the earlier goals of restructuring the economy and for the future recommended continued high priority for development policies based on high technology, research and development, information technology and the services sector – among other policies and sectors. By the end of 1987, the government's short-term measures for stimulating the economy had produced two consecutive years of economic growth; 1986 growth was 1.8 percent and 1987 growth was an unexpectedly high 8.6 percent (*The Wall Street Journal*, 1988, p. 10).

This brief summary of economic policies and results is suggestive of Singapore's style of development. It is characterized by active government planning (Krause, Koh and Lee, 1987) that is premised on long-term goals that have remained relatively constant and on interventions in the short run that are designed to make necessary mid-course corrections that bring development back toward the ultimate imperatives again. In short, Singapore's government-dominated development planning has been elitist, pragmatic, flexible, experimental, farsighted, anticipatory and innovative (Corey, 1987). One of the more innovative policies is the use of information technology in the planning and development of Singapore for the information age.

Singapore and the information age

As a result of the industrialization initiatives of the 1960s, by the early 1970s Singapore was well on its way to becoming an information economy. But what is this type of economy?

The *information economy* consists of functions and activities that are:

> . . . supported by the establishment of industries and markets wherein the necessary technical infrastructure are produced and information as a commodity is sold respectively. As these activities expand in scope and volume, there is a concomitant increase in the share of GNP arising from the value adding which originates from the production and distribution of information goods and services. (Jussawalla and Cheah, 1983, p. 162).

[. . .] Jussawalla and Cheah found that the information sector accounted for 24 percent of Singapore's gross domestic product in 1973. Also, they concluded that

> the public sector appeared to be the most information-intensive with approximately 2/3 of all government services being information-based. Public bureaucracies in administration, planning and development agencies accounted for 49 per cent of total value added. (p. 169).

Singapore's workforce has become dominated by white-collar occupations. In 1921, 27.2 percent of employed persons in Singapore were in white-collar occupations; by 1980, that percentage had risen to 41.5 percent; currently this is the largest occupational category in Singapore's employment structure; blue-collar occupations form the next largest classification at 40.4 percent (Kuo and Chen, 1985).

The 1980 Singapore workforce of 1,077,090 has 34 percent (366,912) of its workers employed in information occupations. In 1921, only 18.6 percent of its workforce was engaged in information occupations (Kuo and Chen, 1985). [. . .]

The national information technology plan

A national computerization effort began in 1981. It was part of Singapore's Economic Development Plan for the 1980s (Goh, 1981). The government's National Computer Board was established then to encourage Singapore's computerization.

With this initiative and the previous history of higher-technology and higher-level services policy experiments and program planning, Singapore by 1985 had positioned itself to be able to formulate a national development plan based on information technology. For Singapore, the operational definition of information technology (IT) 'embraces the use of computer, telecommunication and office systems technologies for the collection, processing, storing, packaging and dissemination of information' (National IT Plan Working Committee, 1985). [. . .]

Information technology manpower

One of the most fundamental of the building blocks in Singapore's national information technology plan is to achieve a workforce that can fully exploit the current and future potential of information technology – both in innovative and economic terms. Singapore has evolved its own philosophy for its manpower policies. These reject Europe's industrial democracy approaches; Singapore has developed a response to its manpower needs that accommodates to its unique ethnic, cultural, economic and locational attributes (Wilkinson, 1986). Through its Economic Development Board, the government of Singapore seeks to attract high value-added industry; this is possible only with a skilled workforce which is computer literate, comfortable in automated workplaces and capable of technological creativity. As a consequence, information technology manpower development is emphasized at all levels, including: 'assembly workers, technicians and technologists, and manufacturing engineering' (Christiansen, 1987).

Manpower development activities are being conducted in an environment of nearly full employment in Singapore (i.e. 3.6 percent unemployment rate in September 1987), and in an environment of continued worker shortage in the industries and services driven by information technologies (Christiansen, 1987; Ministry of Trade and Industry, 1987). To respond to these needs, the Singapore National Computer Board coordinates computer training and education programs; many of these programs are operated by other public-sector and public-funded organizations.

Civil Service Computerization Program

One of the earlier and most important activities of Singapore's information technology initiatives is the Civil Service Computerization Program. This consists of using government agencies as computer pioneers to test the technological and organizational implications of computerizing offices and office staffs. Approximately $50 million (all dollar figures are given in US dollars) have been invested in computerizing the public sector (Khan, 1987, p. 12). Seventeen government ministries and units are having their computer needs supported by the National Computer Board. The lessons of the Civil Service Computerization Program are being disseminated into Singapore's private sector so as to further develop the local economy. [. . .]

These and other programs have surpassed original expectations. 'The original target to have 5,800 computer professionals by 1990 has already been exceeded, three years early' (Khan, 1987, p. 12).

Information technology culture

In order to provide a climate that is supportive and receptive to information technology and to encourage widespread understanding of its benefits, the Singapore government has developed many educational and promotional campaigns. Information technology permeates the daily life of Singapore. For example, October 1987 was designated 'technology month'. At the checkout counter of the supermarket and at the ubiquitous automatic teller machine, the Singaporean is constantly reminded of the country's objectives for cashless shopping and cashless banking. Especially if a Singaporean is a public servant, then he or she has probably been using computers in the office for some time; this is part of the strategy of using the government as the pilot laboratory for computerizing the world of work. Local newspapers routinely publish articles and graphics that educate the citizenry about information technology. An international information technology exhibition has been held three times in recent years. [. . .] The Ministry of Education is implementing information technology education programs throughout Singapore's schools [. . .] (Study Team, 1987).

Another complementary exhibition, designed to inform the populace and to aid in creating the climate for familiarity with information technology, is the Civil Service Productivity Exhibition. It is a means by which the public can be shown the progress being made by the public sector in its drive for improving productivity via information technology. [. . .]

Information communication infrastructure

[. . .]

This infrastructure is the electronic highway of the Information Age: it provides the communication grid of the Singapore information society and the various information-based business and industries. (National Computer Board, 1986b, p. 2).

These telecommunications facilities enable the Telecommunication Authority of Singapore to offer these services to private and corporate customers: (a) international telephone service, including international direct dialing; (b) regional telephone services, including subscriber trunk dialing to Malaysia; (c) private leased circuits, including high-speed international leased circuits for multinational corporations with locations in Singapore; (d) data services, including 'telepac', a remote computer access service with data links to thirty-five overseas destinations; (e) public message services including facsimile (Fax) machines for sale or lease; telefax, a counter service at Telecoms outlets and post offices for the general public; telex, with 6.9 lines per 1,000 population; telemail, which is telex for small businesses; and telebox, which is an electronic mailbox service; (f) an optical fiber network service has been completed that links all of Singapore's twenty-six telephone exchanges enabling data, text and video services; (g) various radio services, including widely used radio pagers, radio memo pagers that can store and display 1,200 characters, landmobile telephone service – i.e. car phones for rent and sale, maritime radio communication services, including radiotelex, radiotelephone and radiotelegram, and maritime satellite communication services, including international telephone traffic and international telex traffic; (h) national telephone service with a telephone network that is 100 percent push-button and Telecoms is proud that 98.4 percent of applications for telephones are met by the date needed by customers; telephone service can be provided within six working days. Telephone services include both business and residential direct exchange lines; at forty-three telephones per 100 population, Singapore, after Japan, has the second-highest telephone density in Asia. Some other telephone services include the 'phone plus service', which consists of abbreviated dialing, absentee message, automatic redial, call transfer, call waiting and three-way calling.

The Telecommunication Authority of Singapore also delivers the more conventional information services of the post office. These complement the telecommunications services in an efficient and low-cost manner.

For the near future, Singapore is taking the steps to provide Integrated Services Digital Network (ISDN) linkages to business by the early 1990s. 'ISDN will essentially be an enhanced telephone network which provides end-to-end digital connectivity. This will allow voice, data, video and text communications simultaneously from one network' (Telecommunication Authority of Singapore, 1987, pp. 18–19). Because separate networks become redundant, one of the major values of ISDN is its one-network efficiency, with resultant cost savings. Teleview is another program for the future; teleview is an interactive system for making electronic transactions that are based on computer-stored information displayed on monitors and home television sets in combination with the use of the telephone network for off-air communication. A design and testing period will be necessary before the full development of Teleview. Other future telecommunications infrastructure and services are under development, including a new cellular radio-telephone system and a consultancy company to provide technical assistance and telecommunications planning services overseas. (The above material on Singapore's information communication infrastructure was abstracted from the 'Operating Review' section of *Telecoms Annual Report 1986/1987*, 1987. Also see Chia, 1986.)

Information technology applications
The government of Singapore continues to promote the widespread and active use of information technology throughout the economy. This is expected to result in greater efficiency in business and a higher quality of life for Singaporeans. Some of the applications mentioned include civil service computerization, widespread availability of automatic teller machines for consumer banking and, among others, the use of credit cards in retail sales so as to move Singapore toward becoming a 'cashless society'. The *National IT Plan* (National IT

Plan Working Committee, 1985) includes the identification of a wide range of information technology applications for Singapore. [. . .]

Information technology industry
The national information technology plan intends that information technology industry will be an increasingly important force in Singapore's economy and its development. The underlying concept is to develop a robust information technology industry and in the process to increase Singapore's industrial productivity and thereby its international competitiveness (National Productivity Board, 1986). The principal sectors of Singapore's information technology industry planning include: (a) computer hardware manufacturing, (b) computer services and (c) telecommunication services.

The manufacturing of high technology products in Singapore has been led by multinational corporations.

> Over the last 20 years, about 600 international manufacturing companies have set up business in Singapore. Most of them are engaged in technologies the government identified as desirable, such as computers, microelectronics, biotechnology and communications. (Cheng, 1987).

[. . .] However, increasingly there will be emphasis on ensuring that Singapore's local industry in information technology will be fully developed. In the manufacturing sector, the following industrial functions have been and will be nurtured and developed in Singapore; computer integrated manufacturing, computer aided engineering, computer aided design, automated product systems and computer numerical control as in machine tool fabrication. These kinds of information technology innovations in manufacturing may be most useful in industries such as consumer electronics and integrated circuit assembly (National IT Plan Working Committee, 1985, p. 39).

The service sector of the Singapore economy will be stimulated by various initiatives and incentives, and 'will concentrate on services which have the potential to be internationally tradable, which make maximal use of our limited brainpower resources, and which generate domestic spin-offs for economic activities' (Singapore Economic Development Board, 1986, p. 53). The Economic Committee recommended the promotion of these services subsectors: sea transport, air transport, land transport, telecommunications, warehousing and distribution, computer services, laboratory and testing services, 'agrotechnology', publishing, and the professional services of: accounting and auditing, legal services, advertising and public relations, management and business consulting, hotel management services, medical services, educational services, and cultural and entertainment services (Economic Committee, 1986, pp. 177–80).

The third sector of Singapore's information technology industrial priorities is the telecommunication services industry. Telecommunications are seen as the major facilitator of both information technology manufacturing and information technology services. The Telecommunication Authority of Singapore is charged with implementing Telecoms infrastructure; its principal activities are listed above (Chia, 1986).

Climate for creativity and entrepreneurship in information technology

The national information technology plan seeks to have Singapore and Singaporeans become increasingly more innovative and developmental in their information technology activities. Toward that end several public-sector research and development institutions have been created. The Institute of Systems Science is researching bilingual systems, intelligent public information systems and office automation (National Computer Board, 1986a, p. 16). The National University of Singapore is conducting applied research in artificial intelligence, local area networks, software engineering, micro-

processor applications and semiconductors. Information technology research also is underway at the Grumman International–Nanyang Technological Institute CAD/CAM (Computer aided design/computer aided manufacturing) Research Center and at the Information Technology Institute of the National Computer Board. As noted above, the Telecommunication Authority of Singapore also conducts its own research and development.

International corporations receive incentives to establish research and development centers in Singapore. The Economic Development Board of Singapore offers overseas companies a variety of tax incentives to stimulate industrial research and development. Several incentives are particularly supportive of information technology development activities. [. . .]

In addition to these incentives, a Science Park has been developed to serve as a locational focal point for research and development activities in Singapore. Both private-sector and public-sector firms may locate in the Science Park, the site of which is near the National University of Singapore; the Park, a 125-hectare estate, permits early start-up of high technology research and incubator activities. The National Computer Board headquarters also is located on the site. [. . .]

Information technology professionals in Singapore are encouraged to be innovative by having access to venture capital financing, incubator facilities, a legal context that provides for copyright protection of software, specialized technical assistance, and a climate for creativity that comes from the synergy that results from the close proximity of others also engaged in related developmental work in information technology (National Computer Board, n.d., p. 10).

Coordination and collaboration of the information technology plan

The actors involved in implementing the above six elements of the *National IT Plan* (1985) require linkage and active attention if they are to realize their objectives. [. . .] The implementation of these functions is shared with the many ministries, boards and organizations noted earlier, with the principal coordinating body composed of the National Computer Board, Telecoms, the Economic Development Board and the educational institutions, with the professional information technology societies and selected private-sector firms in support.

Singapore's information technology planning in sum

[. . .] In just over twenty years Singaporeans have been able to plan and implement themselves from a preindustrial economy and society to one that is importantly industrial and to one that is importantly 'postindustrial' or information-driven.

This has been possible in large part because of an often renewed contract between the elected leaders and the voting electorate. The government and its support bureaucracy have taken up their mandates by formulating policies and engaging in planning that are intended to bring Singaporeans a high quality of life by means, in part, of preparing for and exploiting the potential of information technology. By laying a modern and effective foundation for its development strategies and by being meticulous, tenacious and flexible in its short-term and long-term planning, the government of Singapore has brought its city-state the third highest level of prosperity in Asia – after only Japan and Brunei. [. . .]

Note

This research was made possible through a Fulbright research grant and the support of K. S. Sandhu and the staff of the Institute of Southeast Asian Studies. For the preparation of this chapter special thanks are due to Chia Choon Wei, Loh Chee Meng, Ong Choon Hwa and Martin Anderson

References

CHIA, C.W. (1986) "Economic Aspects of the Information Revolution: The Singapore Experience," in JUSSAWALLA, M., WEDEMEYER, D. and MENON, V. (eds), *The Passing of Remoteness? Information Revolution in the Asia-Pacific*, Singapore, Institute of Southeast Asian Studies, pp. 42–50

CHENG, G. (1987) "The Singapore Way: Acquiring Technology Through MNCs", *The Straits Times*, 3 October, p. 16

CHRISTIANSEN, D. (1987) "An Electronic Nation", *IEEE (Institute of Electrical and Electronics Engineers) Spectrum* 24 (12), p. 30

COREY, K. (1987) "Planning the Information-Age Metropolis: The Case of Singapore", in GUELKE, L. and PRESTON, R. (eds), *Abstract Thoughts: Concrete Solutions: Essays in Honour of Peter Nash*, Waterloo, Ontario, Department of Geography, University of Waterloo, pp. 49–72

DICKEN, P. (1987) "A Tale of Two NICs: Hong Kong and Singapore at the Crossroads", *Geoforum* 18, pp. 151–64

ECONOMIC COMMITTEE (1986) *The Singapore Economy: New Directions*, Singapore, Ministry of Trade and Industry

GOH, C.T. (1981) "Towards Higher Achievement", Budget speech, 6 March 1981, Singapore, Information Division, Ministry of Culture

GRICE, K. and DRAKAKIS-SMITH, D. (1985) "The Role of the State in Shaping Development: Two Decades of Growth in Singapore", *Transactions of the Institute of British Geographers* 10, pp. 347–59

HON, S.S. (1972) *Singapore: Economic Pattern in the Seventies*, Singapore, Ministry of Culture

JUSSAWALLA, M. and CHEAH, C.W. (1983) "Towards an Information Economy: The Case of Singapore", *Information Economics and Policy* 1, pp. 161–76

KHAN, T.L. (1987) "Going for IT", *Mirror* 23 (18), pp. 12–13

KRAUSE, L., KOH AI TEE and LEE (TSAO) YUAN (1987) *The Singapore Economy Reconsidered*, Singapore, Institute of Southeast Asian Studies

KUO, E.C.Y. and CHEN, H.T. (1985) *Towards an Information Society, Changing Occupational Structure in Singapore*, Singapore, Select Books for the Department of Sociology, National University of Singapore

MINISTRY OF TRADE AND INDUSTRY (1987) *Economic Survey of Singapore, Third Quarter 1987* Singapore, Ministry of Trade and Industry

NATIONAL COMPUTER BOARD (1986a) *National Computer Board Yearbook FY85/86*, Singapore, National Computer Board

—— (1986b) *National IT Plan*, Singapore, National Computer Board

—— (n.d.), *The Singapore Information Technology Center*, Singapore, National Computer Board

NATIONAL IT PLAN WORKING COMMITTEE (1985)

National IT Plan: A Strategic Framework, Singapore, National Computer Board

NATIONAL PRODUCTIVITY BOARD (1986) *The Report of the Committee on Productivity in the Manufacturing Sector (1985)*, Singapore, National Productivity Board

SINGAPORE ECONOMIC DEVELOPMENT BOARD (1986) *Economic Development Board Annual Report 1985/86*, Singapore, Singapore Economic Development Board

STUDY TEAM (1987) *Towards Excellence in Schools: A Report to the Minister for Education*, Singapore, Ministry of Education

TAN, T.K.Y. (1985) *Budget Statement 1985*, Singapore, Information Division, Ministry of Communications and Information

TELECOMMUNICATION AUTHORITY OF SINGAPORE (1987) "Operating review", *Telecoms Annual Report 1986/1987*

WALL STREET JOURNAL (1988) "Singapore's growth accelerated in 1987 but is seen slowing", *The Wall Street Journal*, 4 January, p. 10

WILKINSON, B. (1986) "Human Resources in Singapore's Second Industrial Revolution", *Industrial Relations Journal*, 17 (2) pp. 99–114

31

CONTINUITY AND CHANGE IN CONCEPTIONS OF THE WIRED CITY

by William H. Dutton, Jay G. Blumler and Kenneth L. Kraemer

Source: William H. Dutton, Jay G. Blumler and Kenneth L. Kraemer (eds), *Wired Cities: shaping the future of communications*, London, Cassell, 1987, pp. 3–26

The term *wired city* is used in two different ways. At a conceptual level, it refers to a normative forecast of the future of communications – a prescriptive statement about how communications technology should be institutionalized and used. In this context, the wired city has been broadly defined as a community in which all kinds of electronic communication services are available to households and businesses (Martin 1974, 1978). As we shall discuss below, this normative forecast entails a set of values and assumptions. [. . .]

At a concrete level, the wired city concept refers to experiments and projects involving the use of advanced information and communications technologies for the provision of services to households and businesses. Defined broadly, nearly any new development in computing and telecommunications might be called a wired city project if used to provide services to businesses and households of a community. Defined narrowly, wired city projects refer to new developments in cable and telecommunications that affect the public communications systems of a community.

The promotion of cable television during the 1960s generated futuristic visions of what came to be known as the wired city (Martin 1974; National Academy of Engineering 1971; Smith 1970). Beyond the mere introduction of cable technology, however, the wired city concept suggested development of a more integrated and universal system of electronic communications that went further than traditional telephone and television systems to convey a wide range of services to the households of urban communities. This concept caught the imagination of many who anticipated economic, political and social benefits in merging and expanding both traditional and modern communications services (Goldmark 1972; Martin 1978).

Despite revolutionary developments in communications technology over the last several decades, dominant images of the future of telecommunications have remained remarkably stable. In the 1960s, we were said to be moving toward a wired society in which all households and businesses would have access to an integrated array of all kinds of electronic information and communications services (Goldmark 1972). In 1985, we were said to be moving in the same direction. [. . .]

What has changed most dramatically is the economic, social, political and public-policy forces behind these developments. In the 1960s, telecommunications was viewed and later regulated as a vehicle for social and political reform. In the 1980s, it has become viewed and deregulated as a vehicle for international trade and economic development. To understand this shift, it is helpful to briefly trace the history of the wired city.

A history of the wired city

The idea of the wired city

The concept of a wired city was developed in the US in the context of Lyndon Johnson's so-called Great Society. During his six years as president, Johnson launched major governmental programs aimed at solving urban social problems. Many of these initiatives stimulated discussions of the potential of telecommunications. Such a Federal push and a series of Federal agency requests led to the formation of a panel on urban communications by the prestigious National Academy of Engineering. The academy charged the panel, chaired by Peter Goldmark, with advising it on how applications of telecommunications to city functions might (1) improve city living and (2) stimulate valuable patterns of regional development (Goldmark 1972).

Goldmark's panel argued that cable could be central to the emergence of broadband communication networks (BCN), which could support a variety of social-service objectives. Four networks would form the telecommunication infrastructure of the wired city. They were (1) telephone, (2) cable, (3) institutional and (4) community-owned networks (Goldmark 1972). Toward the end of Johnson's administration, a 1968 presidential task force reinforced the academy's work, charging telecommunications with a clear mission to "encourage the growth of communications of all kinds within localities" as an approach to improving urban life (Smith 1970, 37, 88).

Ralph Lee Smith, a journalist who covered these discussions of the new technology, tried to explain the significance of these developments to a broad public in an article in *The Nation* entitled "The Wired Nation," which was later revised to appear as a book by the same title (Smith 1970, 1972). Smith not only popularized these ideas. His work convinced key people at the Federal Communications Commission (FCC) that it was worth known

risks to the vitality of the broadcast industry to knock down regulatory barriers to the development of cable and create regulation aimed at achieving the technological infrastructure of the wired city.

Early experiments and the demise of interactive cable

In the early 1970s, images of the wired city became instrumental to both governmental and corporate investments in interactive cable projects in the United States and Japan. In 1972, the Japanese initiated experiments in Tama New Town and later in Higashi-Ikoma. In 1974, the US National Science Foundation supported a series of experiments with interactive cable. And soon after, in 1977, Warner Communications introduced QUBE, a 30-channel interactive cable television system, on a commercial basis in Columbus, Ohio.

These early ventures had a dual effect. In one respect, they suggested that a variety of public and commercial services could be effectively provided over two-way cable systems (Brownstein 1978; Clarke 1978; Lucas 1979; Moss 1978). But these same experiments also cast doubt over hopes of widespread consumer interest in new information services (Brownstein 1978; Elton 1980; Johnson 1975). So even by the mid-1970s, when these early experiments and trials were first operational, glamorous images of interactive cable had already begun to fade with the weak consumer response and the general "financial malaise throughout much of the cable industry" (Prince 1974). The cable industry refocused on traditional, one-way broadcast services (Meadows 1980; Moss, Warren and Hellebust 1983). It was the commercial success of satellite-linked networks and pay-television in the US that rekindled interest in cable, albeit focused on national networks and entertainment programming rather than on local networks, community-oriented programming or interactive services.

Advanced wired city developments

While discussion of the wired city had ceased by the 1980s, cable and telephone companies of a half-dozen advanced industrial nations launched developments that embodied many elements of the wired city concept. But there were differences. At least four interconnected developments differentiate the inspiration and implementation of advanced from earlier day images of the wired city.

One concerned the more visionary aspect. There was a lessening of utopian social expectations, though to some extent they remained, expressed in different terms. Early visions of wired cities embodied images of electronic democracy, education and huge lists of interactive services, with the public increasingly voting, shopping, working and even producing television from the home. Later versions continued to embody communications technology as an approach to electronic information and transactional services. However, the advanced wired city ventures evoke a more down-to-earth set of concerns over (a) the provision of basic telephone, television and computer services with an already demonstrated value to households and businesses; (b) the growing significance of telecommunications to the economic fortunes of localities and nations; and (c) the furtherance of national interests and cultural values.

This vision is associated, secondly, with a technological shift noted above. Developers of wired city projects in the 1970s aimed at providing an integrated array of communications services via coaxial cable networks. The developers of advanced wired cities still aim at the construction of more integrated communications facilities but now built upon more advanced satellite, microelectronic and fiber-optic technologies.

Third, some changes were implied by a growing awareness that market forces represent a more formidable constraint on the provision of new communications services. The apparent market failures of interactive cable systems, local programming, the videotelephone and videoconferencing services contributed to this perception. Added to this concern was a growing recognition of the high costs of modernizing telecommunications, despite advances in the efficiency of new technologies. These combined to shift attention toward longer-range developments of broadband media and to utilizing narrowband media more effectively in the near term. The telecommunications industry increasingly looked toward "hybrid" systems, such as using the telephone network for the provision of pay-cable services.

Fourth, in the later versions of the wired city, a larger and more diverse set of organizational interests and actors assumed a role. The first wired city experiments were developed by the then relatively new and growing cable companies that had developed primarily to provide clearer reception of broadcast signals into the home and rid new towns of the need for rooftop television antennas. In contrast, the telephone companies and others are prominent among those developing and testing newer networks to deliver an array of both point-to-point and distributed communications and information services to residents and institutions of local communities.

Projects that approach the image of the "advanced wired city" include developments in Milton Keynes, Britain; Biarritz, France; West Berlin, Germany; and Higashi-Ikoma, Japan (Table 31.1). Projects within each of these cities have been oriented toward the use of new communications media, particularly microelectronic and fiber-optic technologies to expand voice, data and visual communications services to households as well as institutions. However, wired city developments represent more than new telecommunication infrastructures. They are new sociotechnical systems in that they represent complex and interdependent social and technological arrangements.

Table 31.1 A profile of selected wired city projects

	United States	Japan	France	Germany	Britain
City	Columbus, OH/ Alameda, CA	Higashi-Ikoma	Biarritz	Berlin (West)	Milton Keynes
Project(s)	QUBE, Mini-Hub 1	Hi-OVIS	Biarritz	BIGFON	Cable, information exchanges and others
Year initiated	1977	1972	1979	1979	1981
Year first operational	1978	1978	1984	1983	1982
Last year operational	1984/ continuing	1985	Continuing	Continuing	Continuing
Major actor(s)	Warner Communications (later Warner-Amex), City of Columbus, City of Alameda, United Cable	MITI; Visual Information System Development Association (later: New Media Development Assoc)	French PTT	Deutsche Bundespost	Dept. of Industry, Home Office, Milton Keynes Development Corp., Eosys Ltd., British Telecom, Selec TV (cable and other projects)
Transmission media	Coaxial cable/ optical fiber and coaxial cable	Optical fiber	Optical fiber	Optical fiber	Coaxial cable and planned optical fiber
Network architecture	Tree network/ star-switched	Star network	Star network	Star network	Tree network and planned star network
Size	31,000 households	156 households 8 institutions	1200 households 300 institutions	106 participants	20,000 cabled households

Source: [Dutton *et al.* 1987], chapters 5, 6, 12, 15, 19 and 24.

They are built in part upon national networks and services but they each have an orientation to a geographically based community, as did earlier wired city projects.[1] [. . .]

Critical assumptions of the wired city concept

At the most general level, the wired city concept blends elements of a utopian society with a strong emphasis on much down-to-earth communications economics and engineering. During a time when there is much skepticism over the value of science and technology for society, the wired city concept reflects an optimistic outlook on technological progress. This is based primarily on a belief that advanced communications technology can provide the basis for more satisfying styles of urban living – politically, economically, socially and pragmatically (Goldmark 1972; Kalba 1975; Martin 1978; Meadows 1980).[2]

Specifically, the wired cities concept seems to embody at least five core organizing principles:

1 Communications is of increasing significance to society.
2 The new media have inherent biases toward more decentralized and democratic modes of communication.
3 Electronic media should emulate and reinforce face-to-face patterns of communication.
4 Communications should be viewed as an electronic highway.

5 Long-range, rational-comprehensive planning should guide development.

[. . .]

Communications of increasing significance to society

Many of the driving forces behind the development of the wired city are common to the electronic media. That is, proponents believe that
communications and information technology,
whether organized as a wired city or in some
other form, will have increasing significance to
the industries, businesses and households of
modern, information-oriented societies.

The invention of a new communications
medium, whether it be the printing press, telephone or computer, poses certain common
prospects for all societies that are in a position
to develop and apply them. Since World War II,
the rapid development and convergence of
computing, telecommunications and management science techniques have been reshaping
advanced industrial nations. Central to this
trend has been the revolution in computing
since the 1950s along with the revolution in
microelectronics since the 1970s. [. . .]

As commonplace as [. . .] expectations of a
more communications-centric society are
becoming, they entail controversial assumptions about the role of electronic communications in human behavior, economic
development and society in general.

A communications-centric consumer?

One assumption is that individuals and organizations act as rational communicators seeking
to increase their access to a greater range of
information. This communications-centric
rationality, however, does not reflect many
behavioral accounts of how individuals and
organizations habitually use information and
communications. Only small percentages of
the public, for example, could be classed as
"information seekers," and much information-
oriented behavior, such as newspaper reading,

seems to be driven as much by habit as by a
need to know (Greenberger 1985). The market
for information is small, fragile and price-sensi-
tive (Komatsuzaki 1982; Hooper 1985). In an
overview of the QUBE system, Edward Meadows noted that a "troublesome difficulty is
that the experimenters may be offering information most families wouldn't want even if it
were cheap. Many of the information services
[. . .] seem designed to satisfy the needs of the
'knowledge workers' who would populate the
'information economy' that some futurists see
coming. But until it arrives, the breadth of the
market for all these data is questionable"
(Meadows 1980, 73).

In addition to doubts about the demand for
information, there are also questions concerning the qualitative implications of providing
faster access to more information. More information and more channels of electronic signals
do not translate necessarily into a qualitatively
better information environment.

But the viability of the wired cities perspective does not turn only on a demand for new
information services. Advanced wired cities
anchor their economics in the growth, convergence and more efficient provision of basic
telephone, television, cable and computing services, most of which have demonstrated a
growing market.

A communications-centric economy?

Another assumption driving this view of the
significance of the electronic media is that
economies are moving toward a greater dependence on information technologies, including
new telecommunications infrastructures. Communities hope to affect business and residential
location decisions by providing telecommunications resources attractive to high-technology
companies. The public officials of major urban
areas like New York City and Los Angeles are
discussing the wisdom of wiring their regions
with wideband networks and developing "teleports." Real estate developers are already using

telecommunications networks and services to keep or attract business and industry, just as an earlier generation of developers competed for industry with tax incentives, industrial parks and physical amenities. [. . .]

Can the problems of a communications-centric society be mitigated?

Finally, the proponents of a wired society are optimistic that the problems confronting a more communications-centric society can be adequately addressed. All societies are aware that the new media pose problems by virtue of the very opportunities they raise (Mesthene 1981). In international trade, all cannot win the race to capture new markets. Likewise, national economic development can be set back rather than furthered if research-and-development dollars are sunk into the wrong ventures or if business and industry expect unrealistic payoffs from investments in new technology. If the new media are springboards for new elites, they may be viewed defensively by more established ones. While the new media are identified with the future, there are many people who are unsettled by change, comfortable with current habits and advantaged by the status quo. If the new media have the potential to be psychologically disturbing to the individual, they can be no less so for the organization concerned over the impacts of new communications systems on travel, budgets, information flows, power shifts and the skills needed to advance within the organization. The new media might permit new elites and industries to develop, but then again, they might reinforce the position of established ones.

The new media have inherent biases

Another driving force behind the new media, whether institutionalized as a wired city or in some other form, is the belief that new communications technologies are, as the late Ithiel de Sola Pool called them, "technologies of freedom" (Pool 1983). Proponents argue that features of the new media promote greater freedom of expression and more democratic modes of communication. Compared to the mass media of broadcast television and even print publishing, the newer electronic media are said to demassify communications. They provide more channels of communication; a greater range and diversity of offerings; easier access to larger volumes of information and entertainment offerings; opportunities for asynchronous communication at the time of one's choosing; and new facilities for more interactive versus one-way communications. Much in the tradition of Harold Innis and Marshall McLuhan, the new media proponents argue that features of the media themselves, independent of their content, are responsible for this democratic bias. [. . .]

Of course, the bias of any medium is a controversial issue. Many claim that media are neutral, albeit malleable, technologies. They are simply tools turned to particular purposes by those who control their adoption and use. Thus, while the new technologies could be technologies of freedom, they could as well be turned into technologies of control that might threaten the privacy of individuals, tailor commercial and political propaganda, and ration the public's media consumption.

Electronic media should emulate face-to-face patterns of communication

[. . .]

Locally based communications

Cities are organized around the primacy of interpersonal communication and geographical communities in which transportation is the basic medium for communication (Meier 1962). The wired city evokes an image of a more communications-centric society in which mediated communications will assume a more central role in enhancing interpersonal communication (Gottman 1983; Pool 1980). For this reason, early views of the wired city were anchored in traditional concepts of the neighborhood and polity as the basic building blocks of wired cities. For example, Peter Goldmark proposed a schematic of a wired city in which neighborhoods and regions within a neighborhood would be linked together and in turn to municipal departments and agencies (Goldmark 1972). This notion that electronic media would support local communities and institutions was an emphasis that carried through experiments of the 1970s in both the US and Japan.

This bias of the new media is frequently dismissed as a holdover of an earlier generation of community antenna cable television systems. The new technology, it is said, will fragment, rather than integrate, communities. First, by raising the costs of information and communication, new media will create "new divisions between those able to pay for information – the information-rich – and those who don't have the financial resources to purchase information – the information-poor" (Demac, Engsberg and Stier 1982). Second, because the new media are becoming distance-insensitive, they will fragment the mass audience into electronic communities quite distinct from the geographical community (Hiltz and Turoff 1980). [. . .]

Communications should be viewed as an electronic highway

The proponents of the wired city, however, are primarily distinguished from other proponents of the new media by their view of communications infrastructure as a public utility rather than a private commodity. The wired cities concept embodies the notion of a universal service to a local community rather than a specialized service to a particular market segment. The wired city label draws an explicit analogy to the telephone system as its model. In 1971, Ralph Lee Smith graphically referred to "an electronic highway" in describing the thrust

of the wired city concept. He noted that just as "the nation provided large federal subsidies for a new interstate highway system to facilitate and modernize the flow of automotive traffic it . . . should make a similar national commitment for an electronic highway system, to facilitate the exchange of information and ideas" (Smith 1972, 83). [. . .]

Long-range rational–comprehensive developments are desirable

Wired city projects are also committed to the relatively comprehensive and interdependent application of advanced communications technology. Whether conceiving development to take place all at once or in measured stages, the developers of wired city projects are likely to have a rational–comprehensive view of technological change. They do not advocate "muddling through," nor do they envision an invisible hand guiding an otherwise uncoordinated evolution of the wired city. In this respect, advanced wired cities are often seen as a logical response to converging technological developments in fiber optics, network architecture and digital communications that create both opportunities and needs for an integration of communications facilities in the home and business. [. . .]

Diverse national receiving grounds

The logic we have described, which underlines the wired city concept, is laden with assumptions of fact and value. Not surprisingly, therefore, there are differences of opinion over the wisdom of the wired city as an approach to new communications technologies. Moreover, the technologies are being introduced, implemented and institutionalized in societies with different cultural, political, institutional and social settings. The new technologies neither spring from the heavens, nor are they injected by some unseen hand. They are developed, adopted, used and discarded by a diverse array of individuals and institutions. In this meeting ground between the universal and the diverse, between technology and societies, the most profound social and political issues of the new technology arise. [. . .]

Summary

The wired city constitutes a highly intriguing juncture of both technological and sociocultural pressures. On the one hand, universal advancements in communications technology have driven the development of wired city projects, presenting all nations with a common set of opportunities. On the other hand, local and national responses to these developments are being shaped by political-administrative traditions, public policies, institutional arrangements and communications cultures that vary from nation to nation.

Nearly every city of modern industrial societies is wired in the literal sense that its residents have access to telephone and broadcast services. However, no city is an "advanced wired city" in the full sense that this concept suggests. The new electronic media have not yet become as centrally important for individuals, businesses or communities as the proponents of wired cities expect they will. New telecommunication services such as electronic mail and teleconferencing might segment, rather than link, the households and businesses of communities. And competing telecommunication networks are as likely to fragment as to integrate communications services and their audiences. Control over the content of media might well be recentralizing rather than decentralizing, despite the emergence of new and competing channels of communications. Finally, legal, economic and political realities continue to frustrate attempts at long-range, comprehensive planning.

Nevertheless, communications technologies and policies are stimulating new approaches to the wired city. There are fundamental assumptions about communications technology and society that underpin interest in the wired city concept that span several decades of experiments and field trials. [. . .] In these assumptions, we see that the wired cities concept is as prescriptive as it is predictive of the future of communications. It provides a perspective on how communications should be developed, not just a forecast of how it will be institutionalized. This is all the more reason it has been a controversial vision of the future of communications.

Beyond experiments and leading-edge developments, the wired city concept has never been institutionalized. It has remained an ideal versus a reality. In fact, new policies of liberalization, privatization and competition might drive the development of communications further toward a market of networks rather than an electronic highway and replace disciplined, long-range, comprehensive planning with muddling through short-term market forces. Likewise, new technologies ranging from communication satellites to microcomputers in the home, are replacing general-purpose local networks of communications with specialized networks that are sometimes confined to an "intelligent building" and sometimes directed to a national and even international community, but seldom oriented around the city. Perhaps the wired city concept presents a normative scenario for the development of communications – one that clarifies current approaches and the values that underpin them.

Notes

1 The wired city also suggests a new city, one that is not necessarily defined by the boundaries of current political jurisdictions. In fact, the wired cities literature of the late 1960s is coincident with an emphasis in political science and urban sociology on defining the social and economic city that has been shaped by modern systems of commerce and transportation as opposed to the legal city. The wired cities concept reinforces this redefinition of the city's boundaries.

2 Kas Kalba's overview of cable technology lists the

following social objectives that could be furthered through an appropriate utilization of cable technology: (1) a reduction in the rate of population growth; (2) control of population density; (3) an increase in the productivity of the region's labor force; (4) protection and preservation of the natural environment and enhancement of the educative and aesthetic qualities of the built environment; (5) more equal access for all; and (6) more direct involvement of all individuals in the administration and decision-making processes affecting the allocation of resources, the delivery of services and the formation of policies (Kalba 1975, 340–42).

References

BAGDIKIAN, B.H. (1983) *The Media Monopoly*, Boston, Beacon Press

BROWNSTEIN, C.N. (1978) "Interactive Cable TV and Social Services", *Journal of Communications* 28(2), pp. 142-7

CLARKE, P. *et al.* (1978) "Rockford, Illinois: In-Service Training for Teachers", *Journal of Communications* (Symposium on Experiments in Interactive Cable TV) 28(2)

DEMAC, D., ENGSBERG, J. and STIER, S. (1982) "Mending or Aggravating Divisions Within Communities?" Office of Communication, United Church of Christ, July (Mimeo)

ELTON, M. (1980) "Educational and Other Two-Way Cable Television Services in the United States", in WILTE, E. (ed.), *Human Aspects of Telecommunication*, Berlin, Springer Verlag, pp. 142-55

GOLDMARK, P.C. (1972) "Communication and the Community", in *Communication, a Scientific American* Book, San Francisco, W.H. Freeman

GOTTMAN, J. (1983) *The Coming of the Transactional City*, College Park, MD, University of Maryland Institute for Urban Studies

GREENBERGER, M. (ed.) (1985) *Electronic Publishing Plus*, White Plains, NY, Knowledge Industry Publications

HILTZ, R.S. and TUROFF, M. (1978) *The Network Nation*, Reading, MA, Addison–Wesley

HOOPER, R. (1985) "Lessons from Overseas: The British Experience", in GREENBERGER, M. (ed.), *Electronic Publishing Plus*, White Plains, NY, Knowledge Industry Publications

JOHNSON, L.L. (1975) "The Social Effects of Cable Televi-sion", Rand Paper Series no. P-5390, Santa Monica, CA, The Rand Corporation

KALBA, K. (1975) "The Wired Future of Urban Communication", in HANNEMAN, G. and MCEWEN, W.J. (eds), *Communication and Behavior*, Reading, MA, Addison–Wesley

KOMATSUZAKI, S. (1982) "Social Impacts of New Communication Media: The Japanese Experience", *Telecommunications Policy* 6, pp. 269-75

LUCAS, W.A. *et al.* (1979) "The Spartanburg Interactive Cable Experiments in Home Education", Santa Monica, CA, The Rand Corporation

MARTIN, J. (1974) *The Future of Telecommunications*, Englewood Cliffs, NJ, Prentice–Hall

—— (1978) *The Wired Society*, Englewood Cliffs, NJ, Prentice–Hall

MEADOWS, E. (1980) "Why TV Sets Do More in Columbus, Ohio", *Fortune*, 6 October, pp. 67-73

MEIER, R.L. (1962) *A Communication Theory of Urban Growth*, Cambridge, MA, MIT Press

MESTHENE, E.M. (1981) "The Role of Technology in Society", in TEICH, A.H. (ed.), *Technology and Man's Future*, 3rd edn, New York, St Martin's Press, pp. 99-129

MOSS, M.L. (1978) "Two-Way Cable Television: An Evaluation of Community Uses in Reading, Pennsylvania", New York, Alternate Media Center, New York University

MOSS, M.L., WARREN, R. and HELLEBUST, K. (1983) "A Third of a Nation Wired", New York, New York University, School of Public Administration

NATIONAL ACADEMY OF ENGINEERING, Committee on Telecommunications (1971) *Communications Technology for Urban Improvement*, Report to the Department of Housing and Urban Development by the Committee on Telecommunications of the National Academy of Engineering, June

POOL, I. DE S. (1980) "Communications Technology and Land Use", *Annals of the American Academy of Political and Social Science* 451 (September) pp. 1-12

—— (1983) *Technologies of Freedom*, Cambridge, MA, Harvard University Press

PRINCE, P. (1974) "The Wired Nation", *Journal of Broadcasting* 18 (3), pp. 375-6

SMITH, R.L. (1970) "The Wired Nation", *The Nation*, 18 May

—— (1972) *The Wired Nation: Cable TV: The Electronic Communications Highway*, New York, Harper & Row

ACKNOWLEDGEMENTS

BUTZER, Karl W. 'The Indian Legacy in the American Landscape', in *The Making of the American Landscape*, edited by M. P. Conzen; 1990. Reprinted by kind permission of Routledge.

HORNBECK, David 'Spanish Legacy in the Borderlands', in *The Making of the American Landscape*, edited by Conzen; 1990. Reprinted by kind permission of Routledge.

REPS, John W. *The Making of Urban America: a history of city planning in the United States*; 1965, Princeton University Press. Reprinted by kind permission of the author.

HARRIS, Cole 'French Landscapes in North America', in *The Making of the American Landscape*, edited by Conzen; 1990. Reprinted by kind permission of Routledge.

LEWIS, Peirce F. 'The Northeast and the Making of American Geographical Habits', in *The Making of the American Landscape*, edited by Conzen; 1990. Reprinted by kind permission of Routledge.

CONDIT, Carl W. *American Building: materials and techniques from the first colonial settlements to the present*, 2nd edition. Copyright © 1968 by The University of Chicago Press. Reprinted by kind permission of the publisher.

CRONON, William *Nature's Metropolis: Chicago and the Great West*. Copyright © 1991 by William Cronon. Reprinted by kind permission of W. W. Norton & Company, Inc.

McKAY, John P. 'Comparative Perspectives on Transit in Europe and the United States 1850-1914', in *Technology and the Rise of the Networked City in Europe and America*, edited by J. A. Tarr and G. Dupuy. Copyright © 1988 by Temple University Press. All rights reserved.

JACKSON, Kenneth T. *Crabgrass Frontier: the suburbanization of the United States*. Copyright © 1985 by Oxford University Press, Inc. Used by kind permission of Oxford University Press, Inc.

McSHANE, Clay 'Transforming the Use of Urban Space: A Look at the Revolution in Street Pavements, 1880-1924', in *Journal of Urban History*, Vol. 5. Copyright © 1979 by Sage Publications, Inc. Reprinted by permission of Sage Publications, Inc.

FOSTER, Mark S. 'The Model-T, the Hard Sell, and Los Angeles's Urban Growth: The Decentralization of Los Angeles during the 1920s', in *Pacific Historical Review*, Vol. 44, No. 4. Copyright © 1975 by American Historical Association, Pacific Coast Branch. Reprinted by permission of University of California Press Journals.

HISE, Greg *Magnetic Los Angeles: planning the twentieth-century metropolis*; 1997. Reprinted by kind permission of The Johns Hopkins University Press.

WILLIS, Carol 'Light, Height and Site: The Skyscraper in Chicago', in *Chicago Architecture* and Design, 1923-1993: reconfiguration of a metropolis, edited by J. Zukowsky; 1993. Reprinted with permission of The Art Institute of Chicago and Prestel Verlag, Munich.

TARR, Joel The Search for the Ultimate Sink: urban pollution in historical perspective; 1996, University of Akron Press.

MELOSI, Martin Pollution and Reform in American Cities, 1870-1930. Copyright © 1980 University of Texas Press. By permission of the University of Texas Press.

BOUMAN, Mark J. '"The Best Lighted City in the World": The Construction of a Nocturnal Landscape in Chicago', in Chicago Architecture and Design, 1923-1993: reconfiguration of a metropolis, edited by Zukowsky; 1993. Reprinted with permission of The Art Institute of Chicago and Prestel Verlag, Munich.

JOHNSON, David A. 'Regional Planning for the Great American Metropolis: New York between the World Wars', in Two Centuries of American Planning, edited by D. Schaffer; 1998, Mansell Publishing Ltd, London.

CLARK, Colin 'Transport: Maker and Breaker of Cities', in Town Planning Review, Vol. 28; 1957, Liverpool University Press. Reproduced by kind permission of the publisher.

WEBBER., Melvin W. 'Order and Diversity: Community without Propinquity', in Cities and Space: the future use of urban land, edited by L. Wingo Jr; 1963. Reprinted by kind permission of The Johns Hopkins University Press.

HALL, Peter 'Squaring the Circle: Can We Resolve the Clarkian Paradox?', in Environment and Planning B: planning and design, Vol. 21; 1994. Copyright Pion Ltd.

MARTIN, James Telematic Society: a challenge for tomorrow; 1981, Prentice-Hall.

WARF, Barney 'Telecommunications and the Changing Geographies of Knowledge Transmission in the Late Twentieth Century', in Urban Studies, Vol. 32; 1995. Reprinted by kind permission of Carfax Publishing Ltd.

BARRETT, Paul 'Cities and their Airports: Policy Formation, 1926-1952', in Journal of Urban History, Vol. 14. Copyright © 1987 by Sage Publications, Inc. Reprinted by permission of Sage Publications, Inc.

BURGESS, Ernest W. 'The Growth of the City: An Introduction to a Research Project', in The City, edited by R. Park, E. Burgess and R. McKenzie; 1967, The University of Chicago Press. Reprinted by kind permission of the publisher.

CASTELLS, Manuel and HALL, Peter Technopoles of the World: the making of twenty-first century industrial complexes; 1994. Reprinted by kind permission of Routledge.

COREY, Kenneth E. 'The Role of Information Technology in the Planning and Development of Singapore', in Collapsing Space and Time: geographic aspects of communications and information, edited by S. Brunn and T. Leinbach; 1991, HarperCollins Academic.

DUTTON, William H., BLUMLER, Jay G. and KRAEMER, Kenneth L. Wired Cities: shaping the future of communications; 1987, Cassell.

The publishers have made every effort to contact copyright holders of material reprinted in The American Cities and Technology Reader. However, this has not been possible in every case and we would welcome correspondence from those individuals or companies we have been unable to trace.

INDEX